Y0-CKL-670

Springer Finance

Editorial Board
M. Avellaneda
G. Barone-Adesi
M. Broadie
M.H.A. Davis
E. Derman
C. Klüppelberg
W. Schachermayer

Springer Finance

Springer Finance is a programme of books addressing students, academics and practitioners working on increasingly technical approaches to the analysis of financial markets. It aims to cover a variety of topics, not only mathematical finance but foreign exchanges, term structure, risk management, portfolio theory, equity derivatives, and financial economics.

For further volumes:
www.springer.com/series/3674

Jakša Cvitanić · Jianfeng Zhang

Contract Theory in Continuous-Time Models

Springer

Jakša Cvitanić
Division of the Humanities
 and Social Sciences
California Institute of Technology
Pasadena, CA, USA
and
EDHEC Business School
Nice, France

Jianfeng Zhang
Department of Mathematics
University of Southern California
Los Angeles, CA, USA

ISBN 978-3-642-14199-7   ISBN 978-3-642-14200-0 (eBook)
DOI 10.1007/978-3-642-14200-0
Springer Heidelberg New York Dordrecht London

Library of Congress Control Number: 2012946750

Mathematics Subject Classification (2010): 91G80, 93E20

JEL Classification: C61, C73, D86, G32, G35, J33, M52

© Springer-Verlag Berlin Heidelberg 2013
This work is subject to copyright. All rights are reserved by the Publisher, whether the whole or part of the material is concerned, specifically the rights of translation, reprinting, reuse of illustrations, recitation, broadcasting, reproduction on microfilms or in any other physical way, and transmission or information storage and retrieval, electronic adaptation, computer software, or by similar or dissimilar methodology now known or hereafter developed. Exempted from this legal reservation are brief excerpts in connection with reviews or scholarly analysis or material supplied specifically for the purpose of being entered and executed on a computer system, for exclusive use by the purchaser of the work. Duplication of this publication or parts thereof is permitted only under the provisions of the Copyright Law of the Publisher's location, in its current version, and permission for use must always be obtained from Springer. Permissions for use may be obtained through RightsLink at the Copyright Clearance Center. Violations are liable to prosecution under the respective Copyright Law.
The use of general descriptive names, registered names, trademarks, service marks, etc. in this publication does not imply, even in the absence of a specific statement, that such names are exempt from the relevant protective laws and regulations and therefore free for general use.
While the advice and information in this book are believed to be true and accurate at the date of publication, neither the authors nor the editors nor the publisher can accept any legal responsibility for any errors or omissions that may be made. The publisher makes no warranty, express or implied, with respect to the material contained herein.

Printed on acid-free paper

Springer is part of Springer Science+Business Media (www.springer.com)

To my parents Antun and Vjera
To my wife Ying and my son Albert

Preface

Why We Wrote This Book In recent years there has been a significant increase in interest in continuous-time Principal–Agent models and their applications. Even though the approach is technical in nature, it often leads to elegant solutions with clear economic predictions. Our monograph sets out to survey some of the literature in a systematic way, using a general theoretical framework. The framework we find natural and general enough to include most of the existing results is the use of the so-called Stochastic Maximum Principle, in models driven by Brownian Motion. It is basically the Stochastic Calculus of Variations, used to find first order conditions for optimality. This leads to the characterization of optimal contracts through a system of Forward-Backward Stochastic Differential Equations (FBSDE's). Even though there is no general existence theory for the FBSDE's that appear in this context, in a number of special cases they can be solved explicitly, thus leading to the analytic form of optimal contracts, and enabling derivation of many qualitative economic conclusions. When assuming Markovian models, we can also identify sufficient conditions via the standard approach of using Hamilton–Jacobi–Bellman Partial Differential Equations (HJB PDE's).

Who Is It For This book is aimed at researchers and graduate students in Economic Theory, Mathematical Economics and Finance, and Mathematics. It provides a general methodological framework, which, hopefully, can be used to develop further advances, both in applications and in theory. It also presents, in its last part, a primer on BSDE's and FBSDE's. We have used the material from the book when teaching PhD courses in contract theory at Caltech and at the University of Zagreb.

Prerequisites A solid knowledge of Stochastic Calculus and the theory of SDE's is required, although the reader not interested in the proofs will need more of an intuitive understanding of the related mathematical concepts, than a familiarity with the technical details of the mathematical theory. A knowledge of Microeconomics is also helpful, although nothing more than a basic understanding of utility functions is required.

Structure of the Book We have divided the book into an introduction, three main middle parts, and the last part. The introduction describes the three main settings: risk sharing, hidden actions and hidden types. It also presents a simple example of each. Then, each middle part presents a general theory for the three settings, with a variety of special cases and applications. The last part presents the basics of the BSDE's theory and the FBSDE's theory.

Web Page for This Book sites.google.com/site/contracttheorycvitaniczhang/. This is a link to the book web page that will be regularly updated with material related to the book, such as corrections of typos.

Acknowledgements Our foremost gratitude goes to our families for the understanding and overall support they provided during the times we spent working on our joint research leading to this book, and for the work on the book itself. We are grateful for the support from the staff of Springer, especially Catriona Byrne, Marina Reizakis and Annika Elting. A number of colleagues and students have made useful comments and suggestions, and pointed out errors in the working manuscript, including Jin Ma, Ajay Subramanian, Xuhu Wan, Xunyu Zhou, Hualei Chang and Nikola Sandrić, and anonymous reviewers.

The research and the writing of this book has been partially supported by the National Science Foundation grants DMS 06-31298, 06-31366, 10-08219 and 10-08873. A great deal of the material for the first draft of the book was written while J.C. was visiting the University of Zagreb in Croatia and teaching a course on contract theory in continuous-time. We are grateful for the hospitality and the support of the university, and the National Foundation for Science, Higher Education and Technological Development of the Republic of Croatia. We are also grateful for the support of our home institutions, California Institute of Technology, and the University of Southern California.

Of course, we are solely responsible for any remaining errors, and the opinions, findings and conclusions or suggestions in this book do not necessarily reflect anyone's opinions but the authors'.

Final Word We hope that you will find the subject of this book interesting in its economic content, and elegant in its mathematical execution. We would be grateful to the careful reader who could inform us of any remaining typos and errors noticed, or any other comments, by sending an e-mail to our current e-mail addresses. Enjoy!

Los Angeles, USA Jakša Cvitanić
April 2012 Jianfeng Zhang

Table of Contents

Part I Introduction

1 Principal–Agent Problem 3
   1.1 Problem Formulation 3
   1.2 Further Reading 6
   References 6

2 Single-Period Examples 7
   2.1 Risk Sharing 7
   2.2 Hidden Action 8
   2.3 Hidden Type 10
   2.4 Further Reading 14
   References 14

Part II First Best: Risk Sharing Under Full Information

3 Linear Models with Project Selection, and Preview of Results 17
   3.1 Linear Dynamics and Control of Volatility 17
      3.1.1 The Model 17
      3.1.2 Risk Sharing, First Best Solution 18
      3.1.3 Implementing the First Best Solution 20
      3.1.4 Optimal Contract as a Function of Output 21
      3.1.5 Examples 22
   3.2 Further Reading 24
   References 24

4 The General Risk Sharing Problem 25
   4.1 The Model and the PA Problem 25
   4.2 Necessary Conditions for Optimality 26
      4.2.1 FBSDE Formulation 27
      4.2.2 Adjoint Processes 28
      4.2.3 Main Result 28

	4.3	Sufficient Conditions for Optimality	30
	4.4	Optimal Contracts	31
		4.4.1 Implementing the First Best Solution	31
		4.4.2 On Uniqueness of Optimal Contracts	32
	4.5	Examples	34
		4.5.1 Linear Dynamics	34
		4.5.2 Nonlinear Volatility Selection with Exponential Utilities	35
		4.5.3 Linear Contracts	37
	4.6	Dual Problem	38
	4.7	A More General Model with Consumption and Recursive Utilities	40
	4.8	Further Reading	43
	References		43

Part III Second Best: Contracting Under Hidden Action—The Case of Moral Hazard

5	**Mathematical Theory for General Moral Hazard Problems**		47
	5.1	The Model and the PA Problem	47
	5.2	Lipschitz Case	51
		5.2.1 Agent's Problem	51
		5.2.2 Principal's Problem	54
		5.2.3 Principal's Problem Based on Principal's Target Actions	57
		5.2.4 Principal's Problem Based on Principal's Target Actions: Another Formulation	60
	5.3	Quadratic Case	64
		5.3.1 Agent's Problem	64
		5.3.2 Principal's Problem	67
	5.4	Special Cases	69
		5.4.1 Participation Constraint at Time Zero	69
		5.4.2 Separable Utility and Participation Constraint at Time Zero	72
		5.4.3 Infinite Horizon	74
		5.4.4 HJB Approach in Markovian Case	76
	5.5	A More General Model with Consumption and Recursive Utilities	77
	5.6	Further Reading	83
	References		84
6	**Special Cases and Applications**		85
	6.1	Exponential Utilities and Lump-Sum Payment	85
		6.1.1 The Model	85
		6.1.2 Necessary Conditions Derived from the General Theory	86
		6.1.3 A Direct Approach	90
		6.1.4 A Solvable Special Case with Quadratic Cost	93
	6.2	General Risk Preferences, Quadratic Cost, and Lump-Sum Payment	94
		6.2.1 The Model	94

	6.2.2	Necessary Conditions Derived from the General Theory	94
	6.2.3	A Direct Approach	98
	6.2.4	Example: Risk-Neutral Principal and Log-Utility Agent	100
6.3	Risk-Neutral Principal and Infinite Horizon		103
	6.3.1	The Model	103
	6.3.2	Necessary Conditions Derived from the General Theory	103
	6.3.3	A Direct Approach	106
	6.3.4	Interpretation and Discussion	109
	6.3.5	Further Economic Conclusions and Extensions	110
6.4	Further Reading		112
References			113

7 An Application to Capital Structure Problems: Optimal Financing of a Company ... 115

7.1	The Model		115
7.2	Agent's Problem		117
7.3	Principal's Problem		121
	7.3.1	Principal's Problem Under Participation Constraint	121
	7.3.2	Properties of the Principal's Value Function	125
	7.3.3	Optimal Contract	126
7.4	Implementation Using Standard Securities		129
7.5	Comparative Statics		130
	7.5.1	Example: Agent Owns the Firm	131
	7.5.2	Computing Parameter Sensitivities	131
	7.5.3	Some Comparative Statics	133
7.6	Further Reading		134
References			134

Part IV Third Best: Contracting Under Hidden Action and Hidden Type—The Case of Moral Hazard and Adverse Selection

8 Adverse Selection ... 137

8.1	The Model and the PA Problem		137
	8.1.1	Constraints Faced by the Principal	138
8.2	Quadratic Cost and Lump-Sum Payment		138
	8.2.1	Technical Assumptions	139
	8.2.2	Solution to the Agent's Problem	140
	8.2.3	Principal's Relaxed Problem	143
	8.2.4	Properties of the Candidate Optimal Contract	144
8.3	Risk-Neutral Agent and Principal		145
8.4	Controlling Volatility		149
	8.4.1	The Model	149
	8.4.2	Main Result: Solving the Relaxed Problem	150
	8.4.3	Comparison with the First Best	152
8.5	Further Reading		153
References			153

Part V Backward SDEs and Forward-Backward SDEs

9 Backward SDEs 157
    9.1 Introduction 157
           9.1.1 Example: Option Pricing and Hedging 158
    9.2 Linear Backward SDEs 159
    9.3 Well-Posedness of BSDEs 160
    9.4 Comparison Theorem and Stability Properties of BSDEs 165
    9.5 Markovian BSDEs and PDEs 170
           9.5.1 Numerical Methods 172
    9.6 BSDEs with Quadratic Growth 173
    9.7 Further Reading 181
    References 181

10 Stochastic Maximum Principle 183
    10.1 Stochastic Control of BSDEs 183
    10.2 Stochastic Control of FBSDEs 188
    10.3 Stochastic Control of High-Dimensional BSDEs 195
    10.4 Stochastic Optimization in Weak Formulation 203
           10.4.1 Weak Formulation Versus Strong Formulation 203
           10.4.2 Sufficient Conditions in Weak Formulation 205
           10.4.3 Necessary Conditions in Weak Formulation 211
           10.4.4 Stochastic Optimization for High-Dimensional BSDEs . . . 215
           10.4.5 Stochastic Optimization for FBSDEs 218
    10.5 Some Technical Proofs 221
           10.5.1 Heuristic Derivation of the Results of Sect. 4.7 221
           10.5.2 Heuristic Derivation of the Results of Sect. 5.5 222
           10.5.3 Sketch of Proof for Theorem 5.2.12 224
    10.6 Further Reading 226
    References 227

11 Forward-Backward SDEs 229
    11.1 FBSDE Definition 229
    11.2 Fixed Point Approach 230
    11.3 Four-Step Scheme—The Decoupling Approach 236
    11.4 Method of Continuation 243
    11.5 Further Reading 247
    References 248

References 249

Index 253

Part I
Introduction

Chapter 1
Principal–Agent Problem

Abstract A Principal–Agent problem is a problem of optimal contracting between two parties, one of which, namely the agent, may be able to influence the value of the outcome process with his actions. What kind of contract is optimal typically depends on whether those actions are observable/contractable or not, and on whether there are characteristics of the agent that are not known to the principal. There are three main types of these problems: (i) the first best case, or risk sharing, in which both parties have the same information; (ii) the second best case, or moral hazard, in which the action of the agent is hidden or not contractable; (iii) the third best case or adverse selection, in which the type of the agent is hidden.

1.1 Problem Formulation

The main topic of this volume is mathematical modeling and analysis of contracting between two parties, **Principal** and **Agent**, in an uncertain environment. As a typical example of a **Principal–Agent problem**, henceforth the **PA problem**, we can think of the principal as an investor (or a group of investors), and of the agent as a portfolio manager who manages the investors' money. Another interesting example from Finance is that of a company (as the principal) and its chief executive (as the agent). As may be guessed, the principal offers a contract to the agent who has to perform a certain task on the principal's behalf (in our model, it's only one type of task).

We will sometimes call the principal P and the agent A, and we will also call the principal "she" and the agent "he".

The economic problem is for the principal to construct a contract in such a way that: (i) the agent will accept the contract; this is called an **individual rationality (IR) constraint**, or a **participation constraint**; (ii) the principal will get the most out of the agent's performance, in terms of expected utility. How this should be done in an optimal way, depends crucially on the amount of information that is available to P and to A. There are three classical cases studied in the literature, and which we also focus on in this volume: **Risk Sharing** (RS) with symmetric information, **Hidden Action** (HA) and **Hidden Type** (HT).

Risk Sharing The case of Risk Sharing, also called the **first best**, is the case in which P and A have the same information. They have to agree how to share the risk between themselves. It is typically assumed that the principal has all the *bargaining power*, in the sense that she offers the contract and also dictates the agent's actions, which the agent has to follow, or otherwise, the principal will penalize him with a severe penalty. Mathematically, the problem becomes a stochastic control problem for a single individual—the principal, who chooses both the contract and the actions, under the IR constraint. Alternatively, it can also be interpreted as a maximization of their joint welfare by a social planner. More precisely, but still in informal notation, if we denote by c the choice of contract and by a the choice of action, and by U_A and U_P the corresponding utility functions, the problem becomes

$$\max_{c,a} \{ E[U_P(c,a)] + \lambda E[U_A(c,a)] \} \tag{1.1}$$

where $\lambda > 0$ is a Lagrange multiplier for the IR constraint, or a parameter which determines the level of risk sharing. The allocations that are obtained in this way are Pareto optimal.

Hidden Action This is the case in which actions of A are not observable by P. Because of this, there will typically be a loss in expected utility for P, and she will only be able to attain the **second best** reward. Many realistic examples do present cases of P not being able to deduce A's actions, either because it may be too costly to monitor A, or quite impossible. For example, it may be costly to monitor which stocks a portfolio manager picks and how much he invests in each, and it may be quite impossible to deduce how much effort he has put into collecting information for selecting those stocks.

It should be mentioned that the problem is of the same type even if the actions are observed, but cannot be contracted upon—the contract payoff cannot depend directly on A's actions.

Due to unobservable or non-contractable actions, P cannot choose directly the actions she would like A to perform. Instead, giving a contract c, she has to be aware which action $a = a(c)$ will be optimal for the agent to choose. Thus, this becomes a problem of *incentives*, in which P indirectly influences A to pick certain actions, by offering an appropriate contract. Because A can undertake actions that are not in the best interest of the principal, this case also goes under the name of **moral hazard**.

Mathematically, we first have to solve the *agent's problem* for a given fixed contract c:

$$V_A(c) := \max_a E[U_A(c,a)]. \tag{1.2}$$

Assuming there is one and only one optimal action $a(c)$ solving this problem, we then have to solve the *principal's problem*:

$$V_P := \max_c \{ E[U_P(c,a(c))] + \lambda E[U_A(c,a(c))] \}. \tag{1.3}$$

Problem (1.2) can be very hard given that c can be chosen in quite an arbitrary way. A standard approach which makes this easier is to assume that the agent does not control the outcome of the task directly by his actions, but that he chooses the distribution of the outcome by choosing specific actions. More precisely, this will be modeled by having A choose probability distributions P^a under which the above expected values will be taken.

Hidden Type In many applications it is reasonable to assume that P does not know some key characteristics of A. For example, she may not know how capable an executive is, in terms of how much return he can produce per unit of effort. Or, P may not know what A's risk aversion is. Or how rich A is. An even more fundamental example is of a buyer (agent) and a seller (principal), in which the buyer may be of a type who cares more or cares less about the quality of the product (wine, for example). Those hidden characteristics, or types, may significantly alter A's behavior, given a certain contract.

It is typically assumed in the HT case, as we also do in this book, that P will offer a **menu of contracts**, one for each type, from which A can choose. Under certain conditions, a so-called **revelation principle** holds, which says that it is sufficient to consider contracts which are **truth-telling**: the agent will reveal his true type by choosing the contract $c(\theta)$ which was meant for his type θ. In particular, the main assumption needed for the revelation principle is that of *full commitment*: once agreed on the contract, the parties cannot change their mind in the future, even if both are willing to renegotiate. This is an assumption that we make throughout.

If the hidden type case is combined with hidden actions, then, generally, the principal gets only her **third best** reward. Since A can pretend to be of a different type than he really is, which can adversely affect P's utility, the hidden type case is also called a case of **adverse selection**. An example is the case of a health insurance company (principal) and an individual (agent) who seeks health insurance, but only if he already has medical problems, and the insurance company may not know about it.

Mathematically, we again first have to solve the agent's problem when he chooses a contract $c(\theta')$ and he is of type θ:

$$V_A(c(\theta'), \theta) := \max_a E^\theta [U_A(c(\theta'), a, \theta)]. \tag{1.4}$$

We assume that the principal's belief about the distribution of types is given by a distribution function $F(\theta)$. Denote by $\mathcal{T}$ the set of truth-telling menus of contracts $c(\theta)$. Assuming there is one and only one optimal action $a(c(\theta'), \theta)$ solving the agent's problem for each pair $(c(\theta'), \theta)$, and denoting $a(c(\theta)) := a(c(\theta), \theta)$ (the action taken when A reveals the truth) we then have to solve the principal's problem

$$V_P := \max_{c \in \mathcal{T}} \int \{E^\theta[U_P(c(\theta), a(c(\theta)))] + \lambda(\theta) E^\theta[U_A(c(\theta), a(c(\theta)))]\} dF(\theta). \tag{1.5}$$

Note that the principal faces now an additional, *truth-telling constraint*, that is, $c \in \mathcal{T}$, which can be written as

$$\max_{\theta'} V_A(c(\theta'), \theta) = V_A(c(\theta), \theta). \tag{1.6}$$

1.2 Further Reading

There are a number of books that have the PA problem as one of the main topics. We mention here Laffont and Martimort (2001), Salanie (2005), and Bolton and Dewatripont (2005), which all contain the general theory in discrete-time, more advanced topics and many applications.

References

Bolton, P., Dewatripont, M.: Contract Theory. MIT Press, Cambridge (2005)
Laffont, J.J., Martimort, D.: The Theory of Incentives: The Principal–Agent Model. Princeton University Press, Princeton (2001)
Salanie, B.: The Economics of Contracts: A Primer, 2nd edn. MIT Press, Cambridge (2005)

Chapter 2
Single-Period Examples

Abstract In this chapter we consider simple examples in one-period models, whose continuous versions will be studied later in the book. Principal–Agent problems in single-period models become more tractable if exponential utility functions are assumed. However, even then, there are cases in which tractability requires considering only linear contracts. Optimal contracts which cannot contract upon the agent's actions are more sensitive to the output than those that can. When the agents' type is unknown to the principal, the agents of "higher" type may have to be paid more to make them reveal their type.

2.1 Risk Sharing

Assume that the contract payment occurs once, at the final time $T = 1$, and we denote it C_1. The principal draws utility from the final value of an *output process* X, given by

$$X_1 = X_0 + a + B_1 \tag{2.1}$$

where B_1 is a fixed random variable. The constant a is the action of the agent.

With full information, the principal maximizes the following case of (1.1), with $g(a)$ denoting a *cost function*:

$$E\big[U_P(X_1 - C_1) + \lambda U_A(C_1 - g(a))\big]. \tag{2.2}$$

Setting the derivative with respect to C_1 inside the expectation equal to zero, we get the first order condition

$$\frac{U'_P(X_1 - C_1)}{U'_A(C_1 - g(a))} = \lambda. \tag{2.3}$$

This is the so-called *Borch rule* for risk-sharing, a classical result that says that *the ratio of marginal utilities of P and A is constant at the risk-sharing optimum*.

We assume now that the utility functions are exponential and the cost of action is quadratic:

$$U_A(C_1 - g(a)) = -\frac{1}{\gamma_A} e^{-\gamma_A[C_1 - ka^2/2]}, \tag{2.4}$$

$$U_P(X_1 - C_1) = -\frac{1}{\gamma_P} e^{-\gamma_P [X_1 - C_1]}. \tag{2.5}$$

Denote

$$\rho := \frac{1}{\gamma_A + \gamma_P}. \tag{2.6}$$

We can compute the optimal C_1 from (2.3), and get

$$C_1 = \rho \big[\gamma_P X_1 + \gamma_A k a^2 / 2 + \log \lambda \big]. \tag{2.7}$$

This is a typical result: *for exponential utility functions the optimal contract is linear* in the output process. We see that the *sensitivity* of the contract with respect to the output is given by $\frac{\gamma_P}{\gamma_A + \gamma_P} \leq 1$, and it gets smaller as the agent's risk aversion gets larger relative to the principal's. A very risk-averse agent should not be exposed much to the uncertainty of the output. In the limit when P is risk-neutral, or A is infinitely risk-averse, that is, $\gamma_P = 0$ or $\gamma_A = \infty$, the agent is paid a fixed cash payment. On the other hand, when A is risk-neutral, that is, $\gamma_A = 0$, the sensitivity is equal to its maximum value of one, and what happens is that *at the end of the period the principal sells the whole firm to the risk-neutral agent in exchange for cash payment*. The risk is completely taken over by the risk-neutral agent.

If we now take a derivative of the objective function with respect to a, and use the first order condition (2.3) for C_1, a simple computation gives us

$$a = 1/k,$$

which is the optimal action. We see another typical feature of exponential utilities: *the optimal action does not depend on the value of the output*. In fact, here, when there is also full information, it does not depend on risk aversions either, and this feature will extend to more general risk-sharing models and other utility functions.

Note that the optimal contract C_1, as given in (2.7), explicitly depends on the action a. Thus, this is not going to be a feasible contract when the action is not observable. Moreover, if, in the hidden action case, the principal replaced a in (2.7) with $1/k$, and offered such a contract, it can be verified that the agent would not choose $1/k$ as the optimal action, and the contract would not attain the first best utility for the principal. We discuss hidden action next.

2.2 Hidden Action

Even though the above example is very simple, it is hard to deal with examples like this in the case of hidden action. We will see that it is actually easier to get more general results in continuous-time models. For example, we will here derive the contract which is optimal among linear contracts, but we will show later that in a continuous-time model the same linear contract is in fact optimal even if we allow general (not just linear) contracts.

Regardless of whether we have a discrete-time or a continuous-time model, for HA models we suppose that the agent can choose the distribution of X_1 by his

2.2 Hidden Action

action, in a way which is unobservable or non-contractable by the principal. More precisely, let us change somewhat the above model by assuming that under some fixed probability $P = P^0$,

$$X_1 = X_0 + \sigma B_1$$

where X_0 is a constant and B_1 is a random variable that has a standard normal distribution. For simplicity of notation set $X_0 = 0$. Given action a we assume that the probability P changes to P^a, under which the distribution of B_1 is normal with mean a/σ and variance one. Thus, under P^a, X_1 has mean a. We see that by choosing action a the agent influences only the distribution and not directly the outcome value of X_1.

Even with that modification, the agent's problem is still hard in this single period model for arbitrary contracts. In fact, Mirrlees (1999) shows that, in general, we cannot expect the existence of an optimal contract in such a setting. For this reason, in this example we restrict ourselves only to the contracts which are linear in X_1, or, equivalently, in B_1:

$$C_1 = k_0 + k_1 B_1.$$

Denoting by E^a the expectation operator under probability P^a, the agent's problem (1.2) then is to minimize

$$E^a\left[e^{-\gamma_A(k_0 + k_1 B_1 - ka^2/2)}\right] = e^{-\gamma_A(k_0 - ka^2/2 + k_1 a/\sigma - \frac{1}{2}k_1^2 \gamma_A)}$$

where we used the fact that

$$E^a\left[e^{cB_1}\right] = e^{ca/\sigma + \frac{1}{2}c^2}. \tag{2.8}$$

We see that the optimal action a is

$$a = \frac{k_1}{k\sigma}. \tag{2.9}$$

That is, it is proportional to the sensitivity k_1 of the contract to the output process, and inversely proportional to the penalty parameter and the uncertainty parameter.

We now use a method which will also prove useful in the continuous-time case. We suppose that the principal decides to give expected utility of R_0 to the agent. This means that, using $C_1 = k_0 + \sigma ka B_1$, the fact that the mean of B_1 under P^a is a/σ, and using (2.8) and (2.9),

$$R_0 = -\frac{1}{\gamma_A} E^a\left[e^{-\gamma_A(C_1 - ka^2/2)}\right] = -\frac{1}{\gamma_A} e^{-\gamma_A(k_0 + ka^2/2 - \frac{1}{2}\gamma_A \sigma^2 k^2 a^2)}. \tag{2.10}$$

Computing $e^{-\gamma_A k_0}$ from this and using $C_1 = k_0 + \sigma ka B_1$ again, we can write

$$-\frac{1}{\gamma_A} e^{-\gamma_A C_1} = R_0 e^{-\gamma_A(-ka^2/2 + \frac{1}{2}\gamma_A \sigma^2 k^2 a^2 + \sigma ka B_1)}. \tag{2.11}$$

This is *a representation of the contract payoff in terms of the agent's promised utility R_0 and the source of uncertainty B_1*, which will be crucial later on, too. Using

$e^{\gamma_P C_1} = (e^{-\gamma_A C_1})^{-\gamma_P/\gamma_A}$, $X_1 = \sigma B_1$ and (2.11), we can write the principal's expected utility as

$$E^a[U_P(X_1 - C_1)] = -\frac{1}{\gamma_P}(-\gamma_A R_0)^{-\gamma_P/\gamma_A} E^a\left[e^{-\gamma_P(\sigma B_1 + ka^2/2 - \frac{1}{2}\gamma_A \sigma^2 k^2 a^2 - \sigma k a B_1)}\right]$$

which can be computed as

$$-\frac{1}{\gamma_P}(-\gamma_A R_0)^{-\gamma_P/\gamma_A} e^{-\gamma_P(-\frac{1}{2}\gamma_P \sigma^2 (ka-1)^2 - a(ka-1) + ka^2/2 - \frac{1}{2}\gamma_A \sigma^2 k^2 a^2)}.$$

Setting the derivative thereof with respect to a to zero, we get the optimal a as

$$a = \frac{1/(k\sigma^2) + \gamma_P}{1/\sigma^2 + k(\gamma_A + \gamma_P)}. \tag{2.12}$$

The sensitivity of the contract is $k_1/\sigma = ka$, that is, we have

$$C_1 = \tilde{k}_0 + kaX_1 = \tilde{k}_0 + \frac{1/(k\sigma^2) + \gamma_P}{1/(k\sigma^2) + \gamma_A + \gamma_P} X_1$$

for some constant $\tilde{k}_0$. Recall that in the risk-sharing, first best case the optimal action is $a = 1/k$ and the sensitivity is $\gamma_P/(\gamma_A + \gamma_P)$, thus both are independent of the level of uncertainty σ, and the action is even independent of the risk aversions. Here, the action and the sensitivity depend on the risk aversions. As the level of risk σ goes to zero, the action approaches the first best action, because then the action becomes, in the limit, fully observable.

It is easy to check that the sensitivity of the above HA contract is decreasing in the level of uncertainty σ and always higher than the sensitivity of the RS contract—when the action is unobservable the principal is forced to try to induce more effort by offering higher incentives, but less so when the risk is higher. In the limit when σ goes to infinity, the two sensitivities become equal.

For fixed σ, the induced action now depends on risk aversions. For the risk-neutral agent, the action is again the first best, and the principal transfers the whole firm to the agent.

However, as the agent's risk aversion increases (relative to the principal's), in the HA case the principal can optimally induce only lower and lower effort from the agent, paying him with lower and lower sensitivity to the output. On the other hand, given A's risk aversion, as P becomes more risk-averse she offers a higher percentage of the output to the agent.

As mentioned at the beginning of this section, we will show later that the above contract is actually optimal among all contracts, linear or not, when we allow continuous actions by the agent.

2.3 Hidden Type

We now add to the above HA model a parameter θ, unknown to the principal, which characterizes the agent. More precisely, for agent of type θ, we assume that, given

2.3 Hidden Type

action a, the mean of the normal random variable B_1 is $(\theta + a)/\sigma$ (the variance is still equal to one). The interpretation is that θ is the "return" that A can produce with no effort, due to his individual-specific skills.

We again restrict ourselves only to the contracts which are linear in X_1, and P offers a menu of contracts depending on type θ, from which A can choose:

$$C_1(\theta) = k_0(\theta) + k_1(\theta) B_1.$$

Denoting by $E^{a,\theta}$ the expectation operator under probability P^a and type θ, the agent's problem (1.4) is

$$-\gamma_A V_A(\theta, \theta') := \min_a E^{a,\theta}\left[e^{-\gamma_A(k_0(\theta') + k_1(\theta')B_1 - ka^2/2)}\right]$$

$$= \min_a e^{-\gamma_A(k_0(\theta') - ka^2/2 + k_1(\theta')(a+\theta)/\sigma - \frac{1}{2}k_1^2(\theta')\gamma_A)}.$$

We see that the optimal action $a = a(\theta')$ is

$$a(\theta') = \frac{k_1(\theta')}{k\sigma} \tag{2.13}$$

and

$$-\gamma_A V_A(\theta, \theta') = e^{-\gamma_A(k_0(\theta') + \frac{1}{2}k_1^2(\theta')(\frac{1}{k\sigma^2} - \gamma_A) + \frac{\theta}{\sigma}k_1(\theta'))}. \tag{2.14}$$

Denote with ∂/∂_θ the derivative with respect to the first argument, and with $\partial/\partial_{\theta'}$ the derivative with respect to the second argument. In order for the contract to be truth-telling, $\max_{\theta'} V_A(\theta, \theta')$ has to be attained at $\theta' = \theta$, which leads to the first order condition

$$0 = \frac{\partial}{\partial \theta'} V_A(\theta, \theta).$$

We denote by $R(\theta)$ the expected utility of the agent of type θ, given that he was offered a truth-telling contract. In other words, we have

$$R(\theta) = V_A(\theta, \theta).$$

Note that then we have, under the above first order condition,

$$R'(\theta) = \frac{d}{d\theta} V_A(\theta, \theta) = \frac{\partial}{\partial \theta} V_A(\theta, \theta) + \frac{\partial}{\partial \theta'} V_A(\theta, \theta) = \frac{\partial}{\partial \theta} V_A(\theta, \theta).$$

Using this and taking the latter derivative in (2.14), we get the following consequence of the first order condition (with a slight abuse of notation introduced by the second term):

$$k_1(\theta) = k_1(R(\theta), R'(\theta)) = -\frac{1}{\gamma_A} \sigma \frac{R'(\theta)}{R(\theta)}. \tag{2.15}$$

Using (2.14) with $\theta = \theta'$, we obtain

$$-\gamma_A R(\theta) = e^{-\gamma_A(k_0(\theta) + \frac{1}{2}k_1^2(\theta)(\frac{1}{k\sigma^2} - \gamma_A) + \frac{\theta}{\sigma}k_1(\theta))}. \tag{2.16}$$

Computing $e^{-\gamma_A k_0(\theta)}$ from this, and using $C_1 = k_0 + k_1 B_1$, we can write

$$-\frac{1}{\gamma_A} e^{-\gamma_A C_1(\theta)} = R(\theta) e^{-\gamma_A(-\frac{1}{2}k_1^2(\theta)(\frac{1}{k\sigma^2} - \gamma_A) - \frac{\theta}{\sigma}k_1(\theta) + k_1(\theta)B_1)}. \quad (2.17)$$

Using $e^{\gamma_P C_1} = (e^{-\gamma_A C_1})^{-\gamma_P/\gamma_A}$, $X_1 = \sigma B_1$ and (2.17), we can write the principal's expected utility as

$$E^{a,\theta}\big[U_P(X_1 - C_1(\theta))\big]$$
$$= -\frac{1}{\gamma_P}(-\gamma_A R(\theta))^{-\gamma_P/\gamma_A} E^{a,\theta}\Big[e^{-\gamma_P(\sigma B_1 + \frac{1}{2}k_1^2(\theta)(\frac{1}{k\sigma^2} - \gamma_A) + \frac{\theta}{\sigma}k_1(\theta) - k_1(\theta)B_1)}\Big].$$

Assume henceforth that the first order condition (2.15) is also sufficient for truth-telling (which has to be verified later when a solution is obtained). Then, the principal's utility can be computed as, abbreviating $k_1 = k_1(R(\theta), R'(\theta))$,

$$v_P(R(\theta), R'(\theta), \theta)$$
$$:= -\frac{1}{\gamma_P}(-\gamma_A R(\theta))^{-\gamma_P/\gamma_A} e^{-\gamma_P(-\frac{1}{2}\gamma_P(\sigma-k_1)^2 + \frac{1}{\sigma}(\frac{k_1}{k\sigma} + \theta)(\sigma-k_1) + \frac{1}{2}k_1^2(\frac{1}{k\sigma^2} - \gamma_A) + \frac{\theta}{\sigma}k_1)}. \quad (2.18)$$

Suppose now that the principal has a prior distribution $F(\theta)$ on the interval $[\theta_L, \theta_H]$ for θ. Also suppose that the agent of type θ needs to be given expected utility of at least $R_0(\theta)$. Then, since we have already taken into account the truth-telling constraint by expressing k_1 in terms of R, R', her problem (1.5) becomes

$$\max_{R(\theta) \geq R_0(\theta)} \int_{\theta_L}^{\theta_U} v_P(R(\theta), R'(\theta), \theta) dF(\theta).$$

This is a calculus of variations problem, which is quite hard in general. We simplify further by assuming the risk-neutral principal,

$$U_P(x) = x.$$

The results are obtained either by repeating the above arguments, or by formally replacing $\frac{1}{\gamma_P}(1 - e^{-\gamma_P x})$ by x (the limit when $\gamma_P = 0$), and noticing that maximizing with utility $\frac{1}{\gamma_P}(1 - e^{-\gamma_P x})$ is the same as maximizing with utility $-\frac{1}{\gamma_P}e^{-\gamma_P x}$. We get

$$v_P(R(\theta), R'(\theta), \theta)$$
$$= \frac{1}{\sigma}(\sigma - k_1)(a + \theta) + \frac{1}{\gamma_A}\log(-\gamma_A R(\theta)) + \frac{1}{2}k_1^2\left(\frac{1}{k\sigma^2} - \gamma_A\right) + \frac{\theta}{\sigma}k_1$$
$$= \frac{1}{\gamma_A}\log(-\gamma_A R(\theta)) - \frac{1}{2}k_1^2\left(\frac{1}{k\sigma^2} + \gamma_A\right) + \frac{k_1}{k\sigma} + \theta. \quad (2.19)$$

2.3 Hidden Type

Simplifying further, we assume that $F(\theta)$ is the uniform distribution on $[\theta_L, \theta_H]$. Introduce a *certainty equivalent*[1] of the agent's utility

$$\tilde{R}(\theta) = -\frac{1}{\gamma_A}\log(-\gamma_A R(\theta))$$

so that, by (2.15),

$$k_1(\theta) = \sigma \tilde{R}'(\theta).$$

Then, the principal's problem is equivalent to

$$\min_{R(\theta) \geq R_0(\theta)} \int_{\theta_L}^{\theta_U} \left[\tilde{R}(\theta) + \frac{1}{2}(\tilde{R}')^2(\theta)(\sigma^2\gamma_A + 1/k) - \tilde{R}'(\theta)/k \right] d\theta. \quad (2.20)$$

This can be solved using standard calculus of variations techniques, as we prove later in an analogous continuous-time model. We state here the results without the proofs. Denote

$$\beta = \frac{1/\sigma^2}{1/(k\sigma^2) + \gamma_A}.$$

We have the following

Theorem 2.3.1 *Assume the above setup and that $R_0(\theta) \equiv R_0$. Then, the principal's problem (2.20) has a unique solution as follows. Denote $\theta^* := \max\{\theta_H - 1/k, \theta_L\}$. The optimal choice of agent's certainty equivalent $\tilde{R}$ by the principal is given by*

$$\tilde{R}(\theta) = \begin{cases} R_0, & \theta_L \leq \theta < \theta^*; \\ R_0 + \beta\theta^2/2 + \beta(1/k - \theta_H)\theta - \beta(\theta^*)^2/2 - \beta(1/k - \theta_H)\theta^*, \\ & \theta^* \leq \theta \leq \theta_H. \end{cases} \quad (2.21)$$

The optimal agent's effort is given by

$$a(\theta) = \tilde{R}'(\theta)/k = \begin{cases} 0, & \theta_L \leq \theta < \theta^*; \\ \frac{\beta}{k}(1/k + \theta - \theta_H), & \theta^* \leq \theta \leq \theta_H. \end{cases} \quad (2.22)$$

The optimal contract is of the form

$$C_1(\theta) = \begin{cases} k_0(\theta), & \theta_L \leq \theta < \theta^*; \\ k_0(\theta) + \beta(1/k + \theta - \theta_H)(X_1 - X_0), & \theta^* \leq \theta \leq \theta_H. \end{cases} \quad (2.23)$$

We see that if the interval of possible type values is large, more precisely, if $\theta_H - \theta_L > 1/k$, a range of lower type agents gets no "rent" above the reservation value R_0, the corresponding contract is not incentive as it does not depend on X_1, and the effort is zero. The higher type agents get certainty equivalent $\tilde{R}(\theta)$ which is quadratically increasing in their type θ. This monotonicity is typical for hidden

---

[1] Given a utility function U, certainty equivalent CE of a random variable X is a real number such that $U(CE) = E[U(X)]$.

action problems: higher type agents may have to be paid an *"informational rent"* above the reservation value R_0 so that they would not pretend to be of lower type and try to shirk.

As the volatility σ, or A's risk aversion γ_A get larger, the contracts for the high type agents get closer to the non-incentive contract for the low type agents, as it gets harder to provide incentives anyway. On the other extreme, as $\sigma^2 \gamma_A$ tends to zero, the incentives and the rent for the high type agents get higher. If the agents are risk-neutral or $\sigma = 0$, the contract for the highest type agent $\theta = \theta_H$ is to sell the whole firm to him.

In the special case when the agent is also risk-neutral, we will show later in a continuous-time setting that the above contract is optimal among all contracts, linear or not.

2.4 Further Reading

Early papers discussing risk sharing are Borch (1962) and Wilson (1968). The hidden action setting with exponential utilities is thoroughly analyzed in Holmström and Milgrom (1987), which is also the first paper that considers the continuous-time setting. The hidden type example is a single-period version of the model in Cvitanić and Zhang (2007).

References

Borch, K.: Equilibrium in a reinsurance market. Econometrica **30**, 424–444 (1962)
Cvitanić, J., Zhang, J.: Optimal compensation with adverse selection and dynamic actions. Math. Financ. Econ. **1**, 21–55 (2007)
Holmström, B., Milgrom, P.: Aggregation and linearity in the provision of intertemporal incentives. Econometrica **55**, 303–328 (1987)
Wilson, R.: The theory of syndicates. Econometrica **36**, 119–132 (1968)

Part II
First Best: Risk Sharing Under Full Information

Chapter 3
Linear Models with Project Selection, and Preview of Results

Abstract The main message of this chapter is that for Principal–Agent problems in which volatility is controlled, as is the case in portfolio management, the first best outcome may be attainable by relatively simple contracts. These may be offered either as those in which the principal "sells" the whole output to the agent for a random "benchmark" amount, and/or as a possibly nonlinear function of the final value of the output. It is not necessary that the agent's actions are observed. Only the final value of the output and, possibly, the final value of the underlying risk process (Brownian motion) need to be observed.

3.1 Linear Dynamics and Control of Volatility

Before we present a general theory for risk sharing in diffusion models, we discuss an example with linear dynamics, and we state many interesting economic conclusions that will be proved later. With a diffusion model, the actions that the agent can naturally perform are the **control of the drift**, or "return", and the **control of the volatility**, or "risk". It turns out that the optimal drift control is of a simple form and not particularly illuminating in the setting of full information. And, it can be argued that the drift/return is not easily observed in practice. Thus, the drift control under full information is mostly interesting only for comparison with the hidden action and hidden type case. On the other hand, if we assume that the output process is observed continuously, then also the volatility (the diffusion coefficient) of the process can be observed fully, and there is no hidden action regarding volatility. Thus, when the volatility is controlled, it is reasonably realistic to consider the first best case. In particular, an important example is the one of a portfolio manager deciding how to invest the money under management. This corresponds to the control of volatility, and, in practice, there is often quite a bit of information available to the investors about the investment strategy used. More generally, control of volatility can be interpreted as the choice of risky projects by the agent.

3.1.1 The Model

Suppose that the output process X is a continuous process satisfying the dynamics

$$dX_t = [r_t X_t + \alpha v_t]dt + v_t dB_t \quad (3.1)$$
$$X_0 = x$$

where B is a Brownian motion process, which generates the information filtration $\{\mathcal{F}_t\}_{0 \leq t \leq T}$ on our probability space. Processes r_t and v_t are adapted to that filtration, with r_t being given and having the interpretation of the interest rate process, and the volatility process v_t is controlled by the agent. Constant α is fixed.

This is exactly the dynamics of a process that represents the wealth process of a portfolio which holds the amount of v_t/σ dollars at time t in a risky asset with volatility σ and with risk premium α, and holds $[X_t - v_t/\sigma]$ dollars in a risk-free asset from which one can borrow and lend at the short rate of interest r_t.

3.1.2 Risk Sharing, First Best Solution

Suppose that the manager is paid once, an amount C_T at time T, and that he maximizes utility $E[U_A(C_T)]$ over the choice of the process v_t, while the investors collectively maximize $E[U_P(X_T - C_T)]$. Under full information the two problems merge into one, that is, we have the risk sharing problem of type (1.1),

$$\max_{C_T, v} E\big[U_P(X_T - C_T) + \lambda U_A(C_T)\big]. \quad (3.2)$$

It turns out that the solution is quite simple and there is an elegant argument, as we demonstrate next. The key step is the following: introduce the "risk-neutral" density process[1]

$$Z_t = e^{-\frac{1}{2}\alpha^2 T - \alpha B_T} \quad (3.3)$$

which satisfies the dynamics

$$dZ_t = -\alpha Z_t dB_t. \quad (3.4)$$

Also denote, for a given random variable Y_t, its discounted version as

$$\bar{Y}(t) = e^{-\int_0^t r_s ds} Y_t. \quad (3.5)$$

The crucial observation is that the process $\bar{Z}X$ satisfies, by Itô's rule,

$$d(\bar{Z}_t X_t) = (\sigma v_t - \alpha X_t)\bar{Z}_t dB_t \quad (3.6)$$

and is thus a local martingale. In fact, we will assume that admissible processes v_t are only those for which $\bar{Z}X$ is a martingale and not just a local martingale, which leads to the **budget constraint**

$$x = E[\bar{Z}_T X_T]. \quad (3.7)$$

---

[1]The terminology comes from the option pricing theory.

3.1 Linear Dynamics and Control of Volatility

Not only that, but *for any $\mathcal{F}_T$-adapted random variable Y_T such that $E[|Y_T|] < \infty$ and such that $E[\bar{Z}_T Y_T] = x$, there exists an adapted portfolio process $v_t = v_t(Y_T)$ such that the corresponding wealth process X satisfies*

$$X_0 = x, \quad \text{and} \quad X_T = Y_T.$$

The processes X and v are found using the **martingale representation theorem** which says that there exists an adapted process $\varphi_t = \varphi_t(Y_T)$ such that

$$E_t[\bar{Z}_T Y_T] = E[\bar{Z}_T Y_T] + \int_0^t \varphi_s dB_s \tag{3.8}$$

where E_t denotes expectation conditional on the information available by time t. We then define

$$\bar{Z}_t X_t := E_t[\bar{Z}_T Y_T]$$

and v_t so that $\sigma v_t - \alpha X_t = \varphi_t$; see (3.6). It is then obvious that X is indeed a wealth process associated with strategy v, and such that $X_0 = x$, $X_T = Y_T$.

Going back to the problem (3.2), let us consider X_T and C_T as the variables over which we maximize, under the budget constraint (3.7). In other words, we are solving the problem

$$\max_{C_T, X_T} E\left[U_P(X_T - C_T) + \lambda U_A(C_T) - z\bar{Z}_T X_T\right] \tag{3.9}$$

where z is the Lagrange multiplier for the budget constraint. For a given utility function U introduce the inverse function of marginal utility as

$$I(z) = (U')^{-1}(z). \tag{3.10}$$

Also denote

$$\lambda' = \frac{1}{\lambda}.$$

Then, maximizing inside the expectation operator, the first order conditions for the above problem are

$$C_T = X_T - I_P(z\bar{Z}_T), \tag{3.11}$$

$$C_T = I_A(z\lambda' \bar{Z}_T). \tag{3.12}$$

In fact, we have the following result, which follows from the above arguments:

Theorem 3.1.1 *Suppose there exists a unique number z, so that for C_T, X_T determined by the first order conditions (3.11) and (3.12) the budget constraint (3.7) is satisfied, and suppose that those first order conditions are sufficient for the problem $\max_{c,x}[U_P(x-c) + \lambda U_A(c) - z\bar{Z}_T x]$ for any positive value of $\bar{Z}_T$. Then, C_T is the optimal contract payoff, and the optimal volatility action v_t is the one obtained from the martingale representation theorem as in (3.8), with*

$$X_T = Y_T = I_A(z\lambda' \bar{Z}_T) + I_P(z\bar{Z}_T). \tag{3.13}$$

Note that the contract (3.11) is a *linear contract with a benchmark*—the final output X_T that is produced is compared to a benchmark value $I_P(z\bar{Z}_T)$. This benchmark value depends on the underlying risk B_T.

Also note that the first order conditions imply the Borch rule (2.3).

3.1.3 Implementing the First Best Solution

From the purely theoretical perspective of risk sharing, the above solution is a complete solution—the principal and the agent share the profits according to C_T and P "forces" A to apply the corresponding process v_t as the action. However, from the practical perspective, it is of interest to study whether there are simple contracts which would induce the agent to apply the first best action without being forced to, and/or without being monitored. This is, in fact, possible, and potentially in more than one way. Here, is the first result of this type:

Theorem 3.1.2 *When offered the contract* $C_T = X_T - I_P(z\bar{Z}_T)$ *from Theorem 3.1.1, and under its assumptions, the agent will optimally choose action* v_t *which will result in the first best wealth* X_T *of* (3.13).

Proof The agent's problem is

$$\max_{X_T} E\left[U_A\left(X_T - I_P(z\bar{Z}_T)\right)\right] - y E[\bar{Z}_T X_T] \tag{3.14}$$

where y is a Lagrange multiplier. The first order condition is

$$X_T = I_A(y\bar{Z}_T) + I_P(z\bar{Z}_T).$$

By the assumptions of Theorem 3.1.1, the only choice of y which satisfies the budget constraint is $y = z\lambda'$, which means that the agent will act to attain the same wealth as in (3.13). □

Note that the benchmark value $I_P(z\bar{Z}_T)$ is of the form which is optimal for the optimal portfolio selection problem or *Merton's problem* $\max_v E[U_P(X_T)]$. The corresponding form of optimal C_T is what drives the agent to apply the controls which are first best for the principal. The principal is able to get the utility of the same form $E[U_P(I_P(z\bar{Z}_T))]$ as if maximizing without the agent's presence, except that the value of z is adjusted to account for the agent's share.

Also note that, except for the constant z, the form of the contract does not require that the principal knows the agent's utility, which may be hard to determine in practice. Moreover, there is no need for the principal to monitor the process X_t or B_t—it's sufficient for her to observe the final value B_T. However, this itself may be an unrealistic assumption, unless the process B_t is fully correlated with some observable process, such as a price of a stock index, for example. In the next section we describe optimal contracts that depend only on the final value X_T of the output.

3.1.4 Optimal Contract as a Function of Output

Suppose the agent is offered a contract $F(X_T)$ for some function F. In practice, a contract like that is at least a part of the compensation package of an executive, in the form of stock shares and call options on the stock. Assuming that this is the only compensation he gets, the agent needs to solve the problem

$$\max_{X_T}\{E[U_A(F(X_T)) - y\bar{Z}_T X_T]\} \tag{3.15}$$

where y is a Lagrange multiplier for the budget constraint. The first order condition is

$$U'_A(F(X_T))F'(X_T) = y\bar{Z}_T. \tag{3.16}$$

Suppose the principal is trying to find a function F so that she gets the first best expected utility $E[U_P(X_T - F(X_T))] = E[U_P(I_P(z\bar{Z}_T))]$ where z is as in Theorem 3.1.1. This will be achieved if $U'_P(X_T - F(X_T)) = z\bar{Z}_T$. If we take (3.16) into account, we see that it is sufficient to have

$$\frac{U'_P(x - F(x))}{U'_A(F(x))} = \frac{z}{y}F'(x). \tag{3.17}$$

Therefore, we get the following result.

Theorem 3.1.3 *Assume the following:*

(i) *there exists a solution $F(x) = F(x; y)$ to the ordinary differential equation (ODE) (3.17) for a constant y such that (ii)–(v) below are satisfied;*
(ii) *the first order condition (3.16) is sufficient for maximizing $U_A(F(X_T)) - y\bar{Z}_T X_T$ a.s.;*
(iii) *there exists a unique random variable $X_T = X_T(y)$ that solves the first order condition;*
(iv) *the constant y can be chosen so that the budget constraint $E[\bar{Z}_T X_T] = x$ is satisfied;*
(v) *agent's expected utility $E[U_A(F(X_T))]$ is the same as in the corresponding first best solution, that is, equal to $E[U_A(I_A(z\lambda'\bar{Z}_T))]$.*

Then, the principal and the agent can attain the first best utilities when the agent is offered the contract $C_T = F(X_T)$.

Note that the action process v chosen by the agent is not necessarily the same as in the first best case of Theorem 3.1.1, but the expected utilities are the same.

Remark 3.1.1 (i) The theorem remains the same if there is a unique maximizer X_T a.s. in (ii), even if it is not determined by the first order conditions.

(ii) The theorem also remains valid in case B, α and v are multidimensional processes. In that case Z_t in the proof has to be replaced by

$$Z_t = e^{-\frac{1}{2}\int_0^t \|\alpha_s^2\|ds - \int_0^t \alpha_s dB_s}$$

where $\|\alpha\|$ is the norm of the vector $\alpha = (\alpha^1, \ldots, \alpha^d)$ and $\alpha \, dB$ denotes the inner product $\sum_i \alpha^i \, dB^i$.

3.1.5 Examples

Example 3.1.1 (Exponential Utilities) Assume

$$U_A(x) = -\frac{1}{\gamma_A} e^{-\gamma_A x} \quad \text{and} \quad U_P(x) = -\frac{1}{\gamma_P} e^{-\gamma_P x}.$$

Then,

$$I_A(z) = -\frac{1}{\gamma_A} \log(z) \quad \text{and} \quad I_P(z) = -\frac{1}{\gamma_P} \log(z).$$

Thus, in the first best solution, the agent will choose volatility v such that

$$X_T = -\frac{1}{\gamma_A} \log(z\lambda' \bar{Z}_T) - \frac{1}{\gamma_P} \log(z\bar{Z}_T).$$

Solving for $\log(\bar{Z}_T)$ in terms of X_T, and using $C_T = I_A(z\lambda' \bar{Z}_T)$, it is easy to check that the first best payoff C_T turns out to be a linear function of X_T, of the form

$$C_T = c + \frac{\gamma_P}{\gamma_A + \gamma_P} X_T.$$

Not only that, it can then easily be verified that the same linear function satisfies the conditions of Theorem 3.1.3, hence this linear contract implements the first best.

Example 3.1.2 (Power Utilities) Assume now, with $0 < \gamma_A < 1, 0 < \gamma_P < 1$,

$$U_A(x) = \frac{1}{\gamma_A} x^{\gamma_A} \quad \text{and} \quad U_P(x) = \frac{1}{\gamma_P} x^{\gamma_P}.$$

Then,

$$I_A(z) = z^{\frac{1}{\gamma_A - 1}} \quad \text{and} \quad I_P(z) = z^{\frac{1}{\gamma_P - 1}}.$$

Thus, in the first best solution, the agent will choose volatility v such that

$$X_T = (z\lambda' \bar{Z}_T)^{\frac{1}{\gamma_A - 1}} + (z\bar{Z}_T)^{\frac{1}{\gamma_P - 1}}.$$

In particular, if the risk aversions are the same, $\gamma_A = \gamma_P$, we get

$$\bar{Z}_T^{\frac{1}{\gamma_A - 1}} = \frac{X_T}{(z\lambda')^{\frac{1}{\gamma_A - 1}} + z^{\frac{1}{\gamma_A - 1}}}$$

and

$$C_T = X_T - I_P(z\bar{Z}_T) = (z\lambda')^{\frac{1}{\gamma_A - 1}} \frac{X_T}{(z\lambda')^{\frac{1}{\gamma_A - 1}} + z^{\frac{1}{\gamma_A - 1}}}.$$

3.1 Linear Dynamics and Control of Volatility

That is, if the principal and agent have the same power utility, and they both behave optimally, the payoff turns out to be linear at time T. Not only that, it can then easily be verified that the same linear function satisfies the conditions of Theorem 3.1.3, hence the linear contract implements the first best.

Problem 3.1.1 (Nonlinear Payoff as the ODE Solution) Assume that

$$U_A(x) = \log(x) \quad \text{and} \quad U_P(x) = -e^{-x}.$$

Recall the special function $Ei(x)$, called *exponential integral*, defined by

$$Ei(x) := -\int_{-x}^{\infty} \frac{e^{-t}}{t} dt. \tag{3.18}$$

This is a well defined function for $x < 0$, and it is continuous and decreases from 0 to $-\infty$.

(i) Show that the solution $F(x)$ to the ODE (3.17) is determined by

$$Ei(-F(x)) = -\frac{y}{z}e^{-x} + c \tag{3.19}$$

where c is a non-positive constant.

(ii) Argue that F is continuous and increasing, with

$$F(-\infty) = 0 \quad \text{and} \quad F(\infty) \in (0, \infty].$$

If $c < 0$ then $F(\infty) < \infty$, if $c = 0$ then $F(\infty) = \infty$.

(iii) Verify the remaining assumptions of Theorem 3.1.3, showing that the contract $F(X_T)$ is optimal. It may be helpful to argue that $Ei(y) > e^y/y$, and that this implies $x\frac{d}{dx}(Ei^{-1})(x) < 1$, which, in turn, implies $F'(x) < 1$.

Problem 3.1.2 (Call Option Contract) Consider numbers K and n such that $0 < K < X_0$, $n < 1$. Suppose that

$$U_A(x) = c\log(x)$$

$$U_P(x) = \begin{cases} b\log(x - K) & \text{if } x > K \\ -\infty & \text{if } x \leq K \end{cases}$$

for some $c > 0, b > 0$, and consider the call option contract

$$C_T = n(X_T - K)^+.$$

(i) Show that the agent will act so that

$$X_T = \frac{c}{y\bar{Z}_T} + K \tag{3.20}$$

for the value of y for which $E[\bar{Z}_T X_T] = X_0$, which is equivalent to

$$\frac{c}{y} = X_0 - KE[\bar{Z}_T]. \tag{3.21}$$

(ii) Find the values of n and K for which the agent and the principal attain their first best expected utilities, that is, the same expected utilities as in Theorem 3.1.1.

We see in this example that we can interpret the strike price K as the lower bound on the wealth value that the principal is willing to tolerate. The higher this bound, the higher the strike price K will be.

3.2 Further Reading

The setting of this section is used in Ou-Yang (2003) to study optimal compensation of portfolio managers. The presentation here follows mostly Cadenillas et al. (2007). Larsen (2005) computes numerically an optimal contract in the context of Theorem 3.1.3, using different methods. In a static setting, Ross (1973) shows that when the agent and the principal have the same power utility, the optimal contract is linear. A very general risk-sharing framework with several agents and recursive utilities is considered in Duffie et al. (1994) and Dumas et al. (2000).

References

Cadenillas, A., Cvitanić, J., Zapatero, F.: Optimal risk-sharing with effort and project choice. J. Econ. Theory **133**, 403–440 (2007)

Duffie, D., Geoffard, P.Y., Skiadas, C.: Efficient and equilibrium allocations with stochastic differential utility. J. Math. Econ. **23**, 133–146 (1994)

Dumas, B., Uppal, R., Wang, T.: Efficient intertemporal allocations with recursive utility. J. Econ. Theory **93**, 240–259 (2000)

Larsen, K.: Optimal portfolio delegation when parties have different coefficients of risk aversion. Quant. Finance **5**, 503–512 (2005)

Ou-Yang, H.: Optimal contracts in a continuous-time delegated portfolio management problem. Rev. Financ. Stud. **16**, 173–208 (2003)

Ross, S.A.: The economic theory of agency: the principal's problem. Am. Econ. Rev. **63**, 134–139 (1973). Papers and Proceedings of the Eighty-fifth Annual Meeting of the American Economic Association

Chapter 4
The General Risk Sharing Problem

Abstract In this chapter we consider general diffusion dynamics for the output process with a general cost function depending on the agent's actions and/or the output values. The main qualitative conclusions from the case of linear drift dynamics of the previous section still hold true with nonlinear drift dynamics—a linear benchmark contract is optimal, and it implements the first best actions. However, the benchmark is now harder to identify, and it may be obtained either as an adjoint process which is a part of a solution to an FBSDE, or as a solution to an appropriate dual problem. The model is also extended to include consumption processes of the principal and the agent, and to the case of so-called recursive utilities, that generalize standard utility functions.

4.1 The Model and the PA Problem

Let $\{B_t\}_{t \geq 0}$ be a d-dimensional Brownian motion on a probability space $(\Omega, \mathcal{F}, P)$ and denote by $\mathbf{F} := \{\mathcal{F}_t\}_{t \leq T}$ its augmented filtration on the interval $[0, T]$. The output process is denoted $X = X^{u,v}$ and its dynamics are given by

$$dX_t = b(t, X_t, u_t, v_t)dt + v_t dB_t \tag{4.1}$$

where (u, v) take values in $A_1 \times A_2 \subset \mathbb{R}^m \times \mathbb{R}^d$, and b is a function taking values in $\mathbb{R}$, possibly random and such that, as a process, it is $\mathbf{F}$-adapted. The notation xy for two vectors $x, y \in \mathbb{R}^d$ indicates the inner product.

The principal offers the agent compensation $C_T = C(\omega, X)$ at time T, where $C : \Omega \times C[0, T] \to A_3 \subset \mathbb{R}$ is a mapping such that C_T is $\mathcal{F}_T$ measurable. Introduce the accumulated cost of the agent,

$$G_T = G_T^{u,v} := \int_0^T g(t, X_t, u_t, v_t)dt.$$

The risk sharing problem is

$$\max_{C,u,v} J(C_T, u, v) := \max_{C,u,v} E\big[U_P(X_T - C_T) + \lambda U_A(C_T, G_T)\big]. \tag{4.2}$$

The functions U_A and U_P are utility functions of the agent and the principal. The function g is a cost function. Typical cases studied in the literature are the *separable* utility case with $U_A(x, y) = U_A(x) - y$, and the *non-separable utility case* with $U_A(x, y) = U_A(x - y)$, where, with a slight abuse of notation, we use the same notation U_A also for the function of one argument only.

Notice that C_T can be optimized ω-wise. For a given $\lambda > 0$, denote

$$U_\lambda(X, G) := \sup_{C \in A_3} \left[U_P(X - C) + \lambda U_A(C, G) \right]. \tag{4.3}$$

Then, (4.2) becomes

$$V_\lambda = \sup_{(u,v) \in \mathcal{A}} E\left[U_\lambda(X_T, G_T) \right]. \tag{4.4}$$

We will specify the admissible set $\mathcal{A}$ for (u, v) in the next subsection.

Remark 4.1.1 (i) If $\hat{C}$ is the optimal contract and takes values in the interior of A_3, then

$$U'_P(X - \hat{C}) = \lambda \partial_C U_A(\hat{C}, G). \tag{4.5}$$

This is simply the Borch rule for risk sharing.

(ii) If A_3 is compact, or if U_P is concave and U_A is concave in C and $U_\lambda < \infty$, then there exists a function $I(X, G)$ such that $\hat{C} := I(X, G)$ is optimal for the problem (4.3). Moreover,

$$U_\lambda(X, G) = U_P(X - I(X, G)) + \lambda U_A(I(X, G), G). \tag{4.6}$$

Assume further that the above function I is differentiable. Then, by (4.6) and (4.5) we obtain

$$\begin{aligned} \partial_X U_\lambda(X, G) &= U'_P(X - I(X, G)), \\ \partial_G U_\lambda(X, G) &= \lambda \partial_G U_A(I(X, G), G). \end{aligned} \tag{4.7}$$

4.2 Necessary Conditions for Optimality

In order to analyze the problem (4.4), we apply the method of the stochastic calculus of variations, which leads to the necessary conditions that go under the name the *stochastic maximum principle*, an analog to Pontryagin's maximum principle in deterministic control. The method basically consists in finding the limits of finite differences of the processes involved, essentially differentiating the objective function, in order to get first order conditions for optimality. Some references for this method are given in the "Further Reading" section. However, we derive general results from scratch in Part V of the book, and apply them here to the problem at hand.

4.2 Necessary Conditions for Optimality

4.2.1 FBSDE Formulation

We can write the problem (4.4) as

$$V_\lambda = \sup_{(u,v)\in\mathcal{A}} Y_0^{u,v}, \qquad (4.8)$$

where

$$Y_0^{u,v} = E\big[U_\lambda\big(X_T^{u,v}, G_T^{u,v}\big)\big].$$

This formulation falls under the framework of Sect. 10.2, with higher dimensions. Indeed, we can consider a quadruple $(X^{u,v}, G^{u,v}, Y^{u,v}, Z^{u,v})$ of adapted processes satisfying the following decoupled Forward-Backward Stochastic Differential Equation (FBSDE):

$$\begin{cases} X_t^{u,v} = x + \int_0^t b(s, X_s^{u,v}, u_s, v_s)ds + \int_0^t v_s dB_s; \\ G_t^{u,v} = \int_0^t g(s, X_s^{u,v}, u_s, v_s)ds; \\ Y_t^{u,v} = U_\lambda(X_T^{u,v}, G_T^{u,v}) - \int_t^T Z_s^{u,v} dB_s. \end{cases} \qquad (4.9)$$

The first two equations are in the form of standard SDEs, and the third one is a BSDE: instead of an initial condition, a terminal condition

$$Y_T^{u,v} = U_\lambda\big(X_T^{u,v}, G_T^{u,v}\big)$$

is imposed, and the solution consists not just of process Y, but of a pair of adapted processes $(Y^{u,v}, Z^{u,v})$.

In light of Assumptions 10.2.1 and 10.2.2, we assume

Assumption 4.2.1

(i) b, g are progressively measurable in all the variables and $\mathbb{F}$-adapted.
(ii) b, g are continuously differentiable in x, and U_λ is continuously differentiable in (X, G), with uniformly bounded derivatives.
(iii) b, g are continuously differentiable in (u, v), and for $\varphi = b, g$,

$$\big|\varphi_u(t, x, u, v)\big| + \big|\varphi_v(t, x, u, v)\big| \le C\big[1 + \big|\varphi_u(t, 0, u, v)\big| \\ + \big|\varphi_v(t, 0, u, v)\big| + |x|\big].$$

Assumption 4.2.2 The admissible set consists of $\mathbb{F}$-progressively measurable $A_1 \times A_2$-valued processes (u, v) satisfying:

(i) For each $(u, v) \in \mathcal{A}$, and $\varphi = b, g$,

$$E\bigg\{\bigg(\int_0^T \big[|\varphi(t, 0, u_t, v_t)| + |\partial_u \varphi(t, 0, u_t, v_t)| + |\partial_v \varphi(t, 0, u_t, v_t)|\big]dt\bigg)^2 \\ + \int_0^T |v_t|^2 dt + \big|U_\lambda(0, 0)\big|^2\bigg\} < \infty. \qquad (4.10)$$

(ii) $\mathcal{A}$ is locally convex, in the sense of Assumption 10.1.7(ii).
(iii) For any (u, v) and $(u^\varepsilon, v^\varepsilon)$ defined in Assumption 10.1.7(ii), $\varphi(t, 0, u^\varepsilon, v^\varepsilon)$, $\partial_u \varphi(t, 0, u^\varepsilon, v^\varepsilon)$, and $\partial_v \varphi(t, 0, u^\varepsilon, v^\varepsilon)$ are integrable uniformly in $\varepsilon \in [0, 1]$ in the sense of (4.10).

4.2.2 Adjoint Processes

It turns out that the necessary conditions of optimality can be more elegantly presented if one introduces appropriate *adjoint processes*, which are defined in terms of BSDEs. One has to look at the proofs in Part V of the book to understand why the adjoint processes are chosen in the way we choose them below, or one can consult standard literature on such problems.

In our case, the appropriate adjoint process corresponding to (10.24) are high-dimensional, and introduced as follows:

$$\Gamma_t^{u,v} = 1;$$

$$\bar{Y}_t^{1,u,v} = \partial_x U_\lambda(X_T^{u,v}, G_T^{u,v}) - \int_t^T \bar{Z}_s^{1,u,v} dB_s$$
$$+ \int_t^T \left[\partial_x b(s, X_s^{u,v}, u_s, v_s) \bar{Y}_s^{1,u,v} + \partial_x g(s, X_s^{u,v}, u_s, v_s) \bar{Y}_s^{2,u,v}\right] ds; \quad (4.11)$$

$$\bar{Y}_t^{2,u,v} = \partial_y U_\lambda(X_T^{u,v}, G_T^{u,v}) - \int_t^T \bar{Z}_s^{2,u,v} dB_s.$$

We can ignore the trivial adjoint process of the first equation, needed in the more general framework analyzed in Part V. The other two equations are BSDEs, whose solution is a pair of adapted processes $(\bar{Y}^{i,u,v}, \bar{Z}^{i,u,v})$, $i = 1, 2$. The first one has as its terminal value the marginal, in output X, utility of the "joint welfare" utility U_λ, and it also accounts for the sensitivities $\partial_x b$, $\partial_x g$ of the output drift to the output value, and of the cost to the output value. The second one is simply the conditional expected value of the marginal, in cost G, utility of the joint welfare utility U_λ.

Note that in the separable case $U_A(x, y) = U_A(x) - y$ we have $\partial_y U_\lambda(x, y) = -\lambda$, $\bar{Y}_t^{2,u,v} \equiv -\lambda$, $\bar{Z}_t^{2,u,v} \equiv 0$, and effectively only one adjoint process is needed.

4.2.3 Main Result

We can now apply Theorem 10.2.5 to get the main result of our general theory for risk sharing:

Theorem 4.2.3 (Necessary Conditions for Optimality) *Let Assumptions* 4.2.1 *and* 4.2.2 *hold.*

4.2 Necessary Conditions for Optimality

(i) *If $(u^*, v^*) \in \mathcal{A}$ is an optimal control of the optimization problem (4.4) and (u^*, v^*) is an interior point of $\mathcal{A}$, then*

$$\bar{Y}_t^{1,u^*,v^*} b_u\big(t, X_t^{u^*,v^*}, u_t^*, v_t^*\big) + \bar{Y}_t^{2,u^*,v^*} g_u\big(t, X_t^{u^*,v^*}, u_t^*, v_t^*\big) = 0;$$

$$\bar{Y}_t^{1,u^*,v^*} b_v\big(t, X_t^{u^*,v^*}, u_t^*, v_t^*\big) + \bar{Y}_t^{2,u^*,v^*} g_v\big(t, X_t^{u^*,v^*}, u_t^*, v_t^*\big) + \bar{Z}_t^{1,u^*,v^*} = 0. \tag{4.12}$$

(ii) *Assume further that there exist unique functions $I_i(t, x, \bar{y}, \bar{z}^1)$, $i = 1, 2$, such that*

$$\bar{y}^1 b_u(t, x, I_1, I_2) + \bar{y}^2 g_u(t, x, I_1, I_2) = 0;$$
$$\bar{y}^1 b_v(t, x, I_1, I_2) + \bar{y}^2 g_v(t, x, I_1, I_2) + \bar{z}^1 = 0. \tag{4.13}$$

Denote

$$\varphi^*(t, x, \bar{y}, \bar{z}^1) := \varphi\big(t, x, I_1(t, x, \bar{y}, \bar{z}^1), I_2(t, x, \bar{y}, \bar{z}^1)\big) \quad \text{for any function } \varphi. \tag{4.14}$$

Then, $(X^*, Y^*, \bar{Y}^*, Z^*, \bar{Z}^*) := (X^{u^*,v^*}, Y^{u^*,v^*}, \bar{Y}^{u^*,v^*}, Z^{u^*,v^*}, \bar{Z}^{u^*,v^*})$ satisfies the following coupled FBSDE:

$$\begin{cases} X_t^* = x + \int_0^t b^*\big(s, X_s^*, \bar{Y}_s^*, \bar{Z}_s^{1,*}\big) ds + \int_0^t I_2\big(s, X_s^*, \bar{Y}_s^*, \bar{Z}_s^{1,*}\big) dB_s; \\[6pt]
G_t^* = \int_0^t g^*\big(s, X_s^*, \bar{Y}_s^*, \bar{Z}_s^{1,*}\big) ds; \\[6pt]
Y_t^* = U_\lambda(X_T^*, G_T^*) - \int_t^T Z_s^* dB_s; \\[6pt]
\bar{Y}_t^{1,*} = \partial_x U_\lambda(X_T^*, G_T^*) - \int_t^T \bar{Z}_s^{1,*} dB_s \\[2pt]
\qquad\quad + \int_t^T \big[(\partial_x b)^*\big(s, X_s^*, \bar{Y}_s^*, \bar{Z}_s^{1,*}\big) \bar{Y}_s^{1,*} + (\partial_x g)^*\big(s, X_s^*, \bar{Y}_s^*, \bar{Z}_s^{1,*}\big) \bar{Y}_s^{2,*}\big] ds; \\[6pt]
\bar{Y}_t^{2,*} = \partial_G U_\lambda(X_T^*, G_T^*) - \int_t^T \bar{Z}_s^{2,*} dB_s, \end{cases} \tag{4.15}$$

and the optimal control satisfies

$$u_t^* = I_1\big(t, X_t^*, \bar{Y}_t^*, \bar{Z}_t^{1,*}\big), \qquad v_t^* = I_2\big(t, X_t^*, \bar{Y}_t^*, \bar{Z}_t^{1,*}\big). \tag{4.16}$$

Remark 4.2.4 (i) The conditions (4.12) set dynamic restrictions on the sensitivities of the cost function and of the drift function with respect to the actions u, v. Together with the static, Borch rule condition (4.5) and the FBSDE system for output process X and the adjoint processes, we call these conditions *necessary conditions* for our problem.

(ii) The second part of the theorem states when the necessary conditions can be written as a coupled FBSDE. This may be useful because it can help find numerical solutions (or analytic solutions, in case they exist). In particular, in Markovian models it is possible to transform an FBSDE into a deterministic PDE.

4.3 Sufficient Conditions for Optimality

We introduce the following Hamiltonian function as in (10.32):

$$H(t, x, \bar{y}, \bar{z}^1, u, v) := \bar{y}^1 b(t, x, u, v) + \bar{y}^2 g(t, x, u, v) + \bar{z}^1 v. \quad (4.17)$$

Assumption 4.3.1

(i) The function U_λ is $\mathcal{F}_T$-measurable and is uniformly Lipschitz continuous and concave in (X, G);
(ii) b and g are $\mathbb{F}$-progressively measurable in all the variables, continuously differentiable in x with uniformly bounded derivatives, and $b, g, \partial_x b, \partial_x g$ are continuous in (u, v).
(iii) The sets A_1, A_2 are convex and the admissible set $\mathcal{A}$ is a set of $\mathbb{F}$-adapted processes (u, v) taking values in $A_1 \times A_2$ satisfying:

$$E\left\{\left(\int_0^T \left[|b(t, 0, u_t, v_t)| + |g(t, 0, u_t, v_t)|\right] dt\right)^2\right\} < \infty.$$

(iv) The Hamiltonian H is concave in (x, u, v) for all $(\bar{y}, \bar{z}^1)$ in the set of possible values the adjoint processes $(\bar{Y}, \bar{Z}^1)$ could take, and there exist functions $I_1(t, x, \bar{y}, \bar{z}^1)$ taking values in A_1 and $I_2(t, x, \bar{y}, \bar{z}^1)$ taking values in A_2 such that

$$H(t, x, \bar{y}, \bar{z}^1, I_1, I_2) = \sup_{(u,v) \in A_1 \times A_2} H(t, x, \bar{y}, \bar{z}^1, u, v). \quad (4.18)$$

Applying Theorem 10.2.9, we obtain

Theorem 4.3.2 *Assume*

(i) *Assumption* 4.3.1 *holds*;
(ii) *the FBSDE* (4.15) *has a solution* $(X^*, G^*, Y^*, \bar{Y}^*, Z^*, \bar{Z}^*)$, *where* φ^* *is defined by* (4.14) *for the functions* I_1, I_2 *in Assumption* 4.3.1(iv);
(iii) *the process* (u^*, v^*) *defined by* (4.16) *is in* $\mathcal{A}$.

Then, $V(\lambda) = Y_0^*$ *and* (u^*, v^*) *is an optimal control*.

Remark 4.3.3 (i) Assume

$$U_P' > 0, \quad \partial_X U_A > 0, \quad \partial_G U_A < 0, \quad \partial_x g < 0. \quad (4.19)$$

Then, $\partial_X U_\lambda > 0, \partial_G U_\lambda < 0$, and one can easily see that

$$\bar{Y}^{1,u,v} > 0, \quad \bar{Y}^{2,u,v} < 0. \quad (4.20)$$

(ii) In addition to (4.19), assume further that b and $-g$ are jointly concave in (x, u, v). Then, H is jointly concave in (x, u, v) for $\bar{y}_1 > 0 > \bar{y}_2$ and thus, in light of (4.20), H satisfies the first condition in Assumption 4.3.1(iv). Moreover, in this case, assuming sufficient differentiability of b and g,

$$\begin{bmatrix} \frac{\partial}{\partial u}(\bar{y}_1 b_u + \bar{y}_2 g_u) & \frac{\partial}{\partial v}(\bar{y}_1 b_u + \bar{y}_2 g_u) \\ \frac{\partial}{\partial u}(\bar{y}_1 b_v + \bar{y}_2 g_v) & \frac{\partial}{\partial v}(\bar{y}_1 b_v + \bar{y}_2 g_v) \end{bmatrix} = \bar{y}_1 \begin{bmatrix} b_{uu} & b_{uv} \\ b_{uv} & b_{vv} \end{bmatrix} + \bar{y}_2 \begin{bmatrix} g_{uu} & g_{uv} \\ g_{uv} & g_{vv} \end{bmatrix}$$

is negative definite for $\bar{y}_1 > 0 > \bar{y}_2$. Thus, under mild additional conditions, the functions I_1, I_2 in (4.13) exist uniquely.

4.4 Optimal Contracts

As in our introductory sections, we now look for contracts that are *incentive compatible*, in the sense that they induce the agent to implement the optimal, first best action (u^*, v^*). We will see that, again, there exists a linear benchmark contract that does that.

4.4.1 Implementing the First Best Solution

Clearly, if the function I in Remark 4.1.1(ii) exists, then the contract $C_T := I(X_T, G_T)$ is an optimal contract. However, optimal contracts are not unique, nor are they necessarily incentive compatible. We next identify an optimal contract which is incentive compatible, and unique among the benchmark contracts that are linear in X_T.

We assume that all the conditions of Theorem 4.3.2 and Remark 4.1.1 hold. Then, by (4.7) we obtain

$$C_T^* = X_T^* - I_P(\bar{Y}_T^{1,*}), \quad \text{where } I_P := (U_P')^{-1}. \tag{4.21}$$

This is of the form of a familiar linear benchmark contract, as in the introductory section on risk sharing. We claim that it is this contract that is incentive compatible. First, we introduce the following

Definition 4.4.1 We say that an admissible triple $(\hat{C}_T, \hat{u}, \hat{v})$ is implementable if there exists a compensation function C such that the agent optimally chooses $\hat{u}, \hat{v}$ and that $C(\omega, X^{\hat{u},\hat{v}}) = \hat{C}_T$.

Here is a result generalizing Theorem 3.1.2.

Proposition 4.4.2 *Assume all the conditions in Theorem 4.3.2 and Remark 4.1.1 hold. Then, the triple (C_T^*, u^*, v^*) is implementable with the compensation function*

$$C(\omega, X) = X_T - I_P(\bar{Y}_T^{1,*}(\omega)). \tag{4.22}$$

Proof Since $\lambda > 0$, we can take the agent's objective function to be given by

$$\sup_{(u,v) \in \mathcal{A}} \lambda E\left[U_A\left(X_T^{u,v} - I_P(Y_T^{1,*}), G_T^{u,v}\right)\right] = \sup_{u,v} \tilde{Y}_0^{u,v}$$

where

$$\tilde{Y}_t^{u,v} = \lambda U_A(X_T^{u,v} - I_P(Y_T^{1,*}), G_T^{u,v}) - \int_t^T \tilde{Z}_s^{u,v} dB_s.$$

This is the same system as (4.9), except that we replace the terminal condition $U_\lambda(X_T^{u,v}, G_T^{u,v})$ of $Y^{u,v}$ with $\lambda U_A(X_T^{u,v} - I_P(Y_T^{1,*}), G_T^{u,v})$. Following exactly the same arguments, we derive a system analogous to (4.15):

$$\begin{cases}
\tilde{X}_t^* = x + \int_0^t b^*(s, \tilde{X}_s^*, \tilde{Y}_s^{1,*}, \tilde{Y}_s^{2,*}, \tilde{Z}_s^{1,*}) ds \\
\qquad + \int_0^t I_2(s, \tilde{X}_s^*, \tilde{Y}_s^{1,*}, \tilde{Y}_s^{2,*}, \tilde{Z}_s^{1,*}) dB_s; \\
\tilde{G}_t^* = \int_0^t g^*(s, \tilde{X}_s^*, \tilde{Y}_s^{1,*}, \tilde{Y}_s^{2,*}, \tilde{Z}_s^{1,*}) ds; \\
\tilde{Y}_t^* = \lambda U_A(\tilde{X}_T^* - I_P(Y_T^{1,*}), \tilde{G}_T^*) - \int_t^T \tilde{Z}_s^* dB_s; \\
\tilde{Y}_t^{1,*} = \lambda \partial_C U_A(\tilde{X}_T^* - I_P(Y_T^{1,*}), \tilde{G}_T^*) - \int_t^T \tilde{Z}_s^{1,*} dB_s \\
\qquad + \int_t^T [(\partial_x b)^*(s, \tilde{X}_s^*, \tilde{Y}_s^{1,*}, \tilde{Y}_s^{2,*}, \tilde{Z}_s^{1,*}) \tilde{Y}_s^{1,*} \\
\qquad + (\partial_x g)^*(s, \tilde{X}_s^*, \tilde{Y}_s^{1,*}, \tilde{Y}_s^{2,*}, \tilde{Z}_s^{1,*}) \tilde{Y}_s^{2,*}] ds; \\
\tilde{Y}_t^{2,*} = \lambda \partial_G U_A(\tilde{X}_T^* - I_P(Y_T^{1,*}), \tilde{G}_T^*) - \int_t^T \tilde{Z}_s^{2,*} dB_s.
\end{cases} \quad (4.23)$$

We emphasize that the argument $Y_T^{1,*}$ of I_P is the random variable from the system (4.15). By (4.5) and (4.7), one can check straightforwardly that

$$(\tilde{X}^*, \tilde{G}^*, \tilde{Y}^{1,*}, \tilde{Y}^{2,*}, \tilde{Z}^{1,*}, \tilde{Z}^{2,*}) = (X^*, G^*, \bar{Y}^{1,*}, \bar{Y}^{2,*}, \bar{Z}^{1,*}, \bar{Z}^{2,*}).$$

Therefore, the optimal contract value of (4.22) is $X_T^* - I_P(\bar{Y}_T^{1,*}) = C_T^*$. □

4.4.2 On Uniqueness of Optimal Contracts

The contract $C(\omega, X) = X_T - I_P(Y_T^{1,*})$ implements the first best solution, but may not be the only contract that does that. We consider now uniqueness in a reduced space of admissible contracts. Ideally, one would like to find optimal contracts in the form $F(X_T)$, for some deterministic function F. For example, recall that we found contracts of such form in Sect. 3.1.4, which attain the first best expected utilities in special cases. In general, it is not possible to do that. We choose to consider the following family of contracts, generalizing the linear benchmark contracts:

4.4 Optimal Contracts

Definition 4.4.3 We say that a contract is of the Increasing State-Contingent Compensation (ISCC) type, if it is of the form $F(X_T) - D_T$ where $F(x)$ is a deterministic function, such that $F'(x) > 0$, and D_T is a given $\mathcal{F}_T$ measurable random variable.

The contract $C(\omega, X) = X_T - I_P(Y_T^{1,*})$ is of the ISCC type, and we will show that it is the only contract which implements the first best solution in that family. The ISCC type contracts are consistent with real-world use, where the agent is often paid based on the performance of an underlying process X compared to a benchmark.

We have the following result:

Proposition 4.4.4 *Under the assumptions of Proposition 4.4.2 and if Range$(X_T^*) = \mathbb{R}$ and $b_u, g_u, \partial_C U_A, \partial_G U_A \neq 0$, the contract $C(\omega, X) = X_T - I_P(Y_T^{1,*})$ is the only ISCC type contract that implements the first best solution.*

Proof Introduce the agent's objective function, given a contract $F(X_T) - D_T$:

$$\tilde{Y}_t^{u,v} := E_t\{U_A(F(X_T^{u,v}) - D_T, G_T^{u,v})\}$$
$$= U_A(F(X_T^{u,v}) - D_T, G_T^{u,v}) - \int_t^T \tilde{Z}_s^{u,v} dB_s.$$

The agent's optimization problem induces the following system analogous to (4.23):

$$\begin{cases} \tilde{X}_t^* = x + \int_0^t b^*(s, \tilde{X}_s^*, \tilde{Y}_s^{1,*}, \tilde{Y}_s^{2,*}, \tilde{Z}_s^{1,*}) ds \\ \qquad + \int_0^t I_2(s, \tilde{X}_s^*, \tilde{Y}_s^{1,*}, \tilde{Y}_s^{2,*}, \tilde{Z}_s^{1,*}) dB_s; \\ \tilde{G}_t^* = \int_0^t g^*(s, \tilde{X}_s^*, \tilde{Y}_s^{1,*}, \tilde{Y}_s^{2,*}, \tilde{Z}_s^{1,*}) ds; \\ \tilde{Y}_t^* = \lambda U_A(\tilde{X}_T^* - I_P(Y_T^{1,*}), \tilde{G}_T^*) - \int_t^T \tilde{Z}_s^* dB_s; \\ \tilde{Y}_t^{1,*} = \partial_C U_A(F(\tilde{X}_T^*) - D_T, \tilde{G}_T^*) F'(\tilde{X}_T^*) - \int_t^T \tilde{Z}_s^{1,*} dB_s \\ \qquad + \int_t^T \big[(\partial_x b)^*(s, \tilde{X}_s^*, \tilde{Y}_s^{1,*}, \tilde{Y}_s^{2,*}, \tilde{Z}_s^{1,*}) \tilde{Y}_s^{1,*} \\ \qquad + (\partial_x g)^*(s, \tilde{X}_s^*, \tilde{Y}_s^{1,*}, \tilde{Y}_s^{2,*}, \tilde{Z}_s^{1,*}) \tilde{Y}_s^{2,*}\big] ds; \\ \tilde{Y}_t^{2,*} = \partial_G U_A(F(\tilde{X}_T^*) - D_T, \tilde{G}_T^*) - \int_t^T \tilde{Z}_s^{2,*} dB_s. \end{cases} \quad (4.24)$$

Assuming the contract $F(X_T) - D_T$ implements the first best solution, we have $(\tilde{u}^*, \tilde{v}^*) = (u^*, v^*)$, and thus $(\tilde{X}^*, \tilde{G}^*) = (X^*, G^*)$. Then, by the necessary condition (10.26) we obtain

$$\tilde{Y}_t^{1,*}b_u(t, X_t^*, u_t^*, v_t^*) + \tilde{Y}_t^{2,*}g_u(t, X_t^*, u_t^*, v_t^*) = 0;$$
$$\bar{Y}_t^{1,*}b_u(t, X_t^*, u_t^*, v_t^*) + \bar{Y}_t^{2,*}g_u(t, X_t^*, u_t^*, v_t^*) = 0.$$

Since $b_u, g_u \neq 0$, we have

$$\tilde{Y}_t^{1,*}\bar{Y}_t^{2,*} = \tilde{Y}_t^{2,*}\bar{Y}_t^{1,*}.$$

In particular, at $t = T$,

$$\partial_C U_A\big(F(\tilde{X}_T^*) - D_T, \tilde{G}_T^*\big)F'(\tilde{X}_T^*)\partial_G U_\lambda(C_T^*, G_T^*)$$
$$= \partial_G U_A\big(F(\tilde{X}_T^*) - D_T, \tilde{G}_T^*\big)\partial_C U_\lambda(C_T^*, G_T^*).$$

If $F(X_T) - D_T$ implements the first best solution, we have $F(\tilde{X}_T^*) - D_T = C_T^*$. This, together with (4.7), implies that

$$\partial_C U_A(C_T^*, \tilde{G}_T^*)F'(\tilde{X}_T^*)\partial_G U_A(C_T^*, G_T^*) = \partial_G U_A(C_T^*, \tilde{G}_T^*)\partial_C U_A(C_T^*, G_T^*).$$

Then, it follows from the assumption $\partial_C U_A, \partial_G U_A \neq 0$ that $F'(X_T^*) = 1$. Since Range$(X_T^*) = \mathbb{R}$, then

$$F(x) = x + \alpha$$

for some constant α. Finally, since $F(X_T^*) - D_T = C_T^*$, we get

$$D_T - \alpha = X_T^* - C_T^* = I_P(Y_T^{1,*}),$$

and therefore, the contract is $F(x) - D_T = x - I_P(Y_T^{1,*})$. □

4.5 Examples

4.5.1 Linear Dynamics

Example 4.5.1 Let's look at a somewhat more general model than (3.1), with $r = 0$ (for simplicity of notation):

$$dX_t = b(u_t)dt + \alpha_t v_t dt + v_t dB_t$$

and $G_T = \int_0^T g(u_t)dt$, where B is a d-dimensional Brownian motion and α_t is a d-dimensional adapted process. The agent's utility is assumed non-separable, $U_A(x, y) = U_A(x - y)$. By (4.5) and (4.7), in this case the adjoint processes in (4.15) become

$$d\bar{Y}_t^{1,*} = \bar{Z}_t^{1,*}dB_t, \qquad d\bar{Y}_t^{2,*} = \bar{Z}_t^{2,*}dB_t;$$
$$\bar{Y}_T^{1,*} = U_P'(X_T^* - C_T^*) = -\lambda U_A'(C_T^* - G_T^*) = -\bar{Y}_T^{2,*}.$$

Then, $\bar{Y}^{1,*} = -\bar{Y}^{2,*}$, $\bar{Z}^{1,*} = -\bar{Z}^{2,*}$, so that the necessary conditions from Theorem 4.2.3 become

4.5 Examples

$$b'(u^*) = g'(u^*), \qquad \alpha_t \bar{Y}_t^{1,*} = -\bar{Z}_t^{1,*}. \tag{4.25}$$

(If there is no control u, then $\bar{Y}^{2,*} = 0$.) We assume that the first equality above has a unique solution u^*, which is then constant. The second equality gives

$$\bar{Y}_t^{1,*} = z \exp\left\{-\frac{1}{2}\int_0^t |\alpha_s|^2 ds - \int_0^t \alpha_s dB_s\right\}$$

for some $z > 0$, to be determined below. Denote

$$I_A(z) := (U_A')^{-1}(z) \quad \text{and} \quad \lambda' = -1/\lambda.$$

The optimal contract should be of the form

$$C_T^* = X_T^* - I_P(\bar{Y}_T^{1,*}) = I_A(\bar{Y}_T^{1,*}\lambda') + G_T^*.$$

Denote $Z_t := \bar{Y}_t^{1,*}/z$. Then, Z is a martingale with $Z_0 = 1$. We have

$$d(Z_t X_t^*) = Z_t\left[b(u^*)dt + (v_t^* - \alpha X_t^*)dB_t\right]$$

so that $Z_t X_t^* - \int_0^t Z_s b(u^*)ds$ has to be a martingale. The above system of necessary and sufficient conditions will have a solution if we can find v^* using the Martingale Representation Theorem, where we now consider the equation for ZX^* as a Backward SDE having a terminal condition

$$Z_T X_T^* = Z_T\{I_P(zZ_T) + I_A(z\lambda' Z_T) + G_T^*\}.$$

This BSDE has a solution which satisfies $X_0^* = x$ if and only if there is a solution z to the budget constraint

$$x = E\left[Z_T X_T^*\right] - b(u^*)T = E\left[Z_T\{I_P(zZ_T) + I_A(z\lambda' Z_T) + G_T^*\}\right] - b(u^*)T.$$

We thus recover results analogous to those of Chap. 3.

4.5.2 Nonlinear Volatility Selection with Exponential Utilities

Example 4.5.2 Consider now this generalization of the previous example, where we allow volatility v to affect the drift in a nonlinear way:

$$dX_t = \left[r_t X_t + b(t, u_t, v_t)\right]dt + v_t dB_t$$

where b is a deterministic function of (t, u, v), and r_t is a given deterministic process. Assume exponential utility functions, $U_i(x) = -\frac{1}{R_i}\exp\{-R_i x\}$. Also assume, with a slight abuse of notation g, that

$$g(t, X_t, u_t, v_t) = \mu_t X_t + g(t, u_t, v_t)$$

where g is a deterministic function of (t, u, v), and μ_t is a given deterministic process. The rest of this example consists of the proof that the optimal pair (u^*, v^*) is the one which solves the system of two equations (4.29) and (4.35) below.

Denote
$$\gamma_{t,s} := e^{\int_s^t r_u du}, \qquad \gamma_t := e^{\int_0^t r_u du}. \qquad (4.26)$$

Then, it is seen by Itô's rule that

$$X_t = X_0 + \int_0^t \gamma_{t,s} b(s, u_s, v_s) ds + \int_0^t \gamma_{t,s} v_s dB_s. \qquad (4.27)$$

The adjoint processes $\bar{Y}^{1,*}$, $\bar{Y}^{2,*}$ satisfy

$$\bar{Y}_t^{1,*} = U_P'(X_T^* - C_T^*) + \int_t^T (\bar{Y}_s^{1,*} r_s + \bar{Y}_s^{2,*} \mu_s) ds$$
$$+ \int_t^T [\bar{Y}_s^{1,*} b_v(s, u_s^*, v_s^*) + \bar{Y}_s^{2,*} g_v(s, u_s^*, v_s^*)] dB_s;$$

$$\bar{Y}_t^{2,*} = -\lambda U_A'(C_T^* - G_T^*) - \int_t^T \bar{Z}_s^{2,*} dB_s.$$

The Borch rule (4.5) gives $\bar{Y}_T^{1,*} = -\bar{Y}_T^{2,*}$.

We conjecture that optimal u^*, v^* are deterministic processes, and that

$$\bar{Y}_t^{2,*} = -h_t \bar{Y}_t^{1,*}, \qquad h_T = 1,$$

where h is a deterministic function of time. Using Itô's rule, it is easily verified that the above equality is satisfied if h is a solution to the ODE

$$h_t(h_t \mu_t - r_t) = -h_t', \qquad h_T = 1. \qquad (4.28)$$

We assume that μ and r are such that a unique solution exists. We get one equation for u^*, v^* from the first condition in (4.12):

$$b_u(t, u_t^*, v_t^*) = h_t g_u(t, u_t^*, v_t^*). \qquad (4.29)$$

Note that $\bar{Y}^{1,*}$ satisfies a linear SDE which is easily solved as

$$\bar{Y}_t^{1,*} = \bar{Y}_0^{1,*} \exp\left\{\int_0^t \left[h_s \mu_s - r_s - \frac{1}{2}|h_s g_v(s, u_s^*, v_s^*) - b_v(s, u_s^* v_s^*)|^2\right] dt \right.$$
$$\left. + \int_0^t [h_s g_v(s, u_s^*, v_s^*) - b_v(s, u_s^*, v_s^*)] dB_s \right\}. \qquad (4.30)$$

Thus, the process

$$A_t = \log(\bar{Y}_t^{1,*}/\bar{Y}_0^{1,*}) - \int_0^t \left[h_s \mu_s - r_s - \frac{1}{2}|h_s g_v(s, u_s^*, v_s^*) - b_v(s, u_s^*, v_s^*)|^2\right] dt$$
$$= \int_0^t [h_s g_v(s, u_s^*, v_s^*) - b_v(s, u_s^*, v_s^*)] dB_s \qquad (4.31)$$

is a local martingale. We conjecture that it is a martingale for the optimal u_s^*, v_s^*.

We will need below this representation for the first term of G_T^*, obtained by integration by parts:

4.5 Examples

$$\int_0^T \mu_s X_s^* ds = \frac{X_T^*}{\gamma_T} \int_0^T \mu_s \gamma_s ds$$
$$- \int_0^T \int_0^s \mu_u \gamma_u du \big[b(s, u_s^*, v_s^*)\gamma_s^{-1} ds + \gamma_s^{-1} v_s^* dB_s\big]. \quad (4.32)$$

From Borch condition (4.5), $\lambda U_A'(C_T^* - G_T^*) = U_P'(X_T^* - C_T^*)$, we can verify that

$$X_T^* - C_T^* = \alpha + \beta X_T^* - \beta G_T^*, \quad \alpha = \frac{\log(\lambda)}{R_A + R_P}, \quad \beta = \frac{R_A}{R_A + R_P}. \quad (4.33)$$

From this and $\bar{Y}_T^{1,*} = U_P'(X_T^* - C_T^*)$ we get

$$\log \bar{Y}_T^{1,*} = -\alpha R_P + \beta R_P (G_T^* - X_T^*). \quad (4.34)$$

Using this, and the fact that v^* is assumed deterministic, denoting by k a generic constant, we can get an alternative representation of A_t, for some constant k, as

$$A_t = E_t[A_T] = k + \beta R_P E_t\big[G_T^* - X_T^*\big].$$

Using (4.32) and that the non-deterministic part of G_T^* is $\int_0^T \mu_u X_u^* du$, we get

$$A_t = k + \beta R_P \left(\frac{\int_0^T \mu_s \gamma_s ds}{\gamma_T} - 1\right) E_t[X_T^*] - \beta R_P \int_0^t \left(\int_0^s \mu_u \gamma_u du\right) \gamma_s^{-1} v_s^* dB_s.$$

Since $E_t[X_T^*] = k + \int_0^t \gamma_{T,s} v_s^* dB_s$, comparing the dB integrand in this last expression to the dB integrand in (4.31), we see that we need to have

$$h_s g_v(s, u_s^*, v_s^*) - b_v(s, u_s^*, v_s^*)$$
$$= \beta R_P \left(\frac{\int_0^T \mu_s \gamma_s ds}{\gamma_T} - 1\right) \gamma_{T,s} v_s^* - \beta R_P \left(\int_0^s \mu_u \gamma_u du\right) \gamma_s^{-1} v_s^*$$
$$= \gamma_s^{-1} v_s^* \beta R_P \left(\int_s^T \mu_u \gamma_u du - \gamma_T\right). \quad (4.35)$$

We assume that b and g are such that this equation together with (4.29) has a unique solution (u_s^*, v_s^*) for all s, and such that in the above calculations the local martingales are, indeed, martingales. Then, we have shown that our conjecture of deterministic optimal controls is correct, and (u^*, v^*) is the first best solution.

4.5.3 Linear Contracts

Example 4.5.3 From (4.33) we get that at time T a linear relationship holds:

$$C_T^* = -\alpha + \frac{R_P}{R_A + R_P} X_T^* - \frac{R_A}{R_A + R_P} G_T^*, \quad \alpha = \frac{\log(\lambda)}{R_A + R_P}. \quad (4.36)$$

The question is whether this contract can be offered without need for monitoring, that is, whether it will induce the agent to apply the first best actions. It turns out that *a linear contract is incentive compatible if there is no control u of the drift, if there is no cost, so that g = 0, and if $\mu \equiv 0$.*[1]

In order to show that, consider the contract

$$C_T = \alpha + kX_T \quad \text{where } k := \frac{R_P}{R_A + R_P}.$$

As in (4.24), with $G_T = 0$, $\mu = 0$, if the agent is given the contract C_T, the corresponding adjoint equations are

$$\tilde{Y}_t^{1,*} = kU_A'(C_T^*) + \int_t^T \tilde{Y}_s^{1,*} r_s ds + \int_t^T \tilde{Y}_s^{1,*} b_v(s, u_s^*, v_s^*) dB_u;$$

$$\tilde{Y}_t^{2,*} = -U_A'(C_T) - \int_t^T \tilde{Z}_s^{2,*} dB_s.$$

We see that $\tilde{Y}_T^{1,*} = -k\tilde{Y}_T^{2,*}$. As before, we conjecture that the agent's optimal v^* is a deterministic process, and we can see that

$$\tilde{Y}_t^{2,*} = -\gamma_t \tilde{Y}_t^{1,*}, \qquad \gamma_T = \frac{1}{k},$$

where γ is a deterministic function of time, with γ being a solution to the ODE (4.28), but with $\gamma_T = \frac{1}{k}$. We can solve for $\tilde{Y}^{1,*}$ as

$$\tilde{Y}_t^{1,*} = \tilde{Y}_0^{1,*} \exp\left\{-\int_0^t \left[r_s + \frac{1}{2}|b_v(s, u_s^*, v_s^*)|^2\right] dt - \int_0^t b_v(s, u_s^*, v_s^*) dB_s\right\}. \tag{4.37}$$

On the other hand,

$$\tilde{Y}_T^{1,*} = kU_A'(C_T^*) = k\exp\{-R_A[\alpha + kX_T^*]\}.$$

Substituting here the expression (4.27) for X_T^*, and comparing the random terms with the one in (4.37), we see that we need to have

$$b_v(s, u_s^*, v_s^*) = R_A k \gamma_{T,s} v_s^* = R_A \frac{v_s^*}{\gamma_s} k\gamma_T. \tag{4.38}$$

We see that this is the same condition as (4.35) with $\mu = G = 0$, therefore the agent will choose the first best action v^*.

4.6 Dual Problem

It is known from the literature on optimal portfolio selection problems that it is often useful to study their dual problems. Our PA problem is similar to portfolio

---

[1] For a reference with a counterexample if u is present see "Further Reading".

4.6 Dual Problem

selection problems insofar as the volatility v can be selected. We illustrate the approach by identifying an appropriate dual problem in the context of model (4.1) and problem (4.2), with $U_A(x, y) = U_A(x - y)$.

We introduce a dual function

$$\tilde{U}(z) := \max_{x,c}\{U_P(x-c) + \lambda U_A(c) - zx\}$$

for which it then holds that

$$U_P(x - c) + \lambda U_A(c) \leq \tilde{U}(z) + zx. \qquad (4.39)$$

Substitute now

$$x = X_T - G_T, \qquad c = C_T - G_T, \qquad z = Z_T$$

for some random variable Z_T. This gives

$$E[U_P(X_T - C_T) + \lambda U_A(C_T - G_T)] \leq E[\tilde{U}(Z_T)] + E[Z_T(X_T - G_T)]. \qquad (4.40)$$

We would like to find random variables Z_T for which $E[Z_T(X_T - G_T)]$ would be bounded by a bound independent of the primal variables u, v, C_T. It turns out that the appropriate dual (random) variables are typically of the form $Z_T^{\kappa,\rho}$, where

$$dZ_t^{\kappa,\rho} = -\rho_t Z_t^{\kappa,\rho} dt - \kappa_t Z_t^{\kappa,\rho} dB_t$$

for some processes κ, ρ belonging to an appropriate family.[2] In order to clarify this further, note that by Itô's rule we get, for $Z = Z^{\kappa,\rho}$,

$$d(Z_t X_t) = Z_t\big[b(t, X_t, u_t, v_t) - \rho_t X_t - \kappa_t v_t\big]dt + (\ldots)dB_t$$

and

$$d\left(Z_t \int_0^t g(s, X_s, u_s, v_s)ds\right) = Z_t g(t, X_t, u_t, v_t)dt + (\ldots)dB_t.$$

Introduce now a dual function

$$\tilde{h}(t, \kappa, \rho) := \max_{x,u,v}\{b(t, x, u, v) - g(t, x, u, v) - \rho x - \kappa v\}.$$

Then, we see that

$$Z_T(X_T - G_T) \leq Z_0 X_0 + \int_0^T Z_s \tilde{h}(s, \kappa_s, \rho_s)ds + \int_0^T (\ldots)dB_s.$$

The process $\int_0^t (\ldots)dB_s$ corresponding to the term on the right-hand side is a local martingale. If it is also a supermartingale, we get a "budget constraint"

$$E[Z_T^{\kappa,\rho}(X_T - G_T)] \leq X_0 Z_0 + E\int_0^T Z_s^{\kappa,\rho}\tilde{h}(s, \kappa_s, \rho_s)ds. \qquad (4.41)$$

Going back to (4.40), we then have

---

[2]Sometimes it is necessary to consider the closure of such a family in an appropriate topology.

$$E[U_P(X_T - C_T) + \lambda U_A(C_T - G_T)]$$
$$\leq E[\tilde{U}(Z_T^{\kappa,\rho})] + Z_0 X_0 + E\int_0^T Z_s^{\kappa,\rho} \tilde{h}(s, \kappa_s, \rho_s) ds. \qquad (4.42)$$

The dual problem is then to minimize the right-hand side over Z_0 and κ, ρ in the domain of the function $\tilde{h}$ (assuming such a function is well defined). Under appropriate conditions, and denoting by Z_T^* the corresponding optimal dual variable, we will have as the optimal contract

$$C_T = X_T - I_P(Z_T^*).$$

Some references with technical details are mentioned in the "Further Reading" section.

4.7 A More General Model with Consumption and Recursive Utilities

In this section we discuss how to extend the above model in two aspects. First, we replace expected utilities with so-called recursive utilities (in continuous-time models also called Stochastic Differential Utilities). These are defined as the initial value of a backward SDE. Second, we allow continuous-time payment or consumption for the principal and the agent.

More specifically, the state process $X := X^{u,v,c,e}$ and the agent's cost function $G_T := G_T^{u,v,c,e}$ are given by

$$X_t = x + \int_0^t b(s, X_s, u_s, v_s, c_s, e_s) ds + \int_0^t v_s dB_s, \qquad (4.43)$$

$$G_t := \int_0^t g(s, X_s, u_s, v_s, c_s, e_s) ds, \qquad (4.44)$$

where u and v are the agent's drift and volatility control, c is the continuous payment and/or the principal's consumption, and e is the agent's consumption. We note that, although for notational simplicity we assume all processes are one-dimensional, by considering high-dimensional c we actually consider both the continuous payment and the principal's consumption. As before, let C_T denote the terminal payment. The agent's and the principal's *recursive utility processes* are defined by the following BSDEs;

$$W_t^A = U_A(X_T, C_T, G_T) + \int_t^T u_A(s, X_s, u_s, v_s, c_s, e_s, W_s^A, Z_s^A) ds$$
$$- \int_t^T Z_s^A dB_s; \qquad (4.45)$$

$$W_t^P = U_P(X_T, C_T) + \int_t^T u_P(s, X_s, v_s, c_s, W_s^P, Z_s^P) ds - \int_t^T Z_s^P dB_s.$$

4.7 A More General Model with Consumption and Recursive Utilities

Thus, the current utility W_t may depend on future values of utility, and on its sensitivity with respect to the underlying uncertainty Z.

The Principal–Agent (or risk-sharing) problem is

$$V := \sup_{C_T, u, v, c, e} W_0^P \quad \text{subject to } W_0^A \geq R_0. \tag{4.46}$$

As before we use the Lagrange multiplier formulation, that is, we consider the following problem: for $\lambda > 0$,

$$V(\lambda) := \sup_{C_T, u, v, c, e} \left[W_0^P + \lambda W_0^A \right]. \tag{4.47}$$

Note that in this case we cannot optimally choose C_T by ω-wise optimization as in Sect. 4.1. Moreover, we cannot apply the results of Sect. 10.2 directly. Nevertheless, we can apply the arguments of that section to derive necessary conditions for the problem (4.47), and we do it in Sect. 10.5.1 in a heuristic way, omitting technical details.

We provide next the main results. Introduce the following adjoint processes:

$$\Gamma_t^A = 1 + \int_0^t \Gamma_s^A \partial_y u_A(s) ds + \int_0^t \Gamma_s^A \partial_z u_A(s) dB_s;$$

$$\Gamma_t^P = 1 + \int_0^t \Gamma_s^P \partial_y u_P(s) ds + \int_0^t \Gamma_s^P \partial_z u_P(s) dB_s;$$

$$\bar{Y}_t^1 = \partial_X U_P(T) \Gamma_T^P + \lambda \partial_X U_A(T) \Gamma_T^A - \int_t^T \bar{Z}_s^1 dB_s \tag{4.48}$$

$$+ \int_t^T \left[\Gamma_s^P \partial_x u_P(s) + \lambda \Gamma_s^A \partial_x u_A(s) + \bar{Y}_s^1 \partial_x b(s) + \bar{Y}_s^2 \partial_x g(s) \right] ds;$$

$$\bar{Y}_t^2 = \lambda \partial_G U_A(T) \Gamma_T^A - \int_t^T \bar{Z}_s^2 dB_s.$$

If $(C_T^*, u^*, v^*, c^*, e^*)$ is an optimal control and is in the interior of the admissible set, then, under technical conditions, we have the following necessary conditions for optimality

$$\Gamma_T^{P,*} \partial_C U_P(X_T^*, C_T^*) + \lambda \Gamma_T^{A,*} \partial_C U_A(X_T^*, C_T^*, G_T^*) = 0;$$

$$\lambda \Gamma_t^{A,*} \partial_u u_A(t, X_t^*, u_t^*, v_t^*, c_t^*, e_t^*, W_t^{A,*}, Z_t^{A,*}) + \bar{Y}_t^{1,*} \partial_u b(t, X_t^*, u_t^*, v_t^*, c_t^*, e_t^*)$$

$$+ \bar{Y}_t^{2,*} \partial_u g(t, X_t^*, u_t^*, v_t^*, c_t^*, e_t^*) = 0;$$

$$\Gamma_t^{P,*} \partial_c u_P(t, X_t^*, v_t^*, c_t^*, W_t^{P,*}, Z_t^{P,*})$$

$$+ \lambda \Gamma_t^{A,*} \partial_v u_A(t, X_t^*, u_t^*, v_t^*, c_t^*, e_t^*, W_t^{A,*}, Z_t^{A,*})$$

$$+ \bar{Y}_t^{1,*} \partial_v b(t, X_t^*, u_t^*, v_t^*, c_t^*, e_t^*) \tag{4.49}$$

$$+ \bar{Y}_t^{2,*} \partial_v g(t, X_t^*, u_t^*, v_t^*, c_t^*, e_t^*) + \bar{Z}_t^{1,*} = 0;$$

$$\Gamma_t^{P,*} \partial_c u_P(t, X_t^*, v_t^*, c_t^*, W_t^{P,*}, Z_t^{P,*})$$

$$+ \lambda \Gamma_t^{A,*} \partial_c u_A(t, X_t^*, u_t^*, v_t^*, c_t^*, e_t^*, W_t^{A,*}, Z_t^{A,*})$$

$$+ \bar{Y}_t^{1,*} \partial_c b(t, X_t^*, u_t^*, v_t^*, c_t^*, e_t^*) + \bar{Y}_t^{2,*} \partial_c g(t, X_t^*, u_t^*, v_t^*, c_t^*, e_t^*) = 0;$$

$$\lambda \Gamma_t^{P,*} \partial_e u_A(t, X_t^*, u_t^*, v_t^*, c_t^*, e_t^*, W_t^{A,*}, Z_t^{A,*}) + \bar{Y}_t^{1,*} \partial_e b(t, X_t^*, u_t^*, v_t^*, c_t^*, e_t^*)$$
$$+ \bar{Y}_t^{2,*} \partial_e g(t, X_t^*, u_t^*, v_t^*, c_t^*, e_t^*) = 0.$$

Note that the first equation reduces to the Borch condition if we have regular expected utilities, rather than recursive utilities.

Assume further that the above conditions determine uniquely the functions

$$C_T^* = I_0(X_T^*, G_T^*, \Gamma_T^{A,*}, \Gamma_T^{P,*});$$
$$u_t^* = I_1(t, \Theta_t^*), \qquad v_t^* = I_2(t, \Theta_t^*), \qquad c_t^* = I_3(t, \Theta_t^*), \qquad e_t^* = I_4(t, \Theta_t^*),$$
$$\text{where } \Theta_t^* := (X_t^*, \Gamma_t^{A,*}, \Gamma_t^{P,*}, W_t^{A,*}, Z_t^{A,*}, W_t^{P,*}, Z_t^{P,*}, \bar{Y}_t^{1,*}, \bar{Z}_t^{1,*}, \bar{Y}_t^{2,*}).$$
(4.50)

Then, we obtain the following FBSDE system for the optimal solution:

$$X_t^* = x + \int_0^t b(s, X_s^*, I_1(s, \Theta_s^*), I_2(s, \Theta_s^*), I_3(s, \Theta_s^*), I_4(s, \Theta_s^*)) ds$$
$$+ \int_0^t I_2(s, \Theta_s^*) dB_s;$$

$$G_t^* = \int_0^t g(s, X_s^*, I_1(s, \Theta_s^*), I_2(s, \Theta_s^*), I_3(s, \Theta_s^*), I_4(s, \Theta_s^*)) ds;$$

$$\Gamma_t^{A,*} = 1 + \int_0^t \Gamma_s^{A,*} \partial_y u_A(s, X_s^*, I_1(s, \Theta_s^*), I_2(s, \Theta_s^*), I_3(s, \Theta_s^*), I_4(s, \Theta_s^*),$$
$$W_s^{A,*}, Z_s^{A,*}) ds$$
$$+ \int_0^t \Gamma_s^{A,*} \partial_z u_A(s, X_s^*, I_1(s, \Theta_s^*), I_2(s, \Theta_s^*), I_3(s, \Theta_s^*), I_4(s, \Theta_s^*),$$
$$W_s^{A,*}, Z_s^{A,*}) dB_s;$$

$$\Gamma_t^{P,*} = 1 + \int_0^t \Gamma_s^{P,*} \partial_y u_P(s, X_s^*, I_3(s, \Theta_s^*), W_s^{P,*}, Z_s^{P,*}) ds$$
$$+ \int_0^t \Gamma_s^{P,*} \partial_z u_P(s, X_s^*, I_3(s, \Theta_s^*), W_s^{P,*}, Z_s^{P,*}) dB_s;$$

$$W_t^{A,*} = U_A(X_T^*, I_0(X_T^*, G_T^*, \Gamma_T^{A,*}, \Gamma_T^{P,*}), G_T^*) - \int_t^T Z_s^{A,*} dB_s$$
$$+ \int_t^T u_A(s, X_s^*, I_1(s, \Theta_s^*), I_2(s, \Theta_s^*), I_3(s, \Theta_s^*), I_4(s, \Theta_s^*),$$
$$W_s^{A,*}, Z_s^{A,*}) ds; \qquad (4.51)$$

$$W_t^{P,*} = U_P(X_T^*, I_0(X_T^*, G_T^*, \Gamma_T^{A,*}, \Gamma_T^{P,*})) - \int_t^T Z_s^{P,*} dB_s$$
$$+ \int_t^T u_P(s, X_s^*, I_2(s, \Theta_s^*), I_3(s, \Theta_s^*), W_s^{P,*}, Z_s^{P,*}) ds;$$

$$\bar{Y}_t^{1,*} = \partial_X U_P\big(X_T^*, I_0\big(X_T^*, G_T^*, \Gamma_T^{A,*}, \Gamma_T^{P,*}\big)\big)\Gamma_T^{P,*}$$
$$+ \lambda \partial_X U_A\big(X_T^*, I_0\big(X_T^*, G_T^*, \Gamma_T^{A,*}, \Gamma_T^{P,*}\big), G_T^*\big)\Gamma_T^{A,*}$$
$$+ \int_t^T \big[\Gamma_s^{P,*} \partial_x u_P u_P\big(s, X_s^*, I_2(s, \Theta_s^*), I_3(s, \Theta_s^*), W_s^{P,*}, Z_s^{P,*}\big)$$
$$+ \lambda \Gamma_s^{A,*} \partial_x u_A\big(s, X_s^*, I_1(s, \Theta_s^*), I_2(s, \Theta_s^*), I_3(s, \Theta_s^*), I_4(s, \Theta_s^*),$$
$$W_s^{A,*}, Z_s^{A,*}\big)$$
$$+ \bar{Y}_t^{1,*} \partial_x b\big(s, X_s^*, I_1(s, \Theta_s^*), I_2(s, \Theta_s^*), I_3(s, \Theta_s^*), I_4(s, \Theta_s^*)\big)$$
$$+ \bar{Y}_t^{2,*} \partial_x g\big(s, X_s^*, I_1(s, \Theta_s^*), I_2(s, \Theta_s^*), I_3(s, \Theta_s^*), I_4(s, \Theta_s^*)\big)\big]ds$$
$$- \int_t^T \bar{Z}_s^{1,*} dB_s;$$
$$\bar{Y}_t^{2,*} = \lambda \partial_G U_A\big(X_T^*, I_0\big(X_T^*, G_T^*, \Gamma_T^{A,*}, \Gamma_T^{P,*}\big), G_T^*\big)\Gamma_T^{A,*} - \int_t^T \bar{Z}_s^{2,*} dB_s.$$

4.8 Further Reading

The presentation here follows mostly Cvitanić et al. (2012) and Cadenillas et al. (2007). In the latter paper one can also find a counterexample in which even with exponential utilities linear contracts cannot induce the first best action, if the control of the drift u is present. Example 4.5.2 presents a somewhat generalized version of a setting analyzed in Ou-Yang (2003). The duality method for portfolio selection problems in Brownian motion models is presented in Karatzas and Shreve (1998). Duffie et al. (1994) and Dumas et al. (2000) solve the first best problem of sharing an endowment process X_t between several agents who consume it at a continuous rate, and who have Stochastic Differential, recursive utilities. In continuous-time model, these were first introduced in Duffie and Epstein (1992).

References

Cadenillas, A., Cvitanić, J., Zapatero, F.: Optimal risk-sharing with effort and project choice. J. Econ. Theory **133**, 403–440 (2007)
Cvitanić, J., Wan, X., Yang, H.: Dynamics of contract design with screening. Manag. Sci. (2012, forthcoming)
Duffie, D., Epstein, L.G.: Stochastic differential utility. Econometrica **60**, 353–394 (1992)
Duffie, D., Geoffard, P.Y., Skiadas, C.: Efficient and equilibrium allocations with stochastic differential utility. J. Math. Econ. **23**, 133–146 (1994)
Dumas, B., Uppal, R., Wang, T.: Efficient intertemporal allocations with recursive utility. J. Econ. Theory **93**, 240–259 (2000)
Karatzas, I., Shreve, S.E.: Methods of Mathematical Finance. Springer, New York (1998)
Ou-Yang, H.: Optimal contracts in a continuous-time delegated portfolio management problem. Rev. Financ. Stud. **16**, 173–208 (2003)

Part III
Second Best: Contracting Under Hidden Action—The Case of Moral Hazard

Chapter 5
Mathematical Theory for General Moral Hazard Problems

Abstract This chapter describes a general theory of optimal contracting with hidden or non-contractable actions in continuous-time, developed by applying the stochastic maximum principle. The main modeling difference with respect to the full information case is that we will now assume that the agent controls the distribution of the output process with his effort. Mathematically, this is modeled using the so-called "weak formulation" and "weak solutions" of the underlying SDEs. Necessary and sufficient conditions are derived in terms of the so-called adjoint processes and corresponding Forward-Backward SDEs. These processes typically include the output process, the agent's expected utility process, the principal's expected utility process, and the ratio of marginal utilities process.

5.1 The Model and the PA Problem

Note to the Reader The reader interested in tractable applications of the general theory of the moral hazard problems and economic conclusions they offer may wish to read only this first section in this chapter, skip the rest, and go then to Chap. 6. The cases discussed in Chap. 6 are the main cases analyzed in the literature, and their results can be derived without knowing all the heavy mathematical machinery of this chapter.

Let B be a standard Brownian motion on a given probability space with probability measure P, and $\mathbb{F}^B = \{\mathcal{F}^B_t\}_{0 \le t \le T}$ be the filtration on $[0, T]$ generated by B. Given an $\mathbb{F}^B$-adapted square integrable strictly positive process v_t:

$$v > 0 \quad \text{and} \quad E\left\{\int_0^T |v_t|^2 dt\right\} < \infty, \tag{5.1}$$

we introduce the state process

$$X_t := x + \int_0^t v_s dB_s. \tag{5.2}$$

Remark 5.1.1 The filtration $\mathbb{F}^X = \{\mathcal{F}^X_t\}_{0 \le t \le T}$ generated by X is the same as the filtration generated by B. In fact, since v is $\mathbb{F}^B$-adapted, then so is X and thus

$\mathbb{F}^X \subset \mathbb{F}^B$. On the other hand, since the quadratic variation $\langle X \rangle_t = \int_0^t |v_s|^2 ds$ is $\mathbb{F}^X$-adapted, then v^2 is $\mathbb{F}^X$-adapted. The assumption that $v > 0$ implies that v is $\mathbb{F}^X$-adapted, and then so is $B_t = \int_0^T v_s^{-1} dX_s$.

For any $\mathbb{F}^B$-adapted effort process u_t representing the agent's action, define

$$B_t^u := B_t - \int_0^t u_s ds,$$
$$M_t^u := \exp\left(\int_0^t u_s dB_s - \frac{1}{2}\int_0^t |u_s|^2 ds\right), \quad (5.3)$$
$$P^u(A) := E[M_T^u \mathbf{1}_A].$$

Then, under certain conditions on the process u, we know by Girsanov Theorem that P^u is a probability measure and B^u is a P^u-Brownian motion. We then have

$$dX_t = u_t v_t dt + v_t dB_t^u. \quad (5.4)$$

In the language of Stochastic Analysis, the triple (X, B^u, P^u) is a weak solution to the SDE:

$$dX_t = u_t v_t dt + v_t dB_t.$$

Unlike the more commonly considered case of the strong solution, the Brownian motion and the probability measure are not fixed here. The choice of u corresponds to the choice of probability measure P^u, thus to the choice of the distribution of process X.

The agent chooses his action based on the output value X which is observable to the principal. However, although u is $\mathbb{F}^X$-adapted, the principal does not know u, and hence does not know the value of B^u either (or, she is not allowed to contract upon the values of u).

Assume the agent receives a terminal payment C_T and continuous payments at the rate c_t. The agent's problem is, given (C_T, c),

$$V_A(C_T, c) := \sup_{u \in \mathcal{U}} V_A(C_T, c, u)$$
$$:= \sup_{u \in \mathcal{U}} E^u \left\{ \int_0^T u_A(t, (X)_t, c_t, u_t) dt + U_A((X)_T, C_T) \right\}$$
$$= \sup_{u \in \mathcal{U}} E \left\{ M_T^u \left[\int_0^T u_A(t, (X)_t, c_t, u_t) dt + U_A((X)_T, C_T) \right] \right\}. \quad (5.5)$$

Here, $(X)_t$ denotes the path of X up to time t, $\mathcal{U}$ is the admissible set of the agent's controls u, which are $\mathbb{F}$-adapted processes taking values in some set $U \subset \mathbb{R}^k$ for some k. We will specify the technical requirements on $\mathcal{U}$ later.

We note that X is a fixed process which does not change its values with the choice of u. For notational simplicity from now on we omit X in all the functions, e.g., we write $u_A(t, c_t, u_t)$ instead of $u_A(t, (X)_t, c_t, u_t)$.

5.1 The Model and the PA Problem

Let $\hat{u} = \hat{u}(C_T, c) \in \mathcal{U}$ denote the agent's optimal action, assumed to exist. Let $W_t^A := W^{A,C_T,c}$ denote the agent's optimal remaining utility:

$$W_t^A := E_t^{\hat{u}}\left\{\int_t^T u_A(s, c_s, \hat{u}_s)ds + U_A(C_T)\right\}. \tag{5.6}$$

We assume (C_T, c) has to satisfy the individual rationality (IR) constraint

$$W_t^A \geq R_t, \tag{5.7}$$

where R_t is a given $\mathbb{F}$-adapted process. Otherwise, the agent would quit and go to work for another principal. The principal wants to prevent that.

The principal's optimization problem is over contracts (C_T, c) in the principal's admissible set $\mathcal{A}$. Here C_T is an $\mathcal{F}_T$-measurable random variable taking values in A_1, c is an $\mathbb{F}$-adapted process taking values in A_2, where A_1 and A_2 are some appropriate sets. We will specify the technical requirements on $\mathcal{A}$ later. In particular, we require that each contract $(C_T, c) \in \mathcal{A}$ is **implementable**, that is, there exists an optimal $\hat{u} = \hat{u}(C_T, c)$ for the agent's problem. If there is more than one such contract, we assume that the agent chooses the one which is best for the principal. Moreover, we denote by $\mathcal{A}_{IR}$ all those contracts $(C_T, c) \in \mathcal{A}$ which satisfy the IR constraint (5.7).

Thus, restricting the optimization over implementable contracts satisfying the IR constraint, we can write the principal's problem as (recalling that we suppress the dependence on X in the utility functions)

$$V_P := \sup_{(C_T,c)\in\mathcal{A}_{IR}} V_P(C_T, c) := \sup_{(C_T,c)\in\mathcal{A}_{IR}} E^{\hat{u}}\left\{\int_0^T u_P(t, c_t)dt + U_P(C_T)\right\}$$

$$= \sup_{(C_T,c)\in\mathcal{A}_{IR}} E\left\{M_T^{\hat{u}}\left[\int_0^T u_P(t, c_t)dt + U_P(C_T)\right]\right\}. \tag{5.8}$$

We deal with the IR constraint by the use of a Lagrange multiplier process λ_t. More precisely, we define the *relaxed principal's problem* to be

$$V_P(\lambda) := \sup_{(C_T,c)\in\mathcal{A}} V_P(C_T, c, \lambda)$$

$$:= \sup_{(C_T,c)\in\mathcal{A}} E\left\{M_T^{\hat{u}}\left[\int_0^T u_P(t, c_t)dt + U_P(C_T) + \int_0^T \lambda_t[W_t^A - R_t]dt\right]\right\}. \tag{5.9}$$

We note that when R is deterministic, and hence also λ is deterministic, clearly

$$V_P(\lambda) := \sup_{(C_T,c)\in\mathcal{A}} E\left\{M_T^{\hat{u}}\left[\int_0^T u_P(t, c_t)dt + U_P(C_T) + \int_0^T \lambda_t W_t^A dt\right]\right\}$$

$$- \int_0^T \lambda_t R_t dt,$$

and, thus, one may drop the last term without changing the problem.

As usual in optimization problems, we have

Proposition 5.1.2 *Assume there exists $\lambda^* > 0$ such that the relaxed principal's problem $V_P(\lambda^*)$ has an optimal control $(C_T^*, c^*) \in \mathcal{A}$ and the agent's corresponding optimal utility satisfies $W_t^{A, C_T^*, c^*} = R_t$, $0 \le t \le T$. Then, $V_P = V_P(\lambda^*)$ and (C_T^*, c^*) is the principal's optimal control.*

Remark 5.1.3 Mathematically, the strong and the weak formulations are in general not equivalent, due to the different requirements on the measurability of the agent's control u. In the weak formulation, u is $\mathbb{F}^X$-adapted, and thus $\mathbb{F}^{B^u}$ may be smaller than $\mathbb{F}^X$, unlike in the strong formulation. In the existing literature, for tractability reasons it is standard to use the weak formulation for the agent's problem and the strong formulation for the principal's problem. This may not be an appropriate approach in general, for a couple of reasons: (i) the optimal action $\hat{u}$ obtained from the agent's problem by using the weak formulation may not be in the admissible set under the strong formulation (if $\mathbb{F}^{B^u}$ is strictly smaller than $\mathbb{F}^X$); (ii) given a principal's desired action (also called *target action*) in the strong formulation, it is not always possible to obtain it as an optimal solution of the agent's problem in the weak formulation, as it may not even be implementable. In the approach of this book that problem is non-existent, as we use the weak formulation for both the agent's problem and the principal's problem. On the other hand, in all the solvable examples in the literature it turns out that the optimal effort u is a functional of output X only (and not of Brownian motion B^u). If that is the case, there is nothing wrong with using the strong formulation for the principal's problem.

Remark 5.1.4 Due to the continuous payment c, a more natural model might be

$$dX_t = v_t dB_t^u + [u_t v_t - c_t]dt. \tag{5.10}$$

However, if we do the transformation

$$\tilde{u} := u - c/v, \qquad \tilde{u}_A(t, c, \tilde{u}) := u_A(t, c, \tilde{u} + c/v),$$

then (5.10) is reduced to the model we introduced above. For simplicity of notation we work with (5.3) instead of (5.10), although one has to keep in mind that the assumptions we impose are on $\tilde{u}$, not on the actual effort u.

Remark 5.1.5 By observing X continuously, the principal observes v and thus can force the agent to choose the process v which the principal prefers. In this sense, the volatility v can be viewed as the principal's control. In this chapter we assume v is given. We will provide some heuristic arguments in Sect. 5.5 in the case in which v is also a control.

Remark 5.1.6 In the full information case, the principal observes both X and the controls (u, v), thus, she also observes the underlying Brownian motion B. It is then reasonable to consider the strong formulation in that case. In the hidden action case that we study here, the principal observes X (and v), but not u. Thus, she does not observe the process B^u. From the principal's point of view, the state process X can be observed, but its distribution is unknown due to the unobservable action u of the agent. This makes it reasonable to use the weak formulation in this case.

From now on we fix v and a Lagrange multiplier $\lambda > 0$:

Assumption 5.1.7

(i) The fixed volatility process v satisfies (5.1) and the Lagrange multiplier $\lambda > 0$ is bounded.
(ii) U_A and U_P are progressively measurable and $\mathcal{F}_T$-measurable, and u_A and u_P are progressively measurable and $\mathbb{F}$-adapted.

We solve, in the following two sections, the relaxed Principal–Agent problem (5.5) and (5.9) in two cases: the Lipschitz case and the quadratic case. We emphasize that these two cases are studied separately mainly for technical reasons: we need to use BSDEs with Lipschitz continuous generator and BSDEs whose generator has quadratic growth, as studied in Sects. 9.3 and 9.6, respectively. We also note that, as long as one can extend the theory of BSDEs, one should also be able to extend our results on Principal–Agent problems to the corresponding more general framework.

5.2 Lipschitz Case

In this section we assume

Assumption 5.2.1 The set U in which u takes values is bounded, and there exists a constant C such that

$$\left|u_A(t,c,u)\right| \leq C\left[1 + \left|u_A(t,c,0)\right|\right].$$

We note that the inequality condition above is merely a technical condition and it is very mild, due to the boundedness of U. As we will see soon, in this case the involved BSDEs have Lipschitz continuous generator, and thus we are able to use the theory introduced in Sect. 9.3 and the corresponding stochastic maximum principle introduced in Sect. 10.1.

5.2.1 Agent's Problem

Let $(C_T, c) \in \mathcal{A}$ be a given contract pair, where the details of the principal's admissible set $\mathcal{A}$ will be provided in Sect. 5.2.2 below. In this section we find optimality conditions for the agent's problem.

For each agent's admissible control u, let $W^{A,u} := W^{A,u,C_T,c}$ denote the agent's remaining utility process:

$$\begin{aligned}W_t^{A,u} &:= E_t^u\left\{\int_t^T u_A(s,c_s,u_s)ds + U_A(C_T)\right\} \\ &= U_A(C_T) + \int_t^T u_A(s,c_s,u_s)ds - \int_t^T Z_s^{A,u}dB_s^u \\ &= U_A(C_T) + \int_t^T \left[u_A(s,c_s,u_s) + u_s Z_s^{A,u}\right]ds - \int_t^T Z_s^{A,u}dB_s, \quad (5.11)\end{aligned}$$

where the existence of $Z^{A,u} := Z^{A,u,C_T,c}$ is due to the extended Martingale Representation Theorem of Lemma 10.4.6. We will apply the results in Sect. 10.1, by using the last equation in (5.11) and Remark 10.4.3. In the notation of Sect. 10.4.1, we have

$$\sigma = v, \quad b = uv, \quad h = u_A(t, c_t, u_t), \quad g = U_A(C_T).$$

Define

$$f_A(t,c,z,u) := u_A(t,c,u) + uz, \quad f_A^*(t,c,z) := \sup_{u \in U} f_A(t,c,z,u). \quad (5.12)$$

By Assumption 5.2.1, $\partial_z f_A = u$ is bounded. Then, the last equation in (5.11) and the following BSDE have Lipschitz continuous generators:

$$W_t^A = U_A(C_T) + \int_t^T f_A^*(t, c_t, Z_t^A)dt - \int_t^T Z_t^A dB_t. \quad (5.13)$$

Assumption 5.2.2 The agent's admissible set $\mathcal{U}$ is a set of $\mathbb{F}^B$-adapted processes u taking values in U.

Applying Theorem 10.1.4, we first get the sufficient conditions.

Theorem 5.2.3 *Assume Assumptions 5.1.7, 5.2.1 and 5.2.2 hold, and*

$$E\left\{|U_A(C_T)|^2 + \left(\int_0^T [|u_A(t,c_t,0)| + |f_A^*(t,c_t,0)|]dt\right)^2\right\} < \infty. \quad (5.14)$$

Let (W^A, Z^A) be the unique solution to the BSDE (5.13). If there exists a progressively measurable and $\mathbb{F}$-adapted function $I(t,c,z)$ such that

$$f_A^*(t,c,z) = f_A(t,c,z,I(t,c,z)) \quad \text{and} \quad u^* := I(\cdot, c, Z^A) \in \mathcal{U}, \quad (5.15)$$

then u^* is an optimal control for the optimization problem (5.5) and $V_A(C_T, c) = W_0^A$.

Remark 5.2.4

(i) As mentioned in Remark 10.1.5(ii), in Theorem 5.2.3 we do not require uniqueness or differentiability of function I.

5.2 Lipschitz Case

(ii) When U is compact (including the discrete case), the function I exists.

(iii) Under the conditions of Theorem 5.2.3, we know in particular that the contract pair (C_T, c) is implementable. In the next section we will check that u^* is indeed in $\mathcal{U}$, a consequence of our assumptions on (C_T, c).

We next provide necessary conditions for the agent's problem.

Assumption 5.2.5 u_A is differentiable in u and $|\partial_u u_A(t, c, u)| \leq C[|\partial_u u_A(t, c, 0)| + 1]$.

Assumption 5.2.6 The agent's admissible set $\mathcal{U}$ is a set of $\mathbb{F}^B$-adapted processes u satisfying:

(i) each $u \in \mathcal{U}$ takes values in U;
(ii) $\mathcal{U}$ is locally convex, in the sense of Assumption 10.1.7(ii).

Applying Theorem 10.1.10 and Proposition 10.1.12 we have

Theorem 5.2.7 *Assume Assumptions 5.1.7, 5.2.1, 5.2.5, and 5.2.6 hold, and*

$$E\left\{|U_A(C_T)|^2 + \left(\int_0^T [|u_A(t, c_t, 0)| + |\partial_u u_A(t, c_t, 0)|] dt\right)^2\right\} < \infty. \quad (5.16)$$

If $u^* \in \mathcal{U}$ is an optimal control of the optimization problem (5.5) and u^* is an interior point of $\mathcal{U}$, then

$$\partial_u u_A(t, c_t, u_t^*) + Z_t^{A, u^*} = 0. \quad (5.17)$$

Assume further that there exists a unique function $I_A(t, c, z)$ taking values in U such that

$$\partial_u u_A(t, c, I_A(t, c, z)) + z = 0 \quad (5.18)$$

and I_A is differentiable in z. Then, BSDE (5.13) with

$$f_A^*(t, c, z) := f_A(t, c, z, I_A(t, c, z)) \quad (5.19)$$

is well-posed and

$$u^* = I_A(t, c_t, Z_t^A) \quad \text{and} \quad V(C_T, c) = W_0^A. \quad (5.20)$$

Remark 5.2.8 The process $W^A = W^{A, u^*}$ is the agent's optimal remaining utility. We can interpret $Z^A = Z^{A, u^*}$ as a "derivative" of the agent's remaining utility with respect to the Brownian motion (actually, it is equal to what is called a Malliavin derivative). In the case when (5.17) holds, at the optimum the agent's local marginal cost of effort has to be equal to the sensitivity of the agent's remaining utility with respect to the underlying uncertainty.

Remark 5.2.9 The result above can be extended to the following case: introduce the cumulative cost

$$G_T = \int_0^T g(s, c_s, u_s) ds$$

and assume that the terminal utility depends on it:

$$U_A = U_A(C_T, G_T).$$

It can be shown then that optimal effort u has to satisfy

$$u^* = \arg\max_u \left[u_A(t, c, u) + u Z_t^A + g(t, c_t, u_t) Y_t^A \right],$$

where

$$Y_t^A = E_t^u \left[\partial_G U_A(C_T, G_T) \right].$$

Similarly, the results we derive below for the principal's problem can also be extended to this case, as well as to the framework in which the principal can also choose the volatility process v. See Sect. 5.5 for some heuristic arguments.

5.2.2 Principal's Problem

We now assume (C_T, c) is implementable with optimal control $u := u(C_T, c) = I_A(t, c_t, Z_t^A)$, where $(W^A, Z^A) := (W^{A, C_T, c}, Z^{A, C_T, c})$ is the solution to BSDE (5.13).

Let $W^{P,\lambda} = W^{P, C_T, c, \lambda}$ denote the principal's relaxed remaining utility:

$$\begin{aligned}
W_t^{P,\lambda} &:= E_t^u \left\{ U_P(C_T) + \int_t^T u_P(s, c_s) ds + \int_t^T \lambda_s W_s^A ds \right\} \\
&= U_P(C_T) + \int_t^T u_P(s, c_s) ds + \int_t^T \lambda_s W_s^A ds - \int_t^T Z_s^{P,\lambda} dB_s^u \\
&= U_P(C_T) + \int_t^T \left[u_P(s, c_s) + \lambda_s W_s^A - \lambda_s R_s + Z_s^{P,\lambda} I_A(s, c_s, Z_s^A) \right] ds \\
&\quad - \int_t^T Z_s^{P,\lambda} dB_s,
\end{aligned} \tag{5.21}$$

where again the process $Z^{P,\lambda} := Z^{P, C_T, c, \lambda}$ is the one whose existence is guaranteed by the extended Martingale Representation Theorem of Lemma 10.4.6. We emphasize that $W^{P,\lambda}$ is the remaining utility for the relaxed principal's problem, which contains the additional term $\lambda[W^A - R]$, and, thus, is not the same as the principal's utility, in general.

Considering (5.13) and the last equation in (5.21) as a two-dimensional BSDE with two-dimensional controls (C_T, c), the relaxed principal's problem (5.9) falls into the framework of Sect. 10.3. Following the arguments in Theorems 10.3.7 and 10.3.10, we can formally establish some sufficient conditions for the problem.

5.2 Lipschitz Case

However, as pointed out in Remarks 10.3.8 and 10.3.11, we are not able to provide tractable sufficient conditions in this case (except for some trivial cases). Instead, we provide only necessary conditions. We will provide some sufficient conditions in Sect. 5.2.4 below, after reformulating the problem slightly.

We now study necessary conditions. Note that the generator of the last equation in (5.21) is in general not Lipschitz-continuous in Z^A, so we adopt some conditions slightly stronger than those in Sect. 10.3; see (5.22) below.

Assumption 5.2.10

(i) Assumptions 5.1.7, 5.2.1, 5.2.5, and 5.2.6 hold.
(ii) U_A, U_P are continuously differentiable; u_P is continuously differentiable in c; u_A is continuously differentiable in c with $|\partial_c u_A(t,c,u)| \leq C[1 + |\partial_c u_A(t,c,0)|]$.
(iii) The equation $\partial_u u_A(t,c,u) + z = 0$ determines uniquely a function $u = I_A(t,c,z)$ and I_A is continuously differentiable in (c,z) with bounded derivatives.

Assumption 5.2.11 The principal's admissible set $\mathcal{A}$ is a set of contract pairs (C_T, c) where C_T is $\mathcal{F}_T$-measurable and c is $\mathbb{F}$-adapted satisfying:

(i) Each $(C_T, c) \in \mathcal{A}$ satisfies

$$E\left\{|\varphi(C_T)|^4 + \left(\int_0^T |\psi(t,c_t)|dt\right)^4\right\} < \infty, \quad (5.22)$$

for $\varphi = U_A, U_A', U_P, U_P'$ and $\psi(t,c) = u_A(t,c,0), \partial_c u_A(t,c,0), u_P(t,c), \partial_c u_P(t,c)$.
(ii) Each $(C_T, c) \in \mathcal{A}$ is implementable and the agent's optimal control $u_t = I_A(t, c_t, Z_t^A)$ is an interior point of $\mathcal{U}$, where Z^A is the solution to BSDE (5.13) with f_A^* defined by (5.19), and $\mathcal{U}$ satisfies Assumption 5.2.6.
(iii) $\mathcal{A}$ is locally convex, in the sense of Assumption 10.1.7(ii).
(iv) For each C_T, C_T^ε, c, c^ε as in Assumption 10.1.7(ii), random variables $\varphi(C_T^\varepsilon)$ and processes $\psi(t, c_t^\varepsilon)$ are integrable uniformly in $\varepsilon \in [0,1]$ in the sense of (5.22).

We remark that, when the agent's optimal control u is an interior point of $\mathcal{U}$, by Theorem 5.2.7 it must take the form $I_A(t, c_t, Z_t^A)$, which is why the condition (ii) above is well formulated.

Notice that f^* is independent of $(W^{P,\lambda}, Z^{P,\lambda})$. In this case, the adjoint process in (10.48) becomes

$$D_t = \int_0^t [\lambda_s - Z_s^P I_A(s, c_s, Z_s^A) \partial_z I_A(s, c_s, Z_s^A)]ds$$

$$+ \int_0^t Z_s^P \partial_z I_A(s, c_s, Z_s^A)dB_s. \quad (5.23)$$

Here we used the fact that, due to (5.12) and (5.18),

$$\partial_z f_A^* = I_A.$$

Our main result in this section is the following theorem, proved in Sect. 10.5.3.

Theorem 5.2.12 *Assume Assumptions 5.2.10 and 5.2.11 hold.*

(i) *Suppose $(C_T^*, c^*) \in \mathcal{A}$ is an optimal control of the principal's relaxed problem (5.9) and it is an interior point of $\mathcal{A}$. Let $W^{A,*}, Z^{A,*}, W^{P,\lambda,*}, Z^{P,\lambda,*}, D^*$ be the corresponding processes. Then,*

$$\begin{aligned} D_T^* U_A'(C_T^*) + U_P'(C_T^*) &= 0, \\ D_t^* \partial_c u_A(t, c_t^*, I_A(t, c^*, Z_t^{A,*})) + \partial_c u_P(t, c_t^*) & \\ + Z_t^{P,\lambda,*} \partial_c I_A(t, c^*, Z_t^{A,*}) &= 0. \end{aligned} \quad (5.24)$$

(ii) *Assume further that there exist unique functions $I_P^1(D)$ and $I_P^2(t, D, z^A, z^P)$ satisfying*

$$\begin{aligned} DU_A'(I_P^1(D)) + U_P'(I_P^1(D)) &= 0, \\ D\partial_c u_A(t, I_P^2(t, D, z^A, z^P), I_A(t, I_P^2(t, D, z^A, z^P), z^A)) & \\ + \partial_c u_P(t, I_P^2(t, D, z^A, z^P)) + z^P \partial_c I_A(t, I_P^2(t, D, z^A, z^P), z^A) &= 0. \end{aligned} \quad (5.25)$$

Then, $(D^*, W^{A,*}, Z^{A,*}, W^{P,\lambda,*}, Z^{P,\lambda,*})$ satisfy the following (high-dimensional) coupled FBSDE:

$$\begin{aligned} D_t^* &= \int_0^t \lambda_s ds + \int_0^t Z_s^{P,\lambda,*} \partial_z I_A(s, I_P^2(s, D_s^*, Z_s^{A,*}, Z_s^{P,\lambda,*}), Z_s^{A,*}) \\ &\quad \times [dB_s - I_A(s, I_P^2(s, D_s^*, Z_s^{A,*}, Z_s^{P,\lambda,*}), Z_s^{A,*}) ds]; \\ W_t^{A,*} &= U_A(I_P^1(D_T^*)) \\ &\quad + \int_t^T u_A(s, I_P^2(s, D_s^*, Z_s^{A,*}, Z_s^{P,\lambda,*}), \\ &\qquad I_A(s, I_P^2(t, D_t^*, Z_t^{A,*}, Z_t^{P,\lambda,*}), Z_s^{A,*})) ds \\ &\quad - \int_t^T Z_s^{A,*} [dB_s - I_A(s, I_P^2(s, D_s^*, Z_s^{A,*}, Z_s^{P,\lambda,*}), Z_s^{A,*}) ds]; \\ W_t^{P,\lambda,*} &= U_P(I_P^1(D_T^*)) + \int_t^T [u_P(s, I_P^2(s, D_s^*, Z_s^{A,*}, Z_s^{P,\lambda,*})) \\ &\quad + \lambda_s W_s^{A,*} - \lambda_s R_s] ds \\ &\quad - \int_t^T Z_s^{P,\lambda,*} [dB_s - I_A(s, I_P^2(s, D_s^*, Z_s^{A,*}, Z_s^{P,\lambda,*}), Z_s^{A,*}) ds], \end{aligned} \quad (5.26)$$

and the principal's optimal control satisfies

$$C_T^* = I_P^1(D_T^*), \quad c_t^* = I_P^2(t, D_t^*, Z_t^{A,*}, Z_t^{P,\lambda,*}). \quad (5.27)$$

5.2 Lipschitz Case   57

Remark 5.2.13 As can be seen from the first equation in (5.24), the negative of process D can be interpreted as the *ratio of marginal utility process* that was constant in the first best case (recall the Borch rule), but that is random and moving in time here, and whose dynamics have to be tracked.

5.2.3 Principal's Problem Based on Principal's Target Actions

In the literature, it is often customary to consider the so-called *principal's target action*, instead of the contract payment, as the principal's control. This makes the technical conditions somewhat easier. We take this approach in this subsection and the following two subsections.

Let (C_T, c) be an implementable contract. For the function I_A of Theorem 5.2.7 (or of Theorem 5.2.3), the agent's optimal action is $u_t = I_A(t, c_t, Z_t^A)$, where Z^A is the solution to BSDE (5.13) with f_A^* defined by (5.19). Note that the principal should either pay a continuous payment c (that is, u_A indeed depends on c), or pay terminal payment C_T, or both. In this subsection we consider the case when $\partial_c u_A \neq 0$ and we replace the continuous payment c by the target action u. In the next subsection we consider the case when $U_A' = \partial_C U_A \neq 0$ and we replace the terminal payment C_T by the target action u.

In the case $\partial_c u_A \neq 0$, we see that I_A indeed depends on c. Recalling (5.17), we assume there exists a unique function $J_A(t, u, z)$ such that

$$\partial_u u_A(t, J_A(t, u, z), z) + z = 0. \tag{5.28}$$

Now, given any $u \in \mathcal{U}$ and terminal payment C_T, consider the following BSDE with $(W^A, Z^A) := (W^{A, C_T, u}, Z^{A, C_T, u})$:

$$W_t^A = U_A(C_T) + \int_t^T \left[u_A(s, J_A(s, u_s, Z_s^A), u_s) + u_s Z_s^A \right] ds$$
$$- \int_t^T Z_s^A dB_s. \tag{5.29}$$

Assume the above BSDE is well-posed and let

$$c_t := J_A(t, u_t, Z_t^A). \tag{5.30}$$

By Theorem 5.2.7, if (C_T, c) is implementable, then u is the corresponding optimal control. So, instead of considering all implementable contracts (C_T, c), in this subsection the principal's control becomes (C_T, u), and the process u is called the principal's target action.

Our relaxed principal's problem becomes

$$V_P(\lambda) := \sup_{(C_T, u) \in \mathcal{A}} W_0^{P, \lambda, C_T, u, \lambda}, \tag{5.31}$$

where $(W^{P, \lambda}, Z^{P, \lambda}) := (W^{P, \lambda, C_T, u, \lambda}, Z^{P, \lambda, C_T, u, \lambda})$ solves the following BSDE:

$$W_t^{P,\lambda} = U_P(C_T) + \int_t^T \left[u_P\left(s, J_A(s, u_s, Z_s^A)\right) + \lambda_s W_s^A - \lambda_s R_s + u_s Z_s^{P,\lambda}\right] ds$$
$$- \int_t^T Z_s^{P,\lambda} dB_s. \tag{5.32}$$

This again is a stochastic optimization problem of a two-dimensional BSDE. As in Sect. 5.2.2, we do not discuss sufficient conditions. To study necessary conditions, we assume

Assumption 5.2.14

(i) Assumptions 5.1.7, 5.2.1, 5.2.5, and 5.2.6 hold.
(ii) U_A, U_P are continuously differentiable; u_A is continuously differentiable in c and u; u_P is continuously differentiable in c.
(iii) The equation $\partial_u u_A(t, c, u) + z = 0$ determines uniquely a function $c = J_A(t, u, z)$ and J_A is continuously differentiable in (u, z).
(iv) Denote

$$u_A^*(t, u, z) := u_A\bigl(t, J_A(t, u, z), u\bigr) \quad \text{and} \quad u_P^*(t, u, z) := u_P\bigl(t, J_A(t, u, z)\bigr). \tag{5.33}$$

For $\varphi = u_A^*$ and u_P^*, we have

$$\left|\partial_z \varphi(t, u, z)\right| \leq C, \qquad \left|\partial_u \varphi(t, u, z)\right| \leq C\bigl[1 + \left|\varphi_u(t, u, 0)\right| + |z|\bigr].$$

Assumption 5.2.15 The principal's admissible set $\mathcal{A}$ is a set of contract pairs (C_T, u), where C_T is $\mathcal{F}_T$-measurable taking values in A_1 and $u \in \mathcal{U}$ with $\mathcal{U}$ satisfying Assumption 5.2.6, such that:

(i) Each $(C_T, u) \in \mathcal{A}$ satisfies

$$E\left\{|\varphi(C_T)|^2 + \left(\int_0^T |\psi(t, u_t, 0)| dt\right)^2\right\} < \infty, \tag{5.34}$$

for $\varphi = U_A, U_A', U_P, U_P'$ and $\psi(t, u) = u_A^*, \partial_u u_A^*, u_P^*, \partial_u u_P^*$.
(ii) For each $(C_T, u) \in \mathcal{A}$, the contract (C_T, c) is implementable and u is the agent's corresponding optimal control, where c is defined by (5.30) and (W^A, Z^A) is the solution to BSDE (5.29).
(iii) $\mathcal{A}$ is locally convex, in the sense of Assumption 10.1.7(ii).
(iv) For each $C_T, C_T^\varepsilon, u, u^\varepsilon$ as in Assumption 10.1.7(ii), the random variables $\varphi(C_T^\varepsilon)$ and processes $\psi(t, u_t^\varepsilon, 0)$ are integrable uniformly in $\varepsilon \in [0, 1]$ in the sense of (5.34).

Now, the problem satisfies all the conditions in Sect. 10.3 with

$$f_1(t, y, z, u) := u_A^*(t, u, z_1) + u z_1, \qquad f_2(t, y, z, u) := u_P^*(t, u, z_1) + \lambda_t y_1 + u z_2.$$

5.2 Lipschitz Case

Introduce the adjoint process:

$$D_t := \int_0^t [\lambda_s - [\partial_z u_A^*(s, u_s, Z_s^A) D_s + \partial_z u_P^*(s, u_s, Z_s^A)] u_s] ds$$
$$+ \int_0^t [\partial_z u_A^*(s, u_s, Z_s^A) D_s + \partial_z u_P^*(s, u_s, Z_s^A)] dB_s. \tag{5.35}$$

Applying Theorem 10.3.5 we have

Theorem 5.2.16 *Assume Assumptions 5.2.14 and 5.2.15 hold.*

(i) *Suppose* $(C_T^*, u^*) \in \mathcal{A}$ *is an optimal control for the principal's relaxed problem (5.31) and it is an interior point of* $\mathcal{A}$. *Let* $W^{A,*}, Z^{A,*}, W^{P,\lambda,*}, Z^{P,\lambda,*}, D^*$ *be the corresponding processes. Then,*

$$D_T^* U_A'(C_T^*) + U_P'(C_T^*) = 0,$$
$$D_t^*[\partial_u u_A^*(t, u_t^*, Z_t^{A,*}) + Z_t^{A,*}] + \partial_c u_P^*(t, u_t^*, Z_t^{A,*}) + Z_t^{P,\lambda,*} = 0.$$

(ii) *Assume further that there exist unique functions* $I_P^1(D)$ *and* $I_P^2(t, D, z^A, z^P)$ *satisfying*

$$DU_A'(I_P^1(D)) + U_P'(I_P^1(D)) = 0,$$
$$D[\partial_c u_A^*(t, I_P^2(t, D, z^A, z^P), z^A) + z^A] + \partial_c u_P^*(t, I_P^2(t, D, z^A, z^P), z^A) + z^P$$
$$= 0. \tag{5.36}$$

Then, $(D^*, W^{A,*}, Z^{A,*}, W^{P,\lambda,*}, Z^{P,\lambda,*})$ *satisfy the following (high-dimensional) coupled FBSDE:*

$$D_t^* = \int_0^t \lambda_s ds + \int_0^t [\partial_z u_P^* + D_s^* \partial_z u_A^*](s, I_P^2(s, D_s^*, Z_s^{A,*}, Z_s^{P,\lambda,*}), Z_s^{A,*})$$
$$\times [dB_s - I_P^2(s, D_s^*, Z_s^{A,*}, Z_s^{P,\lambda,*}) ds];$$

$$W_t^{A,*} = U_A(I_P^1(D_T^*)) + \int_t^T u_A^*(s, I_P^2(s, D_s^*, Z_s^{A,*}, Z_s^{P,\lambda,*}), Z_s^{A,*}) ds$$
$$- \int_t^T Z_s^{A,*}[dB_s - I_P^2(s, D_s^*, Z_s^{A,*}, Z_s^{P,\lambda,*}) ds]; \tag{5.37}$$

$$W_t^{P,\lambda,*} = U_P(I_P^1(D_T^*)) + \int_t^T [u_P^*(s, I_P^2(s, D_s^*, Z_s^{A,*}, Z_s^{P,\lambda,*}), Z_s^{A,*})$$
$$+ \lambda_s W_s^{A,*} - \lambda_s R_s] ds$$
$$- \int_t^T Z_s^{P,\lambda,*}[dB_s - I_P^2(s, D_s^*, Z_s^{A,*}, Z_s^{P,\lambda,*}) ds];$$

and the principal's optimal control satisfies

$$C_T^* = I_P^1(D_T^*), \qquad u_t^* = I_P^2(t, D_t^*, Z_t^{A,*}, Z_t^{P,\lambda,*}). \tag{5.38}$$

5.2.4 Principal's Problem Based on Principal's Target Actions: Another Formulation

We now consider the case when $U'_A \neq 0$. Assume U'_A has an inverse function and denote

$$J_A := [U'_A]^{-1}. \tag{5.39}$$

Let (C_T, c) be an implementable contract, and u be the agent's corresponding optimal action. By Theorem 5.2.7, we have $Z_t^A = -\partial_u u_A(t, c_t, u_t)$. Thus, we may consider W^A as the solution to the following SDE:

$$W_t^A = w^A + \int_0^t \left[u_s \partial_u u_A(s, c_s, u_s) - u_A(s, c_s, u_s)\right] ds$$
$$- \int_0^t \partial_u u_A(s, c_s, u_s) dB_s. \tag{5.40}$$

Here, the agent's initial utility w^A is another parameter to be decided on. Note that

$$C_T = J_A(W_T^A). \tag{5.41}$$

Then, we may consider (w^A, c, u) as the principal's control, instead of (C_T, c). In this case, the process u is again called the principal's target action. It is straightforward to optimize over w^A, it being a real number. So, we fix w^A and then the relaxed principal's problem becomes

$$V_P(\lambda, w^A) := \sup_{(c,u) \in \mathcal{A}} Y_0^P, \tag{5.42}$$

where $(Y^P, Z^{P,\lambda}) := (Y^{P,c,u,\lambda,w_A}, Z^{P,\lambda,c,u,\lambda,w_A})$ is the solution to the BSDE:

$$W_t^{P,\lambda} = U_P(J_A(W_T^A)) + \int_t^T \left[u_P(s, c_s) + \lambda_s W_s^A - \lambda_s R_s + u_s Z_s^{P,\lambda}\right] ds$$
$$- \int_t^T Z_s^{P,\lambda} dB_s. \tag{5.43}$$

Note that (5.40) and (5.43) comprise an FBSDE with control (c, u) and the following coefficients, in the notation used in Sect. 10.2,

$$b(t, c, u) := u \partial_u u_A(t, c, u) - u_A(t, c, u), \qquad \sigma(t, c, u) := -\partial_u u_A(t, c, u),$$
$$f(t, x, z, c, u) := u_P(t, c) + \lambda_t x + uz, \qquad g(x) := U_P^*(x). \tag{5.44}$$

This falls into the framework of Sect. 10.2, except that control (c, u) is two-dimensional here.

As in Sect. 10.2, we start with necessary conditions.

Assumption 5.2.17

(i) Assumptions 5.1.7, 5.2.1, 5.2.5, and 5.2.6 hold.

5.2 Lipschitz Case

(ii) U_A, U_P are continuously differentiable; u_A and $\partial_u u_A$ are continuously differentiable in c and u; u_P is continuously differentiable in c.
(iii) The function U'_A has the inverse function $J_A := [U'_A]^{-1}$, and the function $U_P^*(x) := U_P(J_A(x))$ is continuously differentiable with bounded derivative.

Assumption 5.2.18 The principal's admissible set $\mathcal{A}$ is a set of contract pairs (c, u), where c is an $\mathbb{F}$-adapted process taking values in A_2 and $u \in \mathcal{U}$, with $\mathcal{U}$ satisfying Assumption 5.2.6, such that:

(i) For each $(c, u) \in \mathcal{A}$ and $\varphi = u_A, \partial_u u_A, u_P$, satisfies

$$E\left\{|U_P^*(0)|^2 + \left(\int_0^T |\varphi(t, c_t, u_t)| dt\right)^2\right\} < \infty. \tag{5.45}$$

(ii) For each $(c, u) \in \mathcal{A}$, the contract (C_T, c) is implementable and u is the agent's corresponding optimal control, where C_T is defined by (5.41) and W^A is defined by (5.40).
(iii) $\mathcal{A}$ is locally convex, in the sense of Assumption 10.1.7(ii).
(iv) For each $c, c^\varepsilon, u, u^\varepsilon$ as in Assumption 10.1.7(ii), the processes $\varphi(t, c_t^\varepsilon, u_t^\varepsilon)$ are integrable uniformly in $\varepsilon \in [0, 1]$ in the sense of (5.45).

We introduce the following adjoint processes in analogy to (10.24):

$$\Gamma_t = 1 + \int_0^t \Gamma_s u_s dB_s;$$
$$\bar{Y}_t = \partial_x U_P^*(W_T^A) \Gamma_T + \int_t^T \lambda_s \Gamma_s ds - \int_t^T \bar{Z}_s dB_s. \tag{5.46}$$

Following the arguments of Lemma 10.2.4 and Theorem 10.2.5, one can obtain

Theorem 5.2.19 *Assume Assumptions 5.2.17 and 5.2.18 hold.*

(i) *Suppose $(c^*, u^*) \in \mathcal{A}$ is an optimal control for the principal's relaxed problem (5.42) and it is an interior point of $\mathcal{A}$. Let $W^{A,*}, \Gamma^*, W^{P,*}, Z^{P,*}, \bar{Y}^*, \bar{Z}^*$ be the corresponding processes. Then,*

$$\Gamma_t^* \partial_c u_P(t, c_t^*) + \bar{Y}_t^*[u_t^* \partial_{cu} u_A(t, c_t^*, u_t^*) - \partial_c u_A(t, c_t^*, u_t^*)]$$
$$- \bar{Z}_t^* \partial_{cu} u_A(t, c^*, u^*) = 0, \tag{5.47}$$
$$\Gamma_t^* Z_t^{P,*} + \bar{Y}_t^* u_t^* \partial_{uu} u_A(t, c_t^*, u_t^*) - \bar{Z}_t^* \partial_{uu} u_A(t, c_t^*, u_t^*) = 0.$$

(ii) *Assume further that there exist unique functions $I_P^i(t, \gamma, z, \bar{y}, \bar{z})$, $i = 1, 2$, satisfying*

$$\gamma \partial_c u_P(t, I_P^1) + \bar{y}[I_P^2 \partial_{cu} u_A(t, I_P^1, I_P^2) - \partial_c u_A(t, I_P^1, I_P^2)]$$
$$- \bar{z} \partial_{cu} u_A(t, I_P^1, I_P^2) = 0, \tag{5.48}$$
$$\gamma z + \bar{y} I_P^2 \partial_{uu} u_A(t, I_P^1, I_P^2) - \bar{z} \partial_{uu} u_A(t, I_P^1, I_P^2) = 0.$$

Then, $(W^{A,*}, \Gamma^*, W^{P,*}, Z^{P,*}, \bar{Y}^*, \bar{Z}^*)$ satisfy the following (high-dimensional) coupled FBSDE:

$$W_t^{A,*} = w^A - \int_0^t u_A(s, I_P^1(\cdot), I_P^2(\cdot))(s, \Gamma_s^*, Z_s^{P,*}, \bar{Y}_s^*, \bar{Z}_s^*) ds$$

$$- \int_0^t \partial_u u_A(s, I_P^1(\cdot), I_P^2(\cdot))(s, \Gamma_s^*, Z_s^{P,*}, \bar{Y}_s^*, \bar{Z}_s^*)$$

$$\times [dB_s - I_P^2(s, \Gamma_s^*, Z_s^{P,*}, \bar{Y}_s^*, \bar{Z}_s^*) ds];$$

$$\Gamma_t^* = 1 + \int_0^t \Gamma_s^* I_P^2(s, \Gamma_s^*, Z_s^{P,*}, \bar{Y}_s^*, \bar{Z}_s^*) dB_s; \qquad (5.49)$$

$$W_t^{P,*} = U_P^*(W_T^{A,*}) + \int_t^T [u_P(s, I_P^1(s, \Gamma_s^*, Z_s^{P,*}, \bar{Y}_s^*, \bar{Z}_s^*)) + \lambda_s W_s^{A,*}] ds$$

$$- \int_t^T Z_s^{P,*} [dB_s - I_P^2(s, \Gamma_s^*, Z_s^{P,*}, \bar{Y}_s^*, \bar{Z}_s^*) ds];$$

$$\bar{Y}_t^* = \partial_x U_P^*(W_T^{A,*}) \Gamma_T^* + \int_t^T \lambda_s \Gamma_s^* ds - \int_t^T \bar{Z}_s^* dB_s;$$

and the principal's optimal control satisfies

$$c_t^* = I_P^1(t, \Gamma_t^*, Z_t^{P,*}, \bar{Y}_t^*, \bar{Z}_t^*), \qquad u_t^* = I_P^2(t, \Gamma_t^*, Z_t^{P,*}, \bar{Y}_t^*, \bar{Z}_t^*). \qquad (5.50)$$

Remark 5.2.20 As in Remark 10.2.6, we can simplify the system (5.49) by removing the adjoint process Γ. Indeed, denote

$$\hat{Y}_t^* := \bar{Y}_t^* (\Gamma_t^*)^{-1}, \qquad \hat{Z}_t^* := [\bar{Z}_t^* - \bar{Y}_t^* u_t^*] (\Gamma_t^*)^{-1}. \qquad (5.51)$$

Then, (5.47) becomes

$$\partial_c u_P(t, c_t^*) - \hat{Y}_t^* \partial_c u_A(t, c_t^*, u_t^*) - \hat{Z}_t^* \partial_{cu} u_A(t, c^*, u^*) = 0,$$
$$Z_t^{P,*} - \hat{Z}_t^* \partial_{uu} u_A(t, c_t^*, u_t^*) = 0. \qquad (5.52)$$

Let $\hat{I}_P^i(t, z, \hat{y}, \hat{z})$, $i = 1, 2$ be determined by

$$\partial_c u_P(t, \hat{I}_P^1) - \hat{y} \partial_c u_A(t, \hat{I}_P^1, \hat{I}_P^2) - \hat{z} \partial_{cu} u_A(t, \hat{I}_P^1, \hat{I}_P^2) = 0,$$
$$z - \hat{z} \partial_{uu} u_A(t, \hat{I}_P^1, \hat{I}_P^2) = 0, \qquad (5.53)$$

and assume they are differentiable in z. Then, (5.49) becomes

5.2 Lipschitz Case

$$W_t^{A,*} = w^A - \int_0^t u_A(s, \hat{I}_P^1(\cdot), \hat{I}_P^2(\cdot))(s, Z_s^{P,*}, \hat{Y}_s^*, \hat{Z}_s^*) ds$$

$$- \int_0^t \partial_u u_A(s, \hat{I}_P^1(\cdot), \hat{I}_P^2(\cdot))(s, Z_s^{P,*}, \hat{Y}_s^*, \hat{Z}_s^*)$$

$$\times [dB_s - \hat{I}_P^2(s, Z_s^{P,*}, \hat{Y}_s^*, \hat{Z}_s^*) ds];$$

$$W_t^{P,*} = U_P^*(W_T^{A,*}) + \int_t^T [u_P(s, \hat{I}_P^1(s, Z_s^{P,*}, \hat{Y}_s^*, \hat{Z}_s^*)) + \lambda_s W_s^{A,*}] ds \quad (5.54)$$

$$- \int_t^T Z_s^{P,*} [dB_s - \hat{I}_P^2(s, Z_s^{P,*}, \hat{Y}_s^*, \hat{Z}_s^*) ds];$$

$$\hat{Y}_t^* = \partial_x U_P^*(W_T^{A,*}) + \int_t^T \lambda_s ds - \int_t^T \hat{Z}_s^* [dB_s - \hat{I}_P^2(s, Z_s^{P,*}, \hat{Y}_s^*, \hat{Z}_s^*) ds];$$

and the principal's optimal control satisfies

$$c_t^* = \hat{I}_P^1(t, Z_t^{P,*}, \hat{Y}_t^*, \hat{Z}_t^*), \qquad u_t^* = \hat{I}_P^2(t, Z_t^{P,*}, \hat{Y}_t^*, \hat{Z}_t^*). \quad (5.55)$$

However, as pointed out in Remark 10.2.6, the system (5.49) is more convenient for the sufficiency conditions discussed below.

In the approach of this section we can establish some semi-tractable sufficient conditions, as follows.

Define Hamiltonian in analogy to (10.32):

$$H(t, x, z, \gamma, \bar{y}, \bar{z}, c, u)$$
$$:= \gamma f(t, x, z, c, u) + \bar{y} b(t, c, u) + \bar{z} \sigma(t, c, u)$$
$$= \gamma [u_P(t, c) + \lambda_t x + uz] + \bar{y} [u \partial_u u_A(t, c, u) - u_A] - \bar{z} \partial_u u_A(t, c, u). \quad (5.56)$$

Assumption 5.2.21

(i) Assumptions 5.1.7, 5.2.1, 5.2.5, and 5.2.6 hold.
(ii) u_A is differentiable in u, and u_A, u_P, and $\partial_u u_A$ are continuous in c and u.
(iii) U_A is differentiable, and the derivative U'_A is invertible with $J_A := [U'_A]^{-1}$. Moreover, the function $U_P^*(x) := U_P(J_A(x))$ is uniformly Lipschitz continuous and concave in x and $E\{|U_P^*(0)|^2\} < \infty$.
(iv) The Hamiltonian H defined in (5.56) is concave in (z, c, u) for all $(\gamma, \bar{y}, \bar{z})$ in the set of all possible values that the adjoint processes $(\Gamma, \bar{Y}, \bar{Z})$ could take. Moreover, there exist functions $I_P^i(t, \gamma, z, \bar{y}, \bar{z})$, $i = 1, 2$, taking values in A_2 and U, respectively, such that

$$H(t, x, \gamma, z, \bar{y}, \bar{z}, I_P^1(t, \gamma, z, \bar{y}, \bar{z}), I_P^2(t, \gamma, z, \bar{y}, \bar{z}))$$
$$= \sup_{c \in A_2, u \in U} H(t, x, \gamma, z, \bar{y}, \bar{z}, c, u). \quad (5.57)$$

We remark that we used the fact that H is linear in x in part (iv) of the above assumption.

Assumption 5.2.22 The principal's admissible set $\mathcal{A}$ is a set of contract pairs (c, u), where c is an $\mathbb{F}$-adapted process taking values in A_2 and $u \in \mathcal{U}$, with $\mathcal{U}$ satisfying Assumption 5.2.6, such that:

(i) For each $(c, u) \in \mathcal{A}$ and $\varphi = u_A, \partial_u u_A, u_P$, (5.45) holds.
(ii) For each $(c, u) \in \mathcal{A}$, the contract (C_T, c) is implementable and u is the agent's corresponding optimal control, where C_T is defined by (5.41) and W^A is defined by (5.40).

We then have:

Theorem 5.2.23 *Assume*

(i) *Assumptions 5.2.21 and 5.2.22 hold.*
(ii) *FBSDE (5.49) has a solution* $(W^{A,*}, \Gamma^*, W^{P,*}, Z^{P,*}, \bar{Y}^*, \bar{Z}^*)$, *where the functions* I_P^1, I_P^2 *are given by Assumption 5.2.21(iv).*
(iii) *The processes* (c^*, u^*) *defined by (5.50) are in* $\mathcal{A}$.

Then, $V(\lambda, w^A) = W_0^{P,*}$ *and* (c^*, u^*) *is an optimal control.*

5.3 Quadratic Case

In this section we remove the assumption that U is bounded. The main difference in this case is that BSDE (5.11) is not uniformly Lipschitz continuous in Z^A anymore, and thus we cannot apply the results of Sect. 10.1. We instead use the theory of BSDEs with quadratic growth, introduced in Sect. 9.6, as well as the corresponding Stochastic Maximum Principle introduced in Sect. 10.4. However, we should point out that, although our technical conditions will be different, the equations we are going to derive will be exactly the same as those in Sect. 5.2.

5.3.1 Agent's Problem

Consider the setup of Sect. 5.1 again. Let $(C_T, c) \in \mathcal{A}$ be a given contract pair, where the technical details of the principal's admissible set $\mathcal{A}$ will be provided in the next section. In this section we find optimality conditions for the agent's problem. Unlike in Sect. 5.2 where the agent's admissible set $\mathcal{U}$ is independent of (C_T, c), it is more convenient here to allow $\mathcal{U}$ to depend on (C_T, c), denoted as $\mathcal{U}(C_T, c)$. That is, the agent's problem becomes

$$V_A(C_T, c) := \sup_{u \in \mathcal{U}(C_T, c)} W_0^{A,u}, \tag{5.58}$$

where $W^{A,u}$ is defined by BSDE (5.11). We note that, however, we can still consider a fixed set $\mathcal{U}$, without the dependence on (C_T, c), by imposing stronger technical conditions; see Remark 5.3.7 below.

As in Sect. 10.4, we start with sufficient conditions.

5.3 Quadratic Case

Assumption 5.3.1 The agent's admissible set $\mathcal{U}(C_T, c)$ is a set of $\mathbb{F}^B$-adapted processes u taking values in U such that, for each $u \in \mathcal{U}(C_T, c)$, M^u is a true P-martingale, and

$$E^u\left\{|U_A(C_T)|^2 + \left(\int_0^T |u_A(t, c_t, u_t)| dt\right)^2\right\} < \infty.$$

Clearly, under the above conditions, BSDE (5.11) has a unique solution $(W^{A,u}, Z^{A,u}) \in L^2(\mathbb{F}, P^u) \times L^2(\mathbb{F}, P^u)$.

Recall functions f_A and f_A^* defined in (5.12). Applying Theorems 10.4.7 and 10.4.8, we have

Theorem 5.3.2 *Assume Assumptions 5.1.7 and 5.3.1 hold. Suppose $u^* \in \mathcal{U}(C_T, c)$ satisfies*

$$f_A^*(t, c_t, Z_t^{A,u^*}) = f_A(t, c_t, u_t^*, Z_t^{A,u^*}), \tag{5.59}$$

and for each $u \in \mathcal{U}(C_T, c)$, there exists $\delta > 0$ such that

$$E^u\left\{\left(\int_0^T [|W_t^{A,u^*}|^2 + |Z_t^{A,u^*}|^2] dt\right)^{\frac{1+\delta}{2}}\right\} < \infty. \tag{5.60}$$

Then, u^* is an optimal control for the optimization problem (5.58).

Theorem 5.3.3 *Assume*

(i) *Assumptions 5.1.7 and 5.3.1 hold, and for each $u \in \mathcal{U}(C_T, c)$, there exists $\delta > 0$ such that*

$$E\{|M_T^u|^{1+\delta}\} < \infty. \tag{5.61}$$

(ii) *There exists a progressively measurable and $\mathbb{F}$-adapted function $I_A(t, c, z)$ taking values in U such that*

$$f_A^*(t, c, z) = f_A(t, c, I_A(t, c, z), z). \tag{5.62}$$

(iii) *$U_A(C_T)$ is bounded and*

$$|f^*(t, c, z)| \leq C[1 + |z|^2] \quad \text{and} \quad |I_A(t, c, z)| \leq C[1 + |z|]. \tag{5.63}$$

(iv) *We have*

$$u^* := I_A(\cdot, c, Z^A) \in \mathcal{U}(C_T, c), \tag{5.64}$$

where (W^A, Z^A) is the unique solution to the BSDE (5.13) such that W^A is bounded. Then, u^* is an optimal control for the optimization problem (5.58) and $V_A(C_T, c) = W_0^A$.

We next provide necessary conditions for the agent's problem.

Assumption 5.3.4

(i) u_A is continuously differentiable in u.
(ii) For each $u \in \mathcal{U}(C_T, c)$, (5.61) holds.
(iii) $\mathcal{U}(C_T, c)$ is locally convex, in the sense of Assumption 10.1.7(ii).
(iv) For u, Δu, u^ε as in Assumption 10.1.7(ii), there exists $\delta > 0$ such that

$$\sup_{0 \le \varepsilon \le 1} E^{u^\varepsilon} \left\{ |U_A(C_T)|^{2+\delta} + \left(\int_0^T [|u_A(t, c_t, u_t^\varepsilon)| + |\partial_u u_A(t, c_t, u_t^\varepsilon)|] dt \right)^{2+\delta} \right\}$$
$$< \infty. \tag{5.65}$$

Applying Theorems 10.4.21 and 10.4.22, we have

Theorem 5.3.5 *Assume Assumptions 5.1.7 and 5.3.4 hold. If u^* is an interior point of $\mathcal{U}(C_T, c)$ and is an optimal control for the agent's problem (5.58), then the agent's necessary condition (5.17) holds.*

Theorem 5.3.6 *Assume*

(i) *All the conditions in Theorem 5.3.5 hold.*
(ii) *There exists a unique function $u = I(t, c, z)$ taking values in U such that (5.18) holds and I is differentiable in z.*
(iii) *$U_A(C_T)$ and W^{A,u^*} are bounded, and the above function I and the function f_A^* defined by (5.19) for the above I satisfy (5.63).*

Then, the BSDE (5.13) is well-posed with solution $(W^A, Z^A) := (W^{A,u^*}, Z^{A,u^*})$, and (5.20) holds.

Remark 5.3.7 In the current setting the agent's admissible set $\mathcal{U}(C_T, c)$ depends on the contract pair (C_T, c). One can make $\mathcal{U}$ independent of (C_T, c) by imposing stronger conditions. For example, let us assume $\partial_u u_A$ has linear growth in u and thus u_A has quadratic growth in u, and each $(C_T, c) \in \mathcal{A}$ satisfies

$$E \left\{ |U_A(C_T)|^{4+\delta} + \left(\int_0^T [|u_A(t, c_t, 0)| + |\partial_u u_A(t, c_t, 0)|] dt \right)^{4+\delta} \right\} < \infty$$

for some $\delta > 0$.

Then, we may consider locally convex $\mathcal{U}$ (independent of (C_T, c)) such that, for each $u \in \mathcal{U}$,

$$E \left\{ |M_T^u|^{2+\delta} + \left(\int_0^T |u_t| dt \right)^{4+\delta} \right\} < \infty \quad \text{for some } \delta > 0.$$

One can easily check that the above set $\mathcal{U}$ satisfies Assumption 5.3.4 for all $(C_T, c) \in \mathcal{A}$. However, we should point out that by imposing stronger conditions on $\mathcal{U}$ it will become more difficult in applications to check that process u^* defined by (5.64) is indeed in $\mathcal{U}$.

5.3.2 Principal's Problem

As in Sects. 5.2.2, 5.2.3, 5.2.4, we may take three approaches to the principal's problem. Since the arguments are more or less a direct combination of those in Sects. 5.2.2, 5.2.3, 5.2.4 and those in Sects. 10.4.4 and 10.4.5, we present only one approach to illustrate the main idea. From the technical point of view, it is more convenient to use the principal's target action, so we take the approach corresponding to Sects. 5.2.3 and 10.4.4.

Assume (5.28) determines uniquely a function J_A. The relaxed principal's problem is:

$$V_P(\lambda) := \sup_{(C_T, u) \in \mathcal{A}} W_0^{P,\lambda,C_T,u,\lambda}, \tag{5.66}$$

where $(W^A, Z^A, W^{P,\lambda}, Z^{P,\lambda}) := (W^{A,C_T,u}, Z^{A,C_T,u}, W^{P,\lambda,C_T,u,\lambda}, Z^{P,\lambda,C_T,u,\lambda})$ is the solution to the following BSDE:

$$\begin{aligned}W_t^A &= U_A(C_T) + \int_t^T u_A^*(s, Z_s^A, u_s)\,ds - \int_t^T Z_s^A\,dB_s^u; \\ W_t^{P,\lambda} &= U_P(C_T) + \int_t^T \left[u_P^*(s, Z_s^A, u_s) + \lambda_s W_s^A - \lambda_s R_s\right]ds - \int_t^T Z_d^{P,\lambda}\,dB_s^u,\end{aligned} \tag{5.67}$$

and u_A^*, u_P^* are defined by (5.33).
Our technical conditions are:

Assumption 5.3.8

(i) The functions U_A and U_P are continuously differentiable in C_T, u_A is continuously differentiable in (c, u), and u_P is continuously differentiable in c.
(ii) Equation (5.28) determines uniquely a function J_A, and J_A is continuously differentiable in (z, u).
(iii) The functions u_A^*, u_P^* are continuously differentiable in z with uniformly bounded derivatives, and continuously differentiable in u with

$$\begin{aligned}|\partial_u u_A^*(t, z, u)| &\leq C\left[1 + |\partial_u u_A^*(t, 0, u)| + |z|\right], \\ |\partial_u u_P^*(t, z, u)| &\leq C\left[1 + |\partial_u u_P^*(t, 0, u)| + |z|\right].\end{aligned}$$

Assumption 5.3.9 The principal's admissible set $\mathcal{A}$ is a set of contract pairs (C_T, u), where C_T is $\mathcal{F}_T$-measurable taking values in A_1 and u is $\mathbb{F}$-adapted taking values in U, such that:

(i) For each $(C_T, u) \in \mathcal{A}$, (10.77) holds. Moreover, let (W^A, Z^A) be the solution to the first BSDE in (5.67), and define c by (5.30). Then, $u \in \mathcal{U}(C_T, c)$, where $\mathcal{U}(C_T, c)$ satisfies Assumption 5.3.4, and u is the agent's optimal control corresponding to (C_T, c).
(ii) $\mathcal{A}$ is locally convex, in the sense of Assumption 10.1.7(ii).
(iii) For (C_T, u), $(\Delta C_T, \Delta u)$, and $(C_T^\varepsilon, u^\varepsilon)$ as in Assumption 10.1.7, there exists $\delta > 0$ such that

$$\sup_{0\le\varepsilon\le 1} E^{u^\varepsilon}\left\{|\varphi(C_T^\varepsilon)|^{2+\delta}+\left(\int_0^T|\psi(t,0,u_t^\varepsilon)|dt\right)^{2+\delta}\right\}<\infty,$$

where $\varphi=U_A, U_A', U_P, U_P'$ and $\psi=u_A^*, \partial_u u_A^*, u_P^*, \partial_u u_P^*$.

Adjoint processes of (10.106), corresponding to (5.67), become $\Gamma^{2,C_T,u}=1$ and $\Gamma^{C_T,u}:=\Gamma^{1,C_T,u}$ satisfying

$$\Gamma_t^{C_T,u}=\int_0^t\lambda_s ds+\int_0^t[\partial_z u_A^*(s,Z_s^A,u_s)\Gamma_s^{C_T,u}+\partial_z u_P^*(s,Z_s^A,u_s)]dB_s^u. \quad (5.68)$$

Combining the arguments of Theorem 5.2.16 and those of Sect. 10.4.3, we get

Theorem 5.3.10 *Assume Assumptions* 5.1.7, 5.3.8 *and* 5.3.9 *hold.*

(i) *If* $(C_T^*,u^*)\in\mathcal{A}$ *is an optimal control for the optimization problem* (5.66) *and* (C_T^*,u^*) *is an interior point of* $\mathcal{A}$, *then*

$$\Gamma_T^* U_A'(C_T^*)+U_P'(C_T^*)=0,$$
$$\Gamma_t^*[\partial_u u_A^*(t,W_t^{A,*},Z_t^{A,*},u_t^*)+Z_t^{A,*}] \quad (5.69)$$
$$+[\partial_u u_P^*(t,W_t^{A,*},Z_t^{A,*},u_t^*)+Z_t^{P,\lambda,*}]=0,$$

where

$$(\Gamma^*,W^{A,*},Z^{A,*},W^{P,\lambda,*},Z^{P,\lambda,*})$$
$$:=(\Gamma_T^{C_T^*,u^*},W_t^{A,C_T^*,u^*},Z^{A,C_T^*,u^*},W_t^{P,\lambda,C_T^*,u^*},Z^{P,\lambda,C_T^*,u^*}). \quad (5.70)$$

(ii) *Assume further that there exist unique functions* $I_P^1(\gamma)$ *and* $I_P^2(t,y_1,z,\gamma)$ *such that they are differentiable in* (y,z,γ) *and*

$$\gamma U_A'(I_P^1(\gamma))+U_P'(I_P^1(\gamma))=0,$$
$$\gamma[\partial_u u_A^*(t,y_1,z_1,I_P^2(t,y_1,z,\gamma))+z_1] \quad (5.71)$$
$$+[\partial_u u_P^*(t,y_1,z_1,I_P^2(t,y_1,z,\gamma))+z_2]=0.$$

Define g_i^* *and* φ^* *as in* (10.45). *Then, the multiple* $(\Gamma^*,W^{A,*},Z^{A,*},W^{P,\lambda,*},Z^{P,\lambda,*})$ *defined by* (5.70) *satisfies the following coupled FBSDE:*

$$\Gamma_t^*=\int_0^t\lambda_s ds+\int_0^t[\partial_z u_A^*(s,Z_s^{A,*},I_P^2(s,W_s^{A^*},Z_s^{A,*},Z_s^{P,\lambda,*},\Gamma_s^*))\Gamma_s^*$$
$$+\partial_z u_P^*(s,Z_s^{A,*},I_P^2(s,W_s^{A^*},Z_s^{A,*},Z_s^{P,\lambda,*},\Gamma_s^*))]dB_s^{u^*};$$
$$W_t^{A,*}=U_A(I_P^1(\Gamma_T^*))-\int_t^T Z_s^{A,*}dB_s^{u^*}$$
$$+\int_t^T u_A^*(s,Z_s^{A,*},I_P^2(s,W_s^{A^*},Z_s^{A,*},Z_s^{P,\lambda,*},\Gamma_s^*))ds; \quad (5.72)$$

5.4 Special Cases

$$W_t^{P,\lambda,*} = U_P\left(I_P^1(\Gamma_T^*)\right) - \int_t^T Z_d^{P,\lambda,*} dB_s^{u*}$$
$$+ \int_t^T \left[u_P^*\left(s, Z_s^{A,*}, I_P^2\left(s, W_s^{A*}, Z_s^{A,*}, Z_s^{P,\lambda,*}, \Gamma_s^*\right)\right) \right.$$
$$\left. + \lambda_s W_s^{A,*} - \lambda_s R_s\right] ds,$$

and the optimal control satisfies

$$C_T^* = I_P^1(\Gamma_T^*), \qquad u_t^* = I_P^2(t, W_t^{A*}, Z_t^{A,*}, Z_t^{P,\lambda,*}, \Gamma_t^*). \tag{5.73}$$

5.4 Special Cases

The FBSDEs we have obtained in previous sections are in general difficult to solve. In this section we discuss some special cases and provide the corresponding simplified FBSDEs. In the following sections we will study several examples and applications that can be solved completely.

5.4.1 Participation Constraint at Time Zero

Consider the case in which the participation constraint is only imposed at time zero,

$$V_A(c, C_T) \geq R_0, \tag{5.74}$$

for some constant R_0. Then, for a given constant $\lambda > 0$, the principal's relaxed problem (5.9) becomes

$$V_P(\lambda) := \sup_{(C_T, c) \in \mathcal{A}} V_P(C_T, c, \lambda)$$
$$:= \sup_{(C_T, c) \in \mathcal{A}} E\left\{M_T^{\hat{u}}\left[\int_0^T u_P(t, c_t) dt + U_P(C_T) + \lambda W_0^A - \lambda R_0\right]\right\}. \tag{5.75}$$

Under the same conditions as in Sects. 5.2 or 5.3, the agent's problem is exactly the same. We next present the FBSDEs corresponding to the necessary conditions. All the results can be proved by following the arguments in Sects. 5.2, 5.3 and Chap. 10. Note that (5.75) is equivalent to taking $\lambda_t = \lambda \delta_0(t)$ in (5.9), where $\delta_0(t)$ denotes the Dirac function, that is, for any function $\varphi(t)$,

$$\int_0^t \lambda_s \varphi(s) ds := \lambda \varphi(0). \tag{5.76}$$

We first take the approach of Sect. 5.2.2. In this case, the adjoint process (5.23) becomes

$$D_t := \lambda - \int_0^t Z_s^{P,\lambda} I_A(s, c_s, Z_s^A) \partial_z I_A(s, c_s, Z_s^A) ds + \int_0^t Z_s^{P,\lambda} \partial_z I_A(s, c_s, Z_s^A) dB_s.$$

Denote

$$W_t^{P,*} := W_t^{P,\lambda,*} - \lambda[W_0^{A,*} - R_0]. \tag{5.77}$$

We have

Theorem 5.4.1 *Assume Assumptions 5.2.10 and 5.2.11 hold. Then, for the problem (5.75) all the results in Theorem 5.2.12 hold true by replacing the FBSDE (5.26) with the following FBSDE:*

$$D_t^* = \lambda + \int_0^t Z_s^{P,*} \partial_z I_A(s, I_P^2(s, D_s^*, Z_s^{A,*}, Z_s^{P,*}), Z_s^{A,*})$$

$$\times [dB_s - I_A(s, I_P^2(s, D_s^*, Z_s^{A,*}, Z_s^{P,*}), Z_s^{A,*})ds];$$

$$W_t^{A,*} = U_A(I_P^1(D_T^*))$$

$$+ \int_t^T u_A(s, I_P^2(s, D_s^*, Z_s^{A,*}, Z_s^{P,*}),$$

$$I_A(s, I_P^2(t, D_t^*, Z_t^{A,*}, Z_t^{P,*}), Z_s^{A,*}))ds \tag{5.78}$$

$$- \int_t^T Z_s^{A,*}[dB_s - I_A(s, I_P^2(s, D_s^*, Z_s^{A,*}, Z_s^{P,*}), Z_s^{A,*})ds];$$

$$W_t^{P,*} = U_P(I_P^1(D_T^*)) + \int_t^T u_P(s, I_P^2(s, D_s^*, Z_s^{A,*}, Z_s^{P,*}))ds$$

$$- \int_t^T Z_s^{P,*}[dB_s - I_A(s, I_P^2(s, D_s^*, Z_s^{A,*}, Z_s^{P,*}), Z_s^{A,*})ds].$$

Remark 5.4.2 (i) The process $W^{P,*}$ represents the principal's utility, rather than the utility for the principal's relaxed problem. In the case $W_0^{A,*} = R_0$, we have $W^{P,*} = W^{P,\lambda,0}$ and $V_P = V_P(\lambda^*)$.

(ii) The process $W^{P,\lambda,*}$ does not appear on the right side of the equations in (5.26), so it is legitimate to use $W^{P,*}$ instead of $W^{P,\lambda,*}$. In particular, this modification does not change the value of $Z^{P,\lambda,*}$, that is, we have

$$Z^{P,*} = Z^{P,\lambda,*}. \tag{5.79}$$

Similarly, following the approach of Sect. 5.2.3, we have

Theorem 5.4.3 *Assume Assumptions 5.2.14 and 5.2.15 hold. Then, for the problem (5.75) all the results in Theorem 5.2.16 hold true by replacing FBSDE (5.37) with the following FBSDE:*

5.4 Special Cases

$$D_t^* = \lambda + \int_0^t [\partial_z u_P^* + D_s^* \partial_z u_A^*](s, I_P^2(s, D_s^*, Z_s^{A,*}, Z_s^{P,*}), Z_s^{A,*})$$
$$\times [dB_s - I_P^2(s, D_s^*, Z_s^{A,*}, Z_s^{P,*})ds];$$

$$W_t^{A,*} = U_A(I_P^1(D_T^*)) + \int_t^T u_A^*(s, I_P^2(s, D_s^*, Z_s^{A,*}, Z_s^{P,*}), Z_s^{A,*})ds$$
$$- \int_t^T Z_s^{A,*}[dB_s - I_P^2(s, D_s^*, Z_s^{A,*}, Z_s^{P,*})ds]; \qquad (5.80)$$

$$W_t^{P,*} = U_P(I_P^1(D_T^*)) + \int_t^T u_P^*(s, I_P^2(s, D_s^*, Z_s^{A,*}, Z_s^{P,*}), Z_s^{A,*})ds$$
$$- \int_t^T Z_s^{P,*}[dB_s - I_P^2(s, D_s^*, Z_s^{A,*}, Z_s^{P,*})ds].$$

As for the approach of Sect. 5.2.4, note that in this case $W_0^A = w^A$ is given, so there is no need to introduce the Lagrange multiplier λ, or, we may choose $\lambda_t := 0$. In this case the relaxed principal's problem (5.42) becomes

$$V_P(w^A) := \sup_{(c,u) \in \mathcal{A}} Y_0^P, \qquad (5.81)$$

where, corresponding to (5.43), $(Y^P, Z^P) := (Y^{P,c,u,w_A}, Z^{P,c,u,w_A})$ is the solution to the BSDE:

$$W_t^P = U_P(J_A(W_T^A)) + \int_t^T [u_P(s, c_s) + u_s Z_s^P]ds - \int_t^T Z_s^P dB_s. \qquad (5.82)$$

Our result is then

Theorem 5.4.4 *Assume Assumptions 5.2.17 and 5.2.18 hold. Then, for the problem (5.81) all the results in Theorem 5.2.19 hold true by replacing FBSDE (5.49) with the following FBSDE:*

$$W_t^{A,*} = w^A - \int_0^t u_A(s, I_P^1(\cdot), I_P^2(\cdot))(s, \Gamma_s^*, Z_s^{P,*}, \bar{Y}_s^*, \bar{Z}_s^*)ds$$
$$- \int_0^t \partial_u u_A(s, I_P^1(\cdot), I_P^2(\cdot))(s, \Gamma_s^*, Z_s^{P,*}, \bar{Y}_s^*, \bar{Z}_s^*)$$
$$\times [dB_s - I_P^2(s, \Gamma_s^*, Z_s^{P,*}, \bar{Y}_s^*, \bar{Z}_s^*)ds];$$

$$\Gamma_t^* = 1 + \int_0^t \Gamma_s^* I_P^2(s, \Gamma_s^*, Z_s^{P,*}, \bar{Y}_s^*, \bar{Z}_s^*) dB_s; \qquad (5.83)$$

$$W_t^{P,*} = U_P^*(W_T^{A,*}) + \int_t^T u_P(s, I_P^1(s, \Gamma_s^*, Z_s^{P,*}, \bar{Y}_s^*, \bar{Z}_s^*))ds$$
$$- \int_t^T Z_s^{P,*}[dB_s - I_P^2(s, \Gamma_s^*, Z_s^{P,*}, \bar{Y}_s^*, \bar{Z}_s^*)ds];$$

$$\bar{Y}_t^* = \partial_x U_P^*(W_T^{A,*})\Gamma_T^* - \int_t^T \bar{Z}_s^* dB_s.$$

Remark 5.4.5 (i) As in Remark 5.2.20, we can simplify the system (5.83) by removing the adjoint process Γ:

$$\begin{aligned}
W_t^{A,*} &= w^A - \int_0^t u_A\big(s, \hat{I}_P^1(\cdot), \hat{I}_P^2(\cdot)\big)\big(s, Z_s^{P,*}, \hat{Y}_s^*, \hat{Z}_s^*\big) ds \\
&\quad - \int_0^t \partial_u u_A\big(s, \hat{I}_P^1(\cdot), \hat{I}_P^2(\cdot)\big)\big(s, Z_s^{P,*}, \hat{Y}_s^*, \hat{Z}_s^*\big) \\
&\quad \times \big[dB_s - \hat{I}_P^2\big(s, Z_s^{P,*}, \hat{Y}_s^*, \hat{Z}_s^*\big) ds\big]; \\
W_t^{P,*} &= U_P^*\big(W_T^{A,*}\big) + \int_t^T u_P\big(s, \hat{I}_P^1\big(s, Z_s^{P,*}, \hat{Y}_s^*, \hat{Z}_s^*\big)\big) ds \\
&\quad - \int_t^T Z_s^{P,*}\big[dB_s - \hat{I}_P^2\big(s, Z_s^{P,*}, \hat{Y}_s^*, \hat{Z}_s^*\big) ds\big]; \\
\hat{Y}_t^* &= \partial_x U_P^*\big(W_T^{A,*}\big) - \int_t^T \hat{Z}_s^*\big[dB_s - \hat{I}_P^2\big(s, Z_s^{P,*}, \hat{Y}_s^*, \hat{Z}_s^*\big) ds\big];
\end{aligned} \qquad (5.84)$$

where $\hat{I}_P^i(z, \hat{y}, \hat{z})$, $i = 1, 2$, are determined by (5.53).

(ii) One may expect that the optimal w^A should be R_0. This is indeed true in special cases. However, a general answer to this question involves the comparison principle for high-dimensional FBSDEs, which is a very challenging problem.

Next, in the case of this section the Hamiltonian defined in (5.56) becomes

$$H(t, x, z, \gamma, \bar{y}, \bar{z}, c, u)$$
$$:= \gamma\big[u_P(t, c) + uz\big] + \bar{y}\big[u\partial_u u_A(t, c, u) - u_A\big] - \bar{z}\partial_u u_A(t, c, u) \quad (5.85)$$

and we have the following sufficiency result:

Theorem 5.4.6 *Assume*

(i) *Assumptions 5.2.21 and 5.2.22 hold with H defined by (5.85).*
(ii) *The FBSDE (5.83) has a solution $(W^{A,*}, \Gamma^*, W^{P,*}, Z^{P,*}, \bar{Y}^*, \bar{Z}^*)$, where the functions I_P^1, I_P^2 are given by Assumption 5.2.21(iv).*
(iii) *The pair (c^*, u^*) defined by (5.50) is in $\mathcal{A}$.*

Then, $V(w^A) = W_0^{P,*}$ and (c^*, u^*) is an optimal control.

Let us also note that we can extend similarly the above results to the quadratic case, as in Sect. 5.3.

5.4.2 Separable Utility and Participation Constraint at Time Zero

In addition to assuming that the participation constraint is imposed only at time zero, we now also assume that the agent's utility is separable in effort and consumption:

5.4 Special Cases

$$u_A(t, c_t, u_t) = u_A(t, c_t) - g(t, u_t). \tag{5.86}$$

As before, u_A and g may depend on the path of X. We then have, in self-evident notation,

$$\partial_c u_A = u'_A; \qquad \partial_u u_A = -g'. \tag{5.87}$$

Recall (5.17) and (5.18). For the agent's problem, we have

$$I_A = (g')^{-1}; \qquad \partial_c I_A = 0; \qquad u^*_t = I_A(t, Z^{A,*}_t). \tag{5.88}$$

Then, the optimality conditions (5.24) become

$$D^*_T U'_A(C^*_T) + U'_P(C^*_T) = 0; \qquad D^*_t u'_A(t, c^*_t) + u'_P(t, c^*_t) = 0, \tag{5.89}$$

and thus, the inverse functions in (5.25) are the inverses of the negative of the ratio of marginal utilities:

$$I^1_P = \left[-\frac{U'_P}{U'_A}\right]^{-1}, \qquad I^2_P = \left[-\frac{u'_P}{u'_A}\right]^{-1}, \quad \text{and}$$

$$C^*_T = I^1_P(D^*_T), \qquad c^*_t = I^2_P(t, D^*_t). \tag{5.90}$$

Plugging this into (5.78), we get

$$D^*_t = \lambda + \int_0^t Z^{P,*}_s I'_A(s, Z^{A,*}_s)[dB_s - I_A(s, Z^{A,*}_s)ds];$$

$$W^{A,*}_t = U_A(I^1_P(D^*_T)) + \int_t^T [u_A(s, I^2_P(s, D^*_s)) - g(s, I_A(s, Z^{A,*}_s))]ds$$

$$- \int_t^T Z^{A,*}_s [dB_s - I_A(s, Z^{A,*}_s)ds]; \tag{5.91}$$

$$\tilde{W}^{P,*}_t = U_P(I^1_P(D^*_T)) + \int_t^T u_P(s, I^2_P(s, D^*_s))ds$$

$$- \int_t^T Z^{P,*}_s [dB_s - I_A(s, Z^{A,*}_s)ds].$$

Remark 5.4.7 We see that in order to describe the solution we need to know the agent's utility process $W^{A,*}$, the (adjusted) principal's utility process $\tilde{W}^{P,*}$, their respective volatilities $Z^{A,*}$ and $Z^{P,*}$, and also process D. As seen from (5.89), and already alluded to in Remark 5.2.13, the negative of process D can be interpreted as the *ratio of marginal utilities process*. It is not so surprising that this is an important process, because $D_T = -U'_P(C_T)/U'_A(C_T)$ had an important role in the first best model, via Borch's rule. In that case D was constant, but here it will be changing randomly, and we need to track its value.

Next, in the approach of Sect. 5.2.4, the optimality condition (5.47) becomes

$$\Gamma^*_t u'_P(t, c^*_t) - \bar{Y}^*_t u^*_t g'(t, u^*_t) = 0,$$

$$\Gamma^*_t Z^{P,*}_t - [\bar{Y}^*_t u^*_t - \bar{Z}^*_t]g''(t, u^*_t) = 0. \tag{5.92}$$

Recall (5.48). Assume the above conditions determine uniquely two functions $I_P^i(t, \gamma, z, \bar{y}, \bar{z})$, $i = 1, 2$, such that

$$c_t^* = I_P^1(t, \Gamma_t^*, Z_t^{P,*}, \bar{Y}_t^*, \bar{Z}_t^*), \qquad u_t^* = I_P^2(t, \Gamma_t^*, Z_t^{P,*}, \bar{Y}_t^*, \bar{Z}_t^*). \quad (5.93)$$

Then, FBSDE (5.83) becomes

$$W_t^{A,*} = w^A - \int_0^t [u_A(s, I_P^1(\cdot)) - g(t, I_P^2(\cdot))](s, \Gamma_s^*, Z_s^{P,*}, \bar{Y}_s^*, \bar{Z}_s^*) ds$$

$$- \int_0^t u_A'(s, I_P^1(s, \Gamma_s^*, Z_s^{P,*}, \bar{Y}_s^*, \bar{Z}_s^*))[dB_s - I_P^2(s, \Gamma_s^*, Z_s^{P,*}, \bar{Y}_s^*, \bar{Z}_s^*) ds];$$

$$\Gamma_t^* = 1 + \int_0^t \Gamma_s^* I_P^2(s, \Gamma_s^*, Z_s^{P,*}, \bar{Y}_s^*, \bar{Z}_s^*) dB_s; \quad (5.94)$$

$$W_t^{P,*} = U_P^*(W_T^{A,*}) + \int_t^T u_P(s, I_P^1(s, \Gamma_s^*, Z_s^{P,*}, \bar{Y}_s^*, \bar{Z}_s^*)) ds$$

$$- \int_t^T Z_s^{P,*}[dB_s - I_P^2(s, \Gamma_s^*, Z_s^{P,*}, \bar{Y}_s^*, \bar{Z}_s^*) ds];$$

$$\bar{Y}_t^* = \partial_x U_P^*(W_T^{A,*}) \Gamma_T^* - \int_t^T \bar{Z}_s^* dB_s.$$

Similarly, in this case the simplified system (5.84) becomes

$$W_t^{A,*} = w^A - \int_0^t [u_A(s, \hat{I}_P^1(\cdot)) - g(t, \hat{I}_P^2(\cdot))](s, Z_s^{P,*}, \hat{Y}_s^*, \hat{Z}_s^*) ds$$

$$- \int_0^t u_A'(s, \hat{I}_P^1(s, Z_s^{P,*}, \hat{Y}_s^*, \hat{Z}_s^*))[dB_s - \hat{I}_P^2(s, Z_s^{P,*}, \hat{Y}_s^*, \hat{Z}_s^*) ds];$$

$$W_t^{P,*} = U_P^*(W_T^{A,*}) + \int_t^T u_P(s, \hat{I}_P^1(s, Z_s^{P,*}, \hat{Y}_s^*, \hat{Z}_s^*)) ds \quad (5.95)$$

$$- \int_t^T Z_s^{P,*}[dB_s - \hat{I}_P^2(s, Z_s^{P,*}, \hat{Y}_s^*, \hat{Z}_s^*) ds];$$

$$\hat{Y}_t^* = \partial_x U_P^*(W_T^{A,*}) - \int_t^T \hat{Z}_s^*[dB_s - \hat{I}_P^2(s, Z_s^{P,*}, \hat{Y}_s^*, \hat{Z}_s^*) ds];$$

where $\hat{I}_P^i$ are determined by (5.53), which in this case becomes

$$\partial_c u_P(t, \hat{I}_P^1) + \hat{y} g'(t, \hat{I}_P^2) = 0, \qquad z - \hat{z} u_A''(t, \hat{I}_P^1) = 0. \quad (5.96)$$

Finally, we remark that, by (5.88), in this case I_A does not depend on c, and thus the approach of Sect. 5.2.3 does not work.

5.4.3 Infinite Horizon

Assume now that $T = \infty$. Then, there is no terminal payoff C_T. In this case (5.26) becomes

5.4 Special Cases

$$D_t^* = \int_0^t \lambda_s ds + \int_0^t Z_s^{P,\lambda,*} \partial_z I_A\left(s, I_P^2\left(s, D_s^*, Z_s^{A,*}, Z_s^{P,\lambda,*}\right), Z_s^{A,*}\right)$$
$$\times \left[dB_s - I_A\left(s, I_P^2\left(s, D_s^*, Z_s^{A,*}, Z_s^{P,\lambda,*}\right), Z_s^{A,*}\right)ds\right];$$
$$W_t^{A,*} = \int_t^\infty u_A\left(s, I_P^2\left(s, D_s^*, Z_s^{A,*}, Z_s^{P,\lambda,*}\right),\right.$$
$$\left. I_A\left(s, I_P^2\left(t, D_t^*, Z_t^{A,*}, Z_t^{P,\lambda,*}\right), Z_s^{A,*}\right)\right)ds \qquad (5.97)$$
$$- \int_t^\infty Z_s^{A,*}\left[dB_s - I_A\left(s, I_P^2\left(s, D_s^*, Z_s^{A,*}, Z_s^{P,\lambda,*}\right), Z_s^{A,*}\right)ds\right];$$
$$W_t^{P,\lambda,*} = \int_t^\infty \left[u_P\left(s, I_P^2\left(s, D_s^*, Z_s^{A,*}, Z_s^{P,\lambda,*}\right)\right) + \lambda_s W_s^{A,*} - \lambda_s R_s\right]ds$$
$$- \int_t^\infty Z_s^{P,\lambda,*}\left[dB_s - I_A\left(s, I_P^2\left(s, D_s^*, Z_s^{A,*}, Z_s^{P,\lambda,*}\right), Z_s^{A,*}\right)ds\right],$$

and (5.25) requires only the second equality.

In the case of the IR constraint only at time zero, we use the following FBSDE to replace (5.78):

$$D_t^* = \lambda + \int_0^t Z_s^{P,*} \partial_z I_A\left(s, I_P^2\left(s, D_s^*, Z_s^{A,*}, Z_s^{P,*}\right), Z_s^{A,*}\right)$$
$$\times \left[dB_s - I_A\left(s, I_P^2\left(s, D_s^*, Z_s^{A,*}, Z_s^{P,*}\right), Z_s^{A,*}\right)ds\right];$$
$$W_t^{A,*} = \int_t^\infty u_A\left(s, I_P^2\left(s, D_s^*, Z_s^{A,*}, Z_s^{P,*}\right), I_A\left(s, I_P^2\left(t, D_t^*, Z_t^{A,*}, Z_t^{P,*}\right), Z_s^{A,*}\right)\right)ds$$
$$- \int_t^\infty Z_s^{A,*}\left[dB_s - I_A\left(s, I_P^2\left(s, D_s^*, Z_s^{A,*}, Z_s^{P,*}\right), Z_s^{A,*}\right)ds\right]; \qquad (5.98)$$
$$\tilde{W}_t^{P,*} = \int_t^\infty u_P\left(s, I_P^2\left(s, D_s^*, Z_s^{A,*}, Z_s^{P,*}\right)\right)ds$$
$$- \int_t^\infty Z_s^{P,*}\left[dB_s - I_A\left(s, I_P^2\left(s, D_s^*, Z_s^{A,*}, Z_s^{P,*}\right), Z_s^{A,*}\right)ds\right].$$

In the separable case, (5.91) becomes

$$D_t^* = \lambda + \int_0^t Z_s^{P,*} I_A'\left(s, Z_s^{A,*}\right)\left[dB_s - I_A\left(s, Z_s^{A,*}\right)ds\right];$$
$$W_t^{A,*} = \int_t^\infty \left[u_A\left(s, I_P^2\left(s, D_s^*\right)\right) - g\left(s, I_A\left(s, Z_s^{A,*}\right)\right)\right]ds$$
$$- \int_t^\infty Z_s^{A,*}\left[dB_s - I_A\left(s, Z_s^{A,*}\right)ds\right]; \qquad (5.99)$$
$$\tilde{W}_t^{P,*} = \int_t^\infty u_P\left(s, I_P^2\left(s, D_s^*\right)\right)ds - \int_t^\infty Z_s^{P,*}\left[dB_s - I_A\left(s, Z_s^{A,*}\right)ds\right].$$

5.4.4 HJB Approach in Markovian Case

Consider the principal's problem with finite horizon, the Markovian case and the IR constraint only at zero: $W_0^A = w^A \geq R_0$. We here take the approach of Sect. 5.2.4. That is, we consider the following optimization problem

$$V(w^A) := \sup_{(c,u) \in \mathcal{A}} W_0^{P,c,u}, \qquad (5.100)$$

where, for each (c, u) in some appropriate admissible set $\mathcal{A}$ and for the function $J_A := [U_A']^{-1}$, we have

$$X_t = x + \int_0^t \sigma(s, X_s) dB_s;$$

$$W_t^{A,c,u} = w^A - \int_0^t u_A(s, X_s, c_s, u_s) ds + \int_0^t \partial_u u_A(s, X_s, c_s, u_s) dB_s^u; \quad (5.101)$$

$$W_t^{P,c,u} = U_P(X_T, J_A(X_T, W_T^{A,c,u})) + \int_t^T u_P(s, X_s, c_s) - \int_t^T Z_s^{P,c,u} dB_s^u.$$

Here, as usual, $W_t^{A,c,u}$ is the remaining expected utility of the agent, and $W_t^{P,c,u}$ is the remaining expected utility of the principal.

As is standard in stochastic control problems, for each $(c, u) \in \mathcal{A}$, denote

$$\mathcal{A}(t, c, u) := \{(\tilde{c}, \tilde{u}) \in \mathcal{A} : \tilde{c} = c, \tilde{u} = u \text{ on } [0, t]\}, \qquad (5.102)$$

and introduce the value process

$$V_t^{P,c,u} := \operatorname*{ess\,sup}_{(\tilde{c},\tilde{u}) \in \mathcal{A}(t,c,u)} E_t^{\tilde{u}}\left[U_P(X_T, J_A(X_T, W_T^{A,\tilde{c},\tilde{u}})) + \int_t^T u_P(s, X_s, \tilde{c}_s) ds\right], \tag{5.103}$$

where the essential supremum is under P (or under the equivalent probability measure P^u). Under standard assumptions of the Stochastic Control Theory, this process satisfies the following Dynamical Programming Principle

$$V_t^{P,c,u} := \operatorname*{ess\,sup}_{(\tilde{c},\tilde{u}) \in \mathcal{A}(t,c,u)} E_t^{\tilde{u}}\left[V_{t+\delta}^{P,\tilde{c},\tilde{u}} + \int_t^{t+\delta} u_P(s, X_s, \tilde{c}_s) ds\right], \qquad (5.104)$$

for any $0 < \delta \leq T - t$, and is of the form

$$V_t^{P,c,u} = F(t, X_t, W_t^{A,c,u}) \qquad (5.105)$$

for some deterministic function F (independent of (c, u)), called the value function. Moreover, under appropriate technical conditions, the value function satisfies the Hamilton–Jacobi–Bellman Partial Differential Equation (HJB PDE)

$$\begin{cases} F_t(t, x, w) + \max_{c, u}\left[F_x u - F_w u_A + \frac{1}{2} F_{xx} \sigma^2 + F_{xw} \sigma \partial_u u_A \right. \\ \left. + \frac{1}{2} F_{ww} |\partial_u u_A|^2 + u_P\right](t, x, w, c, u) = 0; \\ F(T, x, w) = U_P(x, J_A(x, w)). \end{cases} \qquad (5.106)$$

Furthermore, if the supremum in HJB PDE (5.106) is attained at $(X, W^{A,c^*,u^*}, c^*, u^*)$ for some (c^*, u^*), then (c^*, u^*) is optimal for optimization problem (5.100).

One way to prove sufficiency of the HJB PDE is to show that

$$\hat{V}_t^{P,c,u} := F(t, X_t, W_t^{A,c,u}) + \int_0^t u_P(s, X_s, c_s) ds$$

is a P^u-supermartingale on $[0, T]$ for all admissible (c, u), and a P^{u^*}-martingale for (c^*, u^*). An analogous approach would work for $T = \infty$, except we may need additional assumptions to guarantee that $\hat{V}_t^{P,c,u}$ is a P^u-supermartingale on the interval including $T = \infty$, and a martingale for (c^*, u^*). This usually boils down to assuming enough conditions to guarantee that $\hat{V}_t^{P,c,u}$ is uniformly integrable.

5.5 A More General Model with Consumption and Recursive Utilities

As in Sect. 4.7, we extend our analysis of second best contracts to a more general model which allows for recursive utilities and consumption. Moreover, we allow the principal to control the volatility v of the state process X that is still defined by (5.2) in weak formulation.

The agent controls (u, e) and the principal controls (C_T, v, c). The agent's cost $G := G^{u,v,c,e}$, the agent's utility $W^A := W^{A,C_T,u,v,c,e}$, and the principal's utility $W^P := W^{P,C_T,u,v,c,e}$ are given by

$$X_t = x + \int_0^t v_s dB_s = x + \int_0^t u_s v_s ds + \int_0^t v_s dB_s^u;$$

$$G_t = \int_0^t g(s, u_s, v_s, c_s, e_s, X_s) ds;$$

$$W_t^A = U_A(X_T, C_T, G_T) + \int_t^T u_A(s, u_s, v_s, c_s, e_s, X_s, W_s^A, Z_s^A) ds \quad (5.107)$$

$$- \int_t^T Z_s^A dB_s^u;$$

$$W_t^P = U_P(X_T, C_T) + \int_t^T u_P(s, v_s, c_s, X_s, W_s^P, Z_s^P) ds - \int_t^T Z_s^P dB_s^u.$$

Here, the coefficients may depend on the path of B and be random, and our controls C_T, u, v, c, e are all $\mathbb{F}^B$-adapted. The agent's problem is

$$V_A(C_T, v, c) := \sup_{u,e} W_0^A, \quad (5.108)$$

and the principal's problem is

$$V_P := \sup_{C_T,v,c} W_0^P \quad \text{subject to} \quad V_A(C_T, v, c) \geq R_0. \quad (5.109)$$

We note that the state process X is not fixed now. While our coefficients and controls are $\mathbb{F}^B$-adapted, in the standard setting it is more natural to assume the principal's controls to be $\mathbb{F}^X$-adapted. However, under the condition $v > 0$, the two settings are equivalent as discussed in the following remark.

Remark 5.5.1 Let $C_T = \alpha(X)$ be a contract, where $\alpha : C[0, T] \to \mathbb{R}$ is a (deterministic) mapping. If $v > 0$, then, as observed in Remark 5.1.1, $\mathbb{F}^X = \mathbb{F}^B$, and there exists a mapping $\alpha_v : C[0, T] \to \mathbb{R}$ such that $C_T = \alpha_v(B.)$. By defining the admissible set carefully, the optimization over all admissible α will be equivalent to the optimization over all α_v. In this sense, we may assume without loss of generality that the principal chooses C_T among all admissible $\mathcal{F}^B_T$-measurable random variables. Similar arguments work for the principal's other controls.

As in Sect. 4.7, we present only necessary conditions and the corresponding FB-SDEs, and provide a heuristic derivation in Sect. 10.5.2.

We start with the agent's problem. Introduce the following adjoint processes:

$$\Gamma_t^A = 1 + \int_0^t \Gamma_s^A \partial_y u_A(s) ds + \int_0^t \Gamma_s^A \partial_z u_A(s) dB_s^u;$$
$$\bar{Y}_t^A = \partial_G U_A(T) \Gamma_T - \int_t^T \bar{Z}_s^A dB_s^u. \tag{5.110}$$

Given (C_T, v, c), if (u, e) is the agent's optimal control and is in the interior of the admissible set, then, under technical conditions, we have the following necessary conditions for optimality

$$\Gamma_t^A \partial_u u_A(t) + \Gamma_t^A Z_t^A + \bar{Y}_t^A \partial_u g(t) = 0 \quad \text{and} \quad \Gamma_t^A \partial_e u_A(t) + \bar{Y}_t^A \partial_e g(t) = 0. \tag{5.111}$$

Assume further that the above conditions determine uniquely the functions

$$u_t = I_1^A(t, \Theta_t^A), \quad e_t = I_2^A(t, \Theta_t^A),$$
$$\text{where } \Theta_t^A := (X_t, \Gamma_t^A, W_t^A, Z_t^A, \bar{Y}_t^A, v_t, c_t). \tag{5.112}$$

Then, we obtain the following FBSDE system for the optimal solution of the agent's problem:

$$X_t = x + \int_0^t v_s dB_s = x + \int_0^t v_s I_1(s, \Theta_s^A) ds + \int_0^t v_s [dB_s - I_1(s, \Theta_s^A) ds];$$
$$G_t = \int_0^t g(s, X_s, I_1(s, \Theta_s^A), v_s, c_s, I_2(s, \Theta_s^A)) ds;$$
$$\Gamma_t^A = 1 + \int_0^t \Gamma_s^A \partial_y u_A(s, X_s, I_1(s, \Theta_s^A), v_s, c_s, I_2(s, \Theta_s^A), W_s^A, Z_s^A) ds$$
$$+ \int_0^t \Gamma_s^A \partial_z u_A(s, X_s, I_1(s, \Theta_s^A), v_s, c_s, I_2(s, \Theta_s^A), W_s^A, Z_s^A)$$
$$\times [dB_s - I_1(s, \Theta_s^A) ds]; \tag{5.113}$$
$$W_t^A = U_A(X_T, C_T, G_T) - \int_t^T Z_s^A [dB_s - I_1(s, \Theta_s^A) ds]$$

5.5 A More General Model with Consumption and Recursive Utilities

$$+ \int_t^T u_A(s, X_s, I_1(s, \Theta_s^A), v_s, c_s, I_2(s, \Theta_s^A), W_s^A, Z_s^A)ds;$$

$$\bar{Y}_t^A = \partial_G U_A(X_T, C_T, G_T)\Gamma_T^A - \int_t^T \bar{Z}_s^A[dB_s - I_1(s, \Theta_s^A)ds].$$

We now turn to the principal's problem. As in Sects. 5.2.2 to 5.2.4, we may take different approaches. We illustrate here the idea following the arguments of Sect. 5.2.3. Let u be the principal's target action. Assume (5.111) uniquely determines two functions J_1^A, J_2^A such that

$$Z_t^A = J_1^A(t, X_t, \Gamma_t^A, \bar{Y}_t^A, u_t, v_t, c_t), \qquad e_t = J_2^A(t, X_t, \Gamma_t^A, \bar{Y}_t^A, u_t, v_t, c_t). \tag{5.114}$$

Moreover, assume

$$W_T^A = U_A(X_T, C_T, G_T)$$

determines uniquely a function J_A such that

$$C_T = J_A(X_T, W_T^A, G_T). \tag{5.115}$$

Then, given the agent's initial utility w_A, (5.113) leads to

$$X_t = x + \int_0^t v_s dB_s = x + \int_0^t v_s u_s ds + \int_0^t v_s dB_s^u;$$

$$G_t = \int_0^t \hat{g}(s, X_s, u_s, v_s, c_s, \Gamma_s^A, \bar{Y}_s^A)ds;$$

$$\Gamma_t^A = 1 + \int_0^t \Gamma_s^A \widehat{\partial_y u_A}(s, X_s, u_s, v_s, c_s, \Gamma_s^A, \bar{Y}_s^A, W_s^A)ds$$

$$+ \int_0^t \Gamma_s^A \widehat{\partial_z u_A}(s, X_s, u_s, v_s, c_s, \Gamma_s^A, \bar{Y}_s^A, W_s^A)dB_s^u; \tag{5.116}$$

$$W_t^A = w_A - \int_0^t \hat{u}_A(s, X_s, u_s, v_s, c_s, \Gamma_s^A, \bar{Y}_s^A, W_s^A)ds$$

$$+ \int_0^t J_1^A(s, X_s, \Gamma_s^A, \bar{Y}_s^A, u_s, v_s, c_s)dB_s^u;$$

$$W_t^P = \hat{U}_P(X_T, W_T^A, G_T) + \int_t^T u_P(s, v_s, c_s, X_s, W_s^P, Z_s^P)ds - \int_t^T Z_s^P dB_s^u;$$

$$\bar{Y}_t^A = \widehat{\partial_G U_A}(X_T, W_T^A, G_T)\Gamma_T^A - \int_t^T \bar{Z}_s^A dB_s^u,$$

where, for $\varphi = u_A, \partial_y u_A, \partial_z u_A$,

$$\hat{g}(s, x, u, v, c, \gamma, \bar{y}) := g(s, x, u, v, c, J_2^A(s, x, \gamma, \bar{y}, u, v, c));$$

$$\hat{\varphi}(s, x, u, v, c, \gamma, \bar{y}, y)$$

$$:= \varphi(s, u, v, c, J_2^A(s, x, \gamma, \bar{y}, u, v, c), x, y, J_1^A(s, x, \gamma, \bar{y}, u, v, c));$$

$$\hat{U}_P(x, y, G) := U_P(x, J_A(x, y, G));$$

$$\widehat{\partial_G U_A}(x, y, G) := \partial_G U_A(x, J_A(x, y, G), G).$$

The principal's problem becomes:

$$V_P := \sup_{w_A \geq R_0} V_P(w_A) \quad \text{where} \quad V_P(w_A) := \sup_{(u,v,c)} W_0^P. \quad (5.117)$$

Remark 5.5.2 (i) In (5.116), W^A is defined by a forward SDE.

(ii) FBSDE (5.116) is a coupled FBSDE, and the backward components are high-dimensional. Thus, one cannot apply the results in Chap. 10 directly. However, following similar arguments we below derive the necessary conditions for the principal's problem in a heuristic way.

(iii) In general, it is not clear whether or not W_0^P is increasing with respect to w_A. Therefore, one cannot assume $w_A = R_0$. See also Remark 5.4.5(ii).

We now fix w_A and consider the optimization problem $V_P(w_A)$ in (5.117). Introduce the following adjoint processes:

$$\Gamma_t^1 = 1 + \int_0^t \Gamma_s^1 \partial_y u_P(s) ds + \int_0^t \Gamma_s^1 \partial_z u_P(s) dB_s^u;$$

$$\Gamma_t^2 = \int_0^t \big[\bar{Y}_s^2 \partial_{\bar{y}} \hat{g}(s) + \bar{Y}_s^3 \Gamma_s^A \partial_{\bar{y}} \widehat{\partial_y u_A}(s) + \bar{Z}_s^3 \Gamma_s^A \partial_{\bar{y}} \widehat{\partial_z u_A}(s) + \bar{Y}_s^4 \partial_{\bar{y}} \hat{u}_A(s)$$
$$+ \bar{Z}_s^4 \partial_{\bar{y}} J_1^A(s)\big] ds;$$

$$\bar{Y}_t^1 = \Gamma_T^1 \partial_x \hat{U}_P(T) + \Gamma_T^A \Gamma_T^2 \partial_x \widehat{\partial_G U_A}(T) + \int_t^T \big[\Gamma_s^1 \partial_x u_P(s) + \bar{Y}_s^2 \partial_x \hat{g}(s)$$
$$+ \bar{Y}_s^3 \Gamma_s^A \partial_x \widehat{\partial_y u_A}(s) + \bar{Z}_s^3 \Gamma_s^A \partial_x \widehat{\partial_z u_A}(s) + \bar{Y}_s^4 \partial_x \hat{u}_A(s) + \bar{Z}_s^4 \partial_x J_1^A(s)\big] ds$$
$$- \int_t^T \bar{Z}_s^1 dB_s^u;$$

$$\bar{Y}_t^2 = \Gamma_T^1 \partial_G \hat{U}_P(T) + \Gamma_T^A \Gamma_T^2 \partial_G \widehat{\partial_G U_A}(T) - \int_t^T \bar{Z}_s^2 dB_s^u; \quad (5.118)$$

$$\bar{Y}_t^3 = \Gamma_T^2 \widehat{\partial_G U_A}(T) + \int_t^T \big[\bar{Y}_s^2 \partial_\Gamma \hat{g}(s) + \bar{Y}_s^3 \widehat{\partial_y u_A}(s) + \bar{Y}_s^3 \Gamma_s^A \partial_\Gamma \widehat{\partial_y u_A}(s)$$
$$+ \bar{Z}_s^3 \widehat{\partial_z u_A}(s) + \bar{Z}_s^3 \Gamma_s^A \partial_\Gamma \widehat{\partial_z u_A}(s) + \bar{Y}_s^4 \partial_\Gamma \hat{u}_A(s) + \bar{Z}_s^4 \partial_\Gamma J_1^A(s)\big] ds$$
$$- \int_t^T \bar{Z}_s^3 dB_s^u;$$

$$\bar{Y}_t^4 = \Gamma_T^1 \partial_y \hat{U}_P(T) + \Gamma_T^A \Gamma_T^2 \partial_y \widehat{\partial_G U_A}(T)$$
$$+ \int_t^T \big[\bar{Y}_s^3 \Gamma_s^A \partial_y \widehat{\partial_y u_A}(s) + \bar{Z}_s^3 \Gamma_s^A \partial_y \widehat{\partial_z u_A}(s) - \bar{Y}_s^4 \partial_y \hat{u}_A(s)\big] ds - \int_t^T \bar{Z}_s^4 dB_s^u.$$

We note that the above system is also coupled. If (u^*, v^*, c^*) is an optimal control of the principal and is in the interior of the admissible set, then, under technical conditions, we have the following necessary conditions for optimality

5.5 A More General Model with Consumption and Recursive Utilities

$$\Gamma^{1,*}Z^{P,*} + \Gamma^{2,*}\bar{Z}^{A,*} + \bar{Y}^{2,*}\partial_u \hat{g}(t, X^*, u^*, v^*, c^*, \Gamma^{A,*}, \bar{Y}^{A,*})$$
$$+ \bar{Y}^{3,*}\Gamma^{A,*}[\widehat{\partial_u \partial_y u_A} - \widehat{\partial_z u_A}](t, X^*, u^*, v^*, c^*, \Gamma^{A,*}, \bar{Y}^{A,*}, W^{A,*})$$
$$+ \bar{Z}^{3,*}\Gamma^{A,*}\widehat{\partial_u \partial_z u_A}(t, X^*, u^*, v^*, c^*, \Gamma^{A,*}, \bar{Y}^{A,*})$$
$$+ \bar{Y}^{4,*}[\partial_u \hat{u}_A(t, X^*, u^*, v^*, c^*, \Gamma^{A,*}, \bar{Y}^{A,*}, W^{A,*})$$
$$+ J_1^A(t, X^*, u^*, v^*, c^*, \Gamma^{A,*}, \bar{Y}^{A,*})]$$
$$+ \bar{Z}^{4,*}\partial_u J_1^A(t, X^*, u^*, v^*, c^*, \Gamma^{A,*}, \bar{Y}^{A,*}) = 0;$$
$$\Gamma^{1,*}\partial_v u_P(t, v^*, c^*, X^*, W^{P,*}, Z^{P,*}) + \bar{Y}^{1,*}u^* + \bar{Z}^{1,*}$$
$$+ \bar{Y}^{2,*}\partial_v \hat{g}(t, X^*, u^*, v^*, c^*, \Gamma^{A,*}, \bar{Y}^{A,*})$$
$$+ \bar{Y}^{3,*}\Gamma^{A,*}\widehat{\partial_v \partial_y u_A}(t, X^*, u^*, v^*, c^*, \Gamma^{A,*}, \bar{Y}^{A,*}, W^{A,*}) \quad (5.119)$$
$$+ \bar{Z}^{3,*}\Gamma^{A,*}\widehat{\partial_v \partial_z u_A}(t, X^*, u^*, v^*, c^*, \Gamma^{A,*}, \bar{Y}^{A,*}, W^{A,*})$$
$$+ \bar{Y}^{4,*}\partial_v \hat{u}_A(t, X^*, u^*, v^*, c^*, \Gamma^{A,*}, \bar{Y}^{A,*}, W^{A,*})$$
$$+ \bar{Z}^{4,*}\partial_v J_1^A(t, X^*, u^*, v^*, c^*, \Gamma^{A,*}, \bar{Y}^{A,*}) = 0;$$
$$\Gamma^{1,*}\partial_c u_P(t, v^*, c^*, X^*, W^{P,*}, Z^{P,*}) + \bar{Y}^{2,*}\partial_c \hat{g}(t, X^*, u^*, v^*, c^*, \Gamma^{A,*}, \bar{Y}^{A,*})$$
$$+ \bar{Y}^{3,*}\Gamma^{A,*}\widehat{\partial_c \partial_y u_A}(t, X^*, u^*, v^*, c^*, \Gamma^{A,*}, \bar{Y}^{A,*}, W^{A,*})$$
$$+ \bar{Z}^{3,*}\Gamma^{A,*}\widehat{\partial_c \partial_z u_A}(t, X^*, u^*, v^*, c^*, \Gamma^{A,*}, \bar{Y}^{A,*}, W^{A,*})$$
$$+ \bar{Y}^{4,*}\partial_c \hat{u}_A(t, X^*, u^*, v^*, c^*, \Gamma^{A,*}, \bar{Y}^{A,*}, W^{A,*})$$
$$+ \bar{Z}^{4,*}\partial_c J_1^A(t, X^*, u^*, v^*, c^*, \Gamma^{A,*}, \bar{Y}^{A,*}) = 0.$$

Assume further that the above conditions determine uniquely the functions

$$u_t^* = I_1(t, \Theta_t^*), \qquad v_t^* = I_2(t, \Theta_t^*), \qquad c_t^* = I_3(t, \Theta_t^*),$$
where $\Theta := (X, \Gamma^A, \Gamma^1, \Gamma^2, W^A, W^P, Z^P, \bar{Y}^A, \bar{Z}^A, \bar{Y}^1, \bar{Z}^1, \bar{Y}^2, \bar{Z}^2, \bar{Y}^3, \bar{Z}^3,$
$$\bar{Y}^4, \bar{Z}^4). \quad (5.120)$$

Then, we obtain the following FBSDE system for the optimal solution:

$$X_t^* = x + \int_0^t I_2(s, \Theta_s^*) dB_s = x + \int_0^t I_1(s, \Theta_s^*) I_2(s, \Theta_s^*) ds$$
$$+ \int_0^t I_2(s, \Theta_s^*)[dB_s - I_1(s, \Theta_s^*) ds];$$
$$G_t^* = \int_0^t \hat{g}(s, X_s^*, I_1(s, \Theta_s^*), I_2(s, \Theta_s^*), I_3(s, \Theta_s^*), \Gamma_s^{A,*}, \bar{Y}_s^{A,*}) ds;$$
$$\Gamma_t^{A,*} = 1 + \int_0^t \Gamma_s^{A,*}\widehat{\partial_y u_A}(s, X_s^*, I_1(s, \Theta_s^*), I_2(s, \Theta_s^*),$$
$$I_3(s, \Theta_s^*), \Gamma_s^{A,*}, \bar{Y}_s^{A,*}, W_s^{A,*}) ds$$

$$+ \int_0^t \Gamma_s^{A,*} \widehat{\partial_z u_A}(s, X_s^*, I_1(s, \Theta_s^*), I_2(s, \Theta_s^*), I_3(s, \Theta_s^*), \Gamma_s^{A,*}, \bar{Y}_s^{A,*}, W_s^{A,*})$$
$$\times [dB_s - I_1(s, \Theta_s^*)ds];$$

$$\Gamma_t^{1,*} = 1 + \int_0^t \Gamma_s^{1,*} \partial_y u_P(s, I_2(s, \Theta_s^*), I_3(s, \Theta_s^*), X_s^*, W_s^{P,*}, Z_s^{P,*})ds$$
$$+ \int_0^t \Gamma_s^{1,*} \partial_z u_P(s, I_2(s, \Theta_s^*), I_3(s, \Theta_s^*), X_s^*, W_s^{P,*}, Z_s^{P,*})$$
$$\times [dB_s - I_1(s, \Theta_s^*)ds];$$

$$\Gamma_t^{2,*} = \int_0^t [\bar{Y}_s^{2,*} \partial_{\bar{y}} \hat{g}(s, X_s^*, I_1(s, \Theta_s^*), I_2(s, \Theta_s^*), I_3(s, \Theta_s^*), \Gamma_s^{A,*}, \bar{Y}_s^{A,*})$$
$$+ \bar{Y}_s^{3,*} \Gamma_s^{A,*} \partial_{\bar{y}} \widehat{\partial_y u_A}(s, X_s^*, I_1(s, \Theta_s^*), I_2(s, \Theta_s^*),$$
$$I_3(s, \Theta_s^*), \Gamma_s^{A,*}, \bar{Y}_s^{A,*}, W_s^{A,*})$$
$$+ \bar{Z}_s^{3,*} \Gamma_s^{A,*} \partial_{\bar{y}} \widehat{\partial_z u_A}(s, X_s^*, I_1(s, \Theta_s^*), I_2(s, \Theta_s^*),$$
$$I_3(s, \Theta_s^*), \Gamma_s^{A,*}, \bar{Y}_s^{A,*}, W_s^{A,*})$$
$$+ \bar{Y}_s^{4,*} \partial_{\bar{y}} \hat{u}_A(s, X_s^*, I_1(s, \Theta_s^*), I_2(s, \Theta_s^*), I_3(s, \Theta_s^*), \Gamma_s^{A,*}, \bar{Y}_s^{A,*}, W_s^{A,*})$$
$$+ \bar{Z}_s^{4,*} \partial_{\bar{y}} J_1^A(s, X_s^*, \Gamma_s^{A,*}, \bar{Y}_s^{A,*}, I_1(s, \Theta_s^*), I_2(s, \Theta_s^*), I_3(s, \Theta_s^*))]ds;$$

$$W_t^{A,*} = w_A - \int_0^t \hat{u}_A(s, X_s^*, I_1(s, \Theta_s^*), I_2(s, \Theta_s^*), I_3(s, \Theta_s^*), \Gamma_s^{A,*}, \bar{Y}_s^{A,*}, W_s^{A,*})ds$$
$$+ \int_0^t J_1^A(s, X_s^*, \Gamma_s^{A,*}, \bar{Y}_s^{A,*}, I_1(s, \Theta_s^*), I_2(s, \Theta_s^*), I_3(s, \Theta_s^*))$$
$$\times [dB_s - I_1(s, \Theta_s^*)ds]; \qquad (5.121)$$

$$W_t^{P,*} = \hat{U}_P(X_T^*, W_T^{A,*}, G_T^*) + \int_t^T u_P(s, I_2(s, \Theta_s^*), I_3(s, \Theta_s^*), X_s^*, W_s^{P,*}, Z_s^{P,*})ds$$
$$- \int_t^T Z_s^{P,*}[dB_s - I_1(s, \Theta_s^*)ds];$$

$$\bar{Y}_t^{A,*} = \widehat{\partial_G U_A}(X_T^*, W_T^{A,*}, G_T^*) \Gamma_T^{A,*} - \int_t^T \bar{Z}_s^{A,*}[dB_s - I_1(s, \Theta_s^*)ds];$$

$$\bar{Y}_t^{1,*} = \Gamma_T^{1,*} \partial_x \hat{U}_P(X_T^*, W_T^{A,*}, G_T^*) + \Gamma_T^{A,*} \Gamma_T^{2,*} \partial_x \widehat{\partial_G U_A}(X_T^*, W_T^{A,*}, G_T^*)$$
$$+ \int_t^T [\Gamma_s^{1,*} \partial_x u_P(s, I_2(s, \Theta_s^*), I_3(s, \Theta_s^*), X_s^*, W_s^{P,*}, Z_s^{P,*})$$
$$+ \bar{Y}_s^{2,*} \partial_x \hat{g}(s, X_s^*, I_1(s, \Theta_s^*), I_2(s, \Theta_s^*), I_3(s, \Theta_s^*), \Gamma_s^{A,*}, \bar{Y}_s^{A,*})$$
$$+ \bar{Y}_s^{3,*} \Gamma_s^{A,*} \partial_x \widehat{\partial_y u_A}(s, X_s^*, I_1(s, \Theta_s^*), I_2(s, \Theta_s^*),$$
$$I_3(s, \Theta_s^*), \Gamma_s^{A,*}, \bar{Y}_s^{A,*}, W_s^{A,*})$$
$$+ \bar{Z}_s^{3,*} \Gamma_s^{A,*} \partial_x \widehat{\partial_z u_A}(s, X_s^*, I_1(s, \Theta_s^*), I_2(s, \Theta_s^*),$$
$$I_3(s, \Theta_s^*), \Gamma_s^{A,*}, \bar{Y}_s^{A,*}, W_s^{A,*})$$

$$+ \bar{Y}_s^{4,*}\partial_x \hat{u}_A\left(s, X_s^*, I_1(s, \Theta_s^*), I_2(s, \Theta_s^*), I_3(s, \Theta_s^*), \Gamma_s^{A,*}, \bar{Y}_s^{A,*}, W_s^{A,*}\right)$$
$$+ \bar{Z}_s^{4,*}\partial_x J_1^A\left(s, X_s^*, \Gamma_s^{A,*}, \bar{Y}_s^{A,*}, I_1(s, \Theta_s^*), I_2(s, \Theta_s^*), I_3(s, \Theta_s^*)\right)]ds$$
$$- \int_t^T \bar{Z}_s^{1,*}\left[dB_s - I_1(s, \Theta_s^*)ds\right];$$
$$\bar{Y}_t^{2,*} = \Gamma_T^{1,*}\partial_G \hat{U}_P\left(X_T^*, W_T^{A,*}, G_T^*\right) + \Gamma_T^{A,*}\Gamma_T^{2,*}\widehat{\partial_G \partial_G U_A}\left(X_T^*, W_T^{A,*}, G_T^*\right)$$
$$- \int_t^T \bar{Z}_s^{2,*}\left[dB_s - I_1(s, \Theta_s^*)ds\right];$$
$$\bar{Y}_t^{3,*} = \Gamma_T^{2,*}\widehat{\partial_G U_A}\left(X_T^*, W_T^{A,*}, G_T^*\right)$$
$$+ \int_t^T \left[\bar{Y}_s^{2,*}\partial_\Gamma \hat{g}\left(s, X_s^*, I_1(s, \Theta_s^*), I_2(s, \Theta_s^*), I_3(s, \Theta_s^*), \Gamma_s^{A,*}, \bar{Y}_s^{A,*}\right)\right.$$
$$+ \bar{Y}_s^{3,*}\widehat{\partial_y u_A}\left(s, X_s^*, I_1(s, \Theta_s^*), I_2(s, \Theta_s^*), I_3(s, \Theta_s^*), \Gamma_s^{A,*}, \bar{Y}_s^{A,*}, W_s^{A,*}\right)$$
$$+ \bar{Y}_s^{3,*}\Gamma_s^{A,*}\partial_\Gamma \widehat{\partial_y u_A}\left(s, X_s^*, I_1(s, \Theta_s^*), I_2(s, \Theta_s^*),\right.$$
$$\left. I_3(s, \Theta_s^*), \Gamma_s^{A,*}, \bar{Y}_s^{A,*}, W_s^{A,*}\right)$$
$$+ \bar{Z}_s^{3,*}\widehat{\partial_z u_A}\left(s, X_s^*, I_1(s, \Theta_s^*), I_2(s, \Theta_s^*), I_3(s, \Theta_s^*), \Gamma_s^{A,*}, \bar{Y}_s^{A,*}, W_s^{A,*}\right)$$
$$+ \bar{Z}_s^{3,*}\Gamma_s^{A,*}\partial_\Gamma \widehat{\partial_z u_A}\left(s, X_s^*, I_1(s, \Theta_s^*), I_2(s, \Theta_s^*),\right.$$
$$\left. I_3(s, \Theta_s^*), \Gamma_s^{A,*}, \bar{Y}_s^{A,*}, W_s^{A,*}\right)$$
$$+ \bar{Y}_s^{4,*}\partial_\Gamma \hat{u}_A\left(s, X_s^*, I_1(s, \Theta_s^*), I_2(s, \Theta_s^*), I_3(s, \Theta_s^*), \Gamma_s^{A,*}, \bar{Y}_s^{A,*}, W_s^{A,*}\right)$$
$$+ \bar{Z}_s^{4,*}\partial_\Gamma J_1^A\left(s, X_s^*, \Gamma_s^{A,*}, \bar{Y}_s^{A,*}, I_1(s, \Theta_s^*), I_2(s, \Theta_s^*), I_3(s, \Theta_s^*)\right)]ds$$
$$- \int_t^T \bar{Z}_s^{3,*}\left[dB_s - I_1(s, \Theta_s^*)ds\right];$$
$$\bar{Y}_t^{4,*} = \Gamma_T^{1,*}\partial_y \hat{U}_P\left(X_T^*, W_T^{A,*}, G_T^*\right) + \Gamma_T^{A,*}\Gamma_T^{2,*}\partial_y \widehat{\partial_G U_A}\left(X_T^*, W_T^{A,*}, G_T^*\right)$$
$$+ \int_t^T \left[\bar{Y}_s^{3,*}\Gamma_s^{A,*}\partial_y \widehat{\partial_y u_A}\left(s, X_s^*, I_1(s, \Theta_s^*), I_2(s, \Theta_s^*), I_3(s, \Theta_s^*),\right.\right.$$
$$\left. \Gamma_s^{A,*}, \bar{Y}_s^{A,*}, W_s^{A,*}\right)$$
$$+ \bar{Z}_s^{3,*}\Gamma_s^{A,*}\partial_y \widehat{\partial_z u_A}\left(s, X_s^*, I_1(s, \Theta_s^*), I_2(s, \Theta_s^*), I_3(s, \Theta_s^*),\right.$$
$$\left. \Gamma_s^{A,*}, \bar{Y}_s^{A,*}, W_s^{A,*}\right)$$
$$+ \bar{Y}_s^{4,*}\partial_y \hat{u}_A\left(s, X_s^*, I_1(s, \Theta_s^*), I_2(s, \Theta_s^*), I_3(s, \Theta_s^*), \Gamma_s^{A,*}, \bar{Y}_s^{A,*}, W_s^{A,*}\right)]ds$$
$$- \int_t^T \bar{Z}_s^{4,*}\left[dB_s - I_1(s, \Theta_s^*)ds\right].$$

5.6 Further Reading

The seminal paper that set up the model and initiated the literature on moral hazard problems in continuous-time is Holmström and Milgrom (1987). Schattler and Sung

(1993) and Sung (1995) generalized those results using a dynamic programming and martingales approach of Stochastic Control Theory. A nice survey of the literature is provided by Sung (2001). The Stochastic Maximum Principle/FBSDE approach that we use is a modification of those employed by Cvitanić et al. (2009) and Williams (2009). The latter paper also models the possibility of hidden savings of the agent. Sannikov (2008) is an important paper that found a tractable way to analyze PA problems in which the payment to the agent is paid at a continuous rate. A nice survey with illuminating discussions on economic implications and many additional references to recent literature is Sannikov (2012).

References

Cvitanić, J., Wan, X., Zhang, J.: Optimal compensation with hidden action and lump-sum payment in a continuous-time model. Appl. Math. Optim. **59**, 99–146 (2009)

Holmström, B., Milgrom, P.: Aggregation and linearity in the provision of intertemporal incentives. Econometrica **55**, 303–328 (1987)

Sannikov, Y.: A continuous-time version of the principal-agent problem. Rev. Econ. Stud. **75**, 957–984 (2008)

Sannikov, Y.: Contracts: the theory of dynamic principal-agent relationships and the continuous-time approach. Working paper, Princeton University (2012)

Schattler, H., Sung, J.: The first-order approach to continuous-time principal-agent problem with exponential utility. J. Econ. Theory **61**, 331–371 (1993)

Sung, J.: Linearity with project selection and controllable diffusion rate in continuous-time principal-agent problems. Rand J. Econ. **26**, 720–743 (1995)

Sung, J.: Lectures on the Theory of Contracts in Corporate Finance: From Discrete-Time to Continuous-Time Models. Com2Mac Lecture Note Series, vol. 4. Pohang University of Science and Technology, Pohang (2001)

Williams, N.: On dynamic principal-agent problems in continuous time. Working paper, University of Wisconsin-Madison (2009)

Chapter 6
Special Cases and Applications

Abstract We present here well-known examples and applications of continuous-time Principal–Agent models. The seminal work of Holmström and Milgrom (Econometrica 55:303–328, 1987) is the first to use a continuous-time model, showing that doing that can, in fact, lead to simple, while realistic optimal contracts. In particular, if the principal and the agent maximize expected utility from terminal output value, and have non-separable cost of effort and exponential utilities, the optimal contract is linear in that value. With other utilities and separable cost of effort, the optimal contract is nonlinear in the terminal output value, obtained as a solution to a nonlinear equation that generalizes the first best Borch condition. In the case of the agent deriving utility from continuous contract payments on an infinite horizon, and if the principal is risk-neutral, the problem reduces to solving an ordinary differential equation for the principal's expected utility process as a function of the agent's expected utility process. That equation can then be solved numerically for various cases, including the case in which the agent can quit, or be replaced by another agent, or be trained and promoted. These cases are analyzed by studying the necessary conditions in terms of an FBSDE system for the agent's problem, and, in Markovian models, by identifying sufficient conditions in terms of the HJB differential equation for the principal's problem.

6.1 Exponential Utilities and Lump-Sum Payment

We now present a model which is an extension of the one from the seminal paper Hölmstrom and Milgrom (1987). For simplicity of notation, as we have done so far, we assume we have a one-dimensional Brownian motion.

6.1.1 The Model

We have, as usual,

$$dX_t = u_t v_t dt + v_t dB_t^u.$$

We assume that the agent is paid only at the final time T in the amount C_T, and the utilities are exponential: the principal maximizes

$$U_P(X_T, C_T) = U_P(X_T - C_T) := -e^{-\gamma_P(X_T - C_T)}$$

and the agent maximizes

$$U_A(C_T - G_T) := -e^{-\gamma_A(C_T - G_T)} \quad \text{with } G_t := \int_0^t [\mu_s X_s + g(s, u_s, v_s)] ds,$$

for some deterministic function of time μ_t. We consider only the participation constraint at time zero:

$$W_0^A \geq R_0.$$

We also allow the principal to choose the volatility process v.

6.1.2 Necessary Conditions Derived from the General Theory

Note to the Reader The reader not interested in the use of general theory of Chap. 5 can skip this section and go to the following section that provides a more direct approach for dealing with the above model.

In this subsection we derive the necessary conditions formally from the general theory established in the previous chapter.

Recalling (5.107), we have

$$g = \mu x + g(t, u, v), \qquad u_A = u_P = 0,$$
$$U_A = -e^{-\gamma_A(C_T - G_T)}, \qquad U_P = -e^{-\gamma_P(X_T - C_T)}. \tag{6.1}$$

We first study the agent's problem. In this case, (5.110) becomes

$$\Gamma^A = 1, \qquad \bar{Y}_t^A = -\gamma_A e^{-\gamma_A(C_T - G_T)} - \int_t^T \bar{Z}_s^A dB_s^u.$$

Comparing this with (5.107), one can easily see that

$$\bar{Y}^A = \gamma_A W^A, \qquad \bar{Z}^A = \gamma_A Z^A.$$

Thus (5.111) becomes

$$Z_t^A + \gamma_A W_t^A \partial_u g(t, u_t, v_t) = 0. \tag{6.2}$$

Note that $W^A < 0$. Denote

$$\tilde{W}_t^A := -\frac{1}{\gamma_A} \ln[-W_t^A] + G_t, \qquad \tilde{Z}^A := -\frac{Z^A}{\gamma_A W^A}. \tag{6.3}$$

Then,

$$\tilde{W}_t^A = C_T - \int_t^T \left[\frac{\gamma_A}{2} (\tilde{Z}_s^A)^2 + \mu_s X_s + g(s, u_s, v_s) \right] ds - \int_t^T \tilde{Z}_s^A dB_s^u, \tag{6.4}$$

6.1 Exponential Utilities and Lump-Sum Payment

and (6.2) becomes

$$\tilde{Z}_t^A = \partial_u g(t, u_t, v_t). \tag{6.5}$$

Assume this uniquely determines a function I^A such that

$$u_t = I^A(t, v_t, \tilde{Z}_t^A). \tag{6.6}$$

Then, FBSDE (5.113) becomes

$$X_t = x + \int_0^t v_s \, dB_s;$$

$$\tilde{W}_t^A = C_T - \int_t^T \left[\frac{\gamma_A}{2} (\tilde{Z}_s^A)^2 + \mu_s X_s + g\left(s, I^A(s, v_s, \tilde{Z}_s^A), v_s\right) \right.$$

$$\left. - \tilde{Z}_s^A I^A(s, v_s, \tilde{Z}_s^A) \right] ds - \int_t^T \tilde{Z}_s^A \, dB_s.$$

This is a decoupled FBSDE that, under certain technical conditions, solves the agent's problem.

We now turn to the principal's problem. Given the principal's target action u, by (6.5) and (6.7) we have

$$\tilde{Z}_t^A = g_u(t, u_t, v_t), \qquad C_T = \tilde{W}_T^A. \tag{6.7}$$

Let w^A denote the agent's initial utility W_0^A, and

$$\tilde{R}_0 := -\frac{1}{\gamma_A} \ln[-R_0], \qquad \tilde{w}^A := -\frac{1}{\gamma_A} \ln[-w^A]. \tag{6.8}$$

Then, (5.116) becomes

$$X_t = x + \int_0^t v_s u_s \, ds + \int_0^t v_s \, dB_s^u;$$

$$\tilde{W}_t^A = \tilde{w}_A + \int_0^t \left[\frac{\gamma_A}{2} [g_u(s, u_s, v_s)]^2 + \mu_s X_s + g(s, u_s, v_s) \right] ds \tag{6.9}$$

$$+ \int_0^t g_u(s, u_s, v_s) \, dB_s^u;$$

$$W_t^P = -\exp(-\gamma_P [X_T - \tilde{W}_T^A]) - \int_t^T Z_s^P \, dB_s^u$$

and the IR constraint is $\tilde{w}^A \geq \tilde{R}_0$. It is clear that $\tilde{W}^A$ is increasing in $\tilde{w}^A$, and thus W^P is decreasing in $\tilde{w}^A$. Therefore, the principal chooses $\tilde{w}^A = \tilde{R}_0$ and then principal's problem (5.117) becomes

$$V_P := \sup_{(u,v)} W_0^P. \tag{6.10}$$

In this case, (5.118) becomes:

$$\Gamma^1 = 1, \quad \Gamma^2 = 0, \quad \bar{Y}^2 = \bar{Z}^2 = 0, \quad \bar{Y}^3 = \bar{Z}^3 = 0,$$

$$\bar{Y}_t^1 = \gamma_P \exp(-\gamma_P[X_T - \tilde{W}_T^A]) + \int_t^T \mu_s \bar{Y}_s^4 ds - \int_t^T \bar{Z}_s^1 dB_s^u, \quad (6.11)$$

$$\bar{Y}_t^4 = -\gamma_P \exp(-\gamma_P[X_T - \tilde{W}_T^A]) - \int_t^T \bar{Z}_s^4 dB_s^u,$$

and (5.119) leads to

$$Z^P + \bar{Y}^4 \left[g_u + \frac{1}{\gamma_A} g_u g_{uu} + \frac{1}{\gamma_A} g_u \right] - \frac{1}{\gamma_A} \bar{Z}^4 g_{uu} = 0;$$

$$\bar{Y}^1 u + \bar{Z}^1 + \bar{Y}^4 \left[g_v + \frac{1}{\gamma_A} g_u g_{uv} \right] - \frac{1}{\gamma_A} \bar{Z}^4 g_{uv} = 0.$$
(6.12)

It is clear that

$$\bar{Y}^4 = \gamma_P W^P, \quad \bar{Z}^4 = \gamma_P Z^P.$$

Moreover, denote

$$\hat{Y} := \frac{\bar{Y}^1}{W^P}, \quad \hat{Z} := \frac{\bar{Z}^1 - \hat{Y} Z^P}{W^P}.$$

We have

$$\hat{Y}_t = -\gamma_P + \int_t^T \left[\gamma_P \mu_s + \frac{Z_s^P}{W_s^P} \hat{Z}_s \right] ds - \int_t^T \hat{Z}_s dB_s^u.$$

Since μ is deterministic, we get

$$\hat{Y}_t = \gamma_P [M_T - M_t - 1], \quad \hat{Z}_t = 0 \quad \text{where } M_t := \int_0^t \mu_s ds. \quad (6.13)$$

Then, (6.12) becomes:

$$Z^P + \gamma_P W^P \left[g_u + \frac{1}{\gamma_A} g_u g_{uu} + \frac{1}{\gamma_A} g_u \right] - \frac{\gamma_P}{\gamma_A} Z^P g_{uu} = 0;$$

$$\hat{Y} W^P u + \hat{Y} Z^P + \gamma_P W^P \left[g_v + \frac{1}{\gamma_A} g_u g_{uv} \right] - \frac{\gamma_P}{\gamma_A} Z^P g_{uv} = 0.$$
(6.14)

We finally solve (6.9), (6.13), and (6.14). Denote

$$\hat{W}_t^A := \tilde{W}_t^A - M_t X_t. \quad (6.15)$$

Then,

$$\hat{W}_t^A = \tilde{R}_0 + \int_0^t \left[\frac{\gamma_A}{2} [g_u(s, u_s, v_s)]^2 + g(s, u_s, v_s) - M_s v_s u_s \right] ds$$

$$+ \int_0^t [g_u(s, u_s, v_s) - M_s v_s] dB_s^u.$$

Also denote

6.1 Exponential Utilities and Lump-Sum Payment

$$\tilde{W}_t^P := -\frac{1}{\gamma_P}\ln(-W_t^P), \quad \tilde{Z}_t^P := -\frac{Z_t^P}{\gamma_P W_t^P}. \quad (6.16)$$

Then,

$$\tilde{W}_t^P = X_T - \tilde{W}_T^A - \int_t^T \frac{\gamma_P}{2}|\tilde{Z}_s^P|^2 ds - \int_t^T \tilde{Z}_s^P dB_s^u,$$

and (6.14) becomes

$$[\gamma_A - \gamma_P g_{uu}]\tilde{Z}^P = \gamma_A g_u + g_u g_{uu} + g_u;$$
$$[\gamma_A \hat{Y} - \gamma_P g_{uv}]\tilde{Z}^P = \frac{\gamma_A}{\gamma_P}\hat{Y}u + \gamma_A g_v + g_u g_{uv}. \quad (6.17)$$

Note that

$$X_T - \tilde{W}_T^A = (1 - M_T)X_T - \hat{W}_T^A$$

and that M is deterministic. Denote

$$\hat{W}_t^P := \tilde{W}_t^P - (1 - M_T)X_t + \hat{W}_t^A. \quad (6.18)$$

Then,

$$\hat{W}_t^P = -\int_t^T \left[\frac{\gamma_P}{2}|\tilde{Z}_s^P|^2 - (1 - M_T)v_s u_s + \frac{\gamma_A}{2}[g_u(s, u_s, v_s)]^2\right.$$
$$\left. + g(s, u_s, v_s) - M_s v_s u_s\right]ds$$
$$- \int_t^T \left[\tilde{Z}_s^P - (1 - M_T)v_s + g_u(s, u_s, v_s) - M_s v_s\right]dB_s^u.$$

One solution for this and (6.17) is

$$\tilde{Z}_s^P = (1 - M_T)v_s - g_u(s, u_s, v_s) + M_s v_s;$$
$$\hat{W}_t^P = -\int_t^T \left[\frac{\gamma_P}{2}|\tilde{Z}_s^P|^2 - (1 - M_T)v_s u_s + \frac{\gamma_A}{2}[g_u(s, u_s, v_s)]^2\right. \quad (6.19)$$
$$\left. + g(s, u_s, v_s) - M_s v_s u_s\right]ds;$$

where u, v are deterministic and satisfy

$$[\gamma_A - \gamma_P g_{uu}][(1 - M_T)v - g_u + Mv] = \gamma_A g_u + g_u g_{uu} + g_u;$$
$$[\gamma_A \hat{Y} - \gamma_P g_{uv}][(1 - M_T)v - g_u + Mv] = \frac{\gamma_A}{\gamma_P}\hat{Y}u + \gamma_A g_v + g_u g_{uv}. \quad (6.20)$$

Assume (6.20) determines uniquely deterministic functions (u^*, v^*). Solving (6.19) we obtain $\hat{W}_0^{P,*}$. Then, the principal's optimal utility is

$$W_0^{P,*} := -\exp(-\gamma_P \tilde{W}_0^{P,*}) = -\exp(-\gamma_P[\hat{W}_0^{P,*} + (1 - M_T)x - \tilde{R}_0]),$$

and the optimal contract is $C_T^* = \tilde{W}_T^{A,*}$, where the latter is defined by (6.9) with optimal control (u^*, v^*). In the following section we prove rigorously that the above solution is indeed optimal for the problem.

6.1.3 A Direct Approach

In this section we provide a direct approach for solving the problem, without using the results of Chap. 5. However, we use additional results from the BSDE theory. We start with the agent's problem. Given a pair (C_T, v), the agent's utility process is given by

$$W_t^A = U_A(C_T - G_T) - \int_t^T Z_s^A dB_s^u.$$

Recall the agent's certainty equivalent process $\tilde{W}^A$ and the corresponding process $\tilde{Z}^A$ as defined in (6.3):

$$\tilde{W}_t^A := -\frac{1}{\gamma_A} \ln[-W_t^A] + G_t, \qquad \tilde{Z}^A := -\frac{Z^A}{\gamma_A W^A}. \qquad (6.21)$$

We have the following result.

Proposition 6.1.1 *Assume, for a given pair (C_T, v), that the admissible set for u is such that BSDE (6.23) below is well-posed and satisfies the BSDE comparison principle (as stated in Part V of the book). Then, the necessary and sufficient condition for the agent's optimal effort is*

$$u_t = I_A(t, v_t, \tilde{Z}^A) := \underset{u}{\mathrm{argmin}}\big[g(t, u, v_t) - u\tilde{Z}_t^A\big]. \qquad (6.22)$$

Proof Note that $\tilde{W}_0^A = -\frac{1}{\gamma_A} \ln(-W_0^A)$. Then, the optimization of the agent's utility W_0^A is equivalent to the optimization of $\tilde{W}_0^A$. By (6.4), or by applying Itô's rule directly, we get

$$\tilde{W}_t^A = C_T - \int_t^T \left[\frac{\gamma_A}{2}(\tilde{Z}_s^A)^2 + \mu_s X_s + g(s, u_s, v_s) - u_s \tilde{Z}_s^A\right] ds - \int_t^T \tilde{Z}_s^A dB_s. \qquad (6.23)$$

By the comparison principle for BSDEs we see that the optimal u is obtained by minimizing the integrand in the first integral in the previous expression, which completes the proof. □

Remark 6.1.2 BSDE (6.23) has quadratic growth in $\tilde{Z}^A$. When C_T is bounded, we prove the well-posedness and the comparison principle for such BSDEs in Sect. 9.6. However, C_T corresponding to the optimal contract in Theorem 6.1.3 below is in general not bounded. Instead we can use the comparison theorem from Briand and Hu (2008). In order to apply that theorem, we need to assume that we only allow actions (u, v) and contracts C_T such that

$$E\big[e^{\lambda \sup_{0 \le t \le T} |\tilde{W}_t^A|}\big] < \infty, \quad \forall \lambda > 0$$

and the random variable

$$\int_0^T \big|\mu_t X_t + g(t, u_t, v_t) - u_t \tilde{Z}_t^A\big| dt$$

has exponential moments of all orders.

6.1 Exponential Utilities and Lump-Sum Payment

We now turn to the principal's problem. Assume g is differentiable in u and the optimal u is in the interior of the admissible set. Then, (6.22) leads to

$$Z_t^A + \gamma_A W_t^A \partial_u g(t, u_t, v_t) = 0 \qquad (6.24)$$

and thus we get, also using (6.23),

$$\tilde{Z}_t^A = g_u(t, u_t, v_t), \qquad C_T = \tilde{W}_T^A. \qquad (6.25)$$

Denote

$$\tilde{R}_0 := -\frac{1}{\gamma_A} \ln[-R_0], \qquad W_0^A = w^A, \qquad \tilde{w}^A := -\frac{1}{\gamma_A} \ln[-w^A]. \qquad (6.26)$$

Then, we can write

$$\begin{aligned}
X_t &= x + \int_0^t v_s u_s \, ds + \int_0^t v_s \, dB_s^u; \\
\tilde{W}_t^A &= \tilde{w}_A + \int_0^t \left[\frac{\gamma_A}{2} [g_u(s, u_s, v_s)]^2 + \mu_s X_s + g(s, u_s, v_s) \right] ds \\
&\quad + \int_0^t g_u(s, u_s, v_s) \, dB_s^u; \\
W_t^P &= -\exp(-\gamma_P [X_T - \tilde{W}_T^A]) - \int_t^T Z_s^P \, dB_s^u.
\end{aligned} \qquad (6.27)$$

As in Sect. 5.2.3, instead of using contract payment C_T as the principal's control, we use the corresponding agent's optimal action u as the principal's control. Given a principal's "target action" u, the volatility control v, and the agent's initial utility $w^A \geq R_0$, the corresponding contract is $C_T = \tilde{W}_T^A$. Clearly, $\tilde{W}^A$ is increasing in $\tilde{w}^A$, and so W^P is decreasing in $\tilde{w}^A$. Thus, the principal chooses $\tilde{w}^A = \tilde{R}_0$ and faces the problem

$$V_P := \sup_{(u,v)} W_0^P. \qquad (6.28)$$

The solution is given by the following result:

Theorem 6.1.3 *Consider the function*

$$\begin{aligned}
L(t, u_t, v_t) &:= \frac{\gamma_P}{2} \left[(1 - M_T + M_t) v_t - g_u(t, u_t, v_t) \right]^2 - (1 - M_T + M_t) u_t v_t \\
&\quad + \frac{\gamma_A}{2} |g_u(t, u_t, v_t)|^2 + g(t, u_t, v_t),
\end{aligned} \qquad (6.29)$$

where M is defined by

$$M_t := \int_0^t \mu_s \, ds.$$

Assume that, for every t, there exists a pair (u_t^*, v_t^*) minimizing this expression, and such that $\int_0^T L(t, u_t^*, v_t^*) \, dt$ is finite. Then, the deterministic controls (u_t^*, v_t^*) are optimal for the principal's problem. The optimal contract payoff is given by

$$C_T^* = c + \int_0^T \left[M_T - M_t + \frac{g_u(t, u_t^*, v_t^*)}{v_t^*} \right] dX_t^* \qquad (6.30)$$

for a constant c chosen so that the agent's expected utility is equal to his reservation value R_0. In particular, if $g_u(t, u_t^*, v_t^*)/v_t^* + M_T - M_t$ is a constant, then contract C_T^* is linear in X_T^*.

Proof Doing integration by parts we get the following representation for the first part of the cost G_T:

$$\int_0^T \mu_t X_t dt = X_T M_T - \int_0^T M_t \left[u_t v_t dt + v_t dB_t^u \right]. \qquad (6.31)$$

Then, by (6.27) and $\tilde{W}_0^A = \tilde{R}_0$, we see that we need to minimize

$$-W_0^P = E^u \left[\exp\left(-\gamma_P [X_T - \tilde{W}_T^A]\right) \right]$$
$$= E^u \left[\exp\left(-\gamma_P \left[(1 - M_T)x - \tilde{R}_0 \right. \right. \right.$$
$$+ \int_0^T \left[(1 - M_T + M_t) u_t v_t - \left(\frac{\gamma_A}{2} |g_u|^2 + g \right) \right] dt$$
$$+ \left. \left. \left. \int_0^T \left[(1 - M_T + M_t) v_t - g_u \right] dB_t^u \right] \right) \right]. \qquad (6.32)$$

This is a standard stochastic control problem, for which the solution, when it exists, turns out to be a pair of deterministic processes (u^*, v^*). (This can be verified, once the solution is found, by verifying the corresponding HJB equation.) Assuming that u, v are deterministic, the expectation above can be computed by using the fact that

$$E^u \left[\exp\left(\int_0^T f_s dB_s^u \right) \right] = \exp\left(\frac{1}{2} \int_0^T f_s^2 ds \right)$$

for a given square-integrable deterministic function f. Then,

$$-W_0^P = \exp\left(-\gamma_P \left[(1 - M_T)x - \tilde{R}_0 + \int_0^T \left[(1 - M_T + M_t) u_t v_t \right. \right. \right.$$
$$\left. \left. - \left(\frac{\gamma_A}{2} |g_u|^2 + g \right) \right] dt \right] + \frac{1}{2} \gamma_P^2 \int_0^T \left[(1 - M_T + M_t) v_t - g_u \right]^2 dt \right)$$
$$= \exp\left(-\gamma_P [(1 - M_T)x - \tilde{R}_0] + \gamma_P \int_0^T L(t, u_t, v_t) dt \right).$$

Thus, the minimization can be done inside the integral in the exponent, and boils down to minimizing $L(t, u_t, v_t)$ over (u_t, v_t), which proves the first part of the theorem.

The optimal contract is found from $C_T^* = \tilde{W}_T^{A,*}$. Note that (6.31) is equivalent to

$$\int_0^T \mu_t X_t dt = \int_0^T [M_T - M_t] dX_t$$

6.1 Exponential Utilities and Lump-Sum Payment

and that

$$\int_0^T g_u(t, u_t^*, v_t^*) dB_t^{u^*} = \int_0^T \frac{g_u(t, u_t^*, v_t^*)}{v_t^*} dX_t^* - \int_0^T g_u(t, u_t^*, v_t^*) u_t^* dt.$$

Plugging these into (6.27) we obtain (6.30). □

Remark 6.1.4 Assume the functions below are smooth enough and the optimal controls (u^*, v^*) are in the interior of the admissible set. Then, the minimization of $L(t, u, v)$ leads to

$$-\gamma_P \big[(1 - M_T + M_t) v_t^* - g_u \big] g_{uu} - (1 - M_T + M_t) v_t^* + \gamma_A g_u g_{uu} + g_u = 0;$$
$$\gamma_P \big[(1 - M_T + M_t) v_t^* - g_u \big] [1 - M_T + M_t - g_{uv}] - (1 - M_T + M_t) u_t^*$$
$$+ \gamma_A g_u g_{uv} + g_v = 0.$$

One can check straightforwardly that this is equivalent to (6.20).

6.1.4 A Solvable Special Case with Quadratic Cost

Consider now the special case of Holmström–Milgrom (1987), with

$$\mu_t \equiv 0, \qquad v_t \equiv v, \qquad g(t, x, u, v) = (uv)^2/2.$$

Then, $g_u = v^2 u$ and the expression (6.29) becomes

$$L(t, u_t) := \frac{\gamma_P}{2} \big[v - v^2 u_t \big]^2 - u_t v + \frac{\gamma_A}{2} |v^2 u_t|^2 + \frac{1}{2} |v u_t|^2. \qquad (6.33)$$

Minimizing this we get constant optimal u^* of Holmström–Milgrom (1987), given by

$$u^* = \frac{\frac{1}{v} + \gamma_P v}{1 + (\gamma_A + \gamma_P) v^2}.$$

The optimal contract is linear, and given by

$$C_T^* = c + \frac{1 + \gamma_P v^2}{1 + (\gamma_A + \gamma_P) v^2} X_T,$$

where c is such that the IR constraint is satisfied,

$$c = -\frac{1}{\gamma_A} \log(-R_0) - u^* v x + \frac{|u^* v|^2 T}{2} (\gamma_A - 1). \qquad (6.34)$$

In particular, one prediction is that with lower uncertainty v, the "pay-per-performance" sensitivity (the slope) of the contract is higher; in fact, it is equal to 1 when $v = 0$: the principal turns over the whole firm to the agent when there is no risk.

6.2 General Risk Preferences, Quadratic Cost, and Lump-Sum Payment

6.2.1 The Model

Consider now the setting in which the participation constraint is imposed only at time zero, there is no intermediate consumption, just the lump sum payment C_T at the end, no volatility control, and the cost is quadratic:

$$u_A = u_P = 0, \qquad g = ku^2/2 \quad \text{for some constant } k. \tag{6.35}$$

Moreover, the agent's utility is separable in effort and contract payment, so that the model becomes

$$\begin{aligned} W_t^A &= U_A(C_T) - \int_t^T \frac{ku_s^2}{2} ds - \int_t^T Z_s^A dB_s^u, \\ W_t^P &= U_P(C_T) - \int_t^T Z_s^P dB_s^u \end{aligned} \tag{6.36}$$

and the IR constraint is

$$W_0^A \geq R_0. \tag{6.37}$$

6.2.2 Necessary Conditions Derived from the General Theory

Note to the Reader The reader not interested in the use of general theory of Chap. 5 can skip this section and go to the following section that provides a more direct approach for dealing with the above model.

As usual, we start with the agent's problem. In this case, by (5.88) we have

$$I_A(z) = \frac{z}{k}, \quad \text{and the agent's optimal control satisfies } u = \frac{1}{k} Z^A. \tag{6.38}$$

Consequently, given C_T, the agent's optimal utility process satisfies

$$\begin{aligned} W_t^A &= U_A(C_T) - \int_t^T \left[\frac{ku_s^2}{2} - u_s Z_s^A \right] ds - \int_t^T Z_s^A dB_s \\ &= U_A(C_T) + \int_t^T \frac{1}{2k} |Z_s^A|^2 ds - \int_t^T Z_s^A dB_s. \end{aligned} \tag{6.39}$$

Denote

$$\tilde{W}_t^A := e^{W_t^A/k}, \qquad \tilde{Z}_t^A := \frac{1}{k} \tilde{W}^A Z^A. \tag{6.40}$$

Applying Itô's rule, we get

6.2 General Risk Preferences, Quadratic Cost, and Lump-Sum Payment

$$\tilde{W}_t^A = e^{U_A(C_T)/k} - \int_t^T \tilde{Z}_s^A dB_s. \tag{6.41}$$

If $E[e^{2U_A(C_T)/k}] < \infty$, the above BSDE is well-posed, with the solution $\tilde{W}_t^A = E_t[e^{U_A(C_T)/k}]$, and we obtain the agent's optimal utility and optimal control:

$$W_t^A = k \ln(\tilde{W}_t^A) = k \ln\bigl(E_t[e^{U_A(C_T)/k}]\bigr), \qquad u_t = Z_t^A/k = \frac{\tilde{Z}_t^A}{\tilde{W}_t^A}. \tag{6.42}$$

We now turn to the principal's problem. As in Sect. 5.4.2, we take two different approaches, corresponding to Sects. 5.2.2 and 5.2.4, respectively. For the first approach, we consider the relaxed principal's problem

$$V_P(\lambda) := \sup_{C_T}\bigl[W_0^P + \lambda W_0^A\bigr], \quad \text{with } u = Z^A/k \text{ in (6.36)}. \tag{6.43}$$

The first equation in (5.89) gives us the optimality condition for C_T, that translates into $C_T = I_P(D_T)$, assuming the following inverse function exists:

$$I_P := \bigl[-U_P'/U_A'\bigr]^{-1}. \tag{6.44}$$

Recall (6.38) and, as in Theorem 5.4.1,

$$\tilde{W}_t^P := W_t^P - \lambda W_0^A. \tag{6.45}$$

Then, (5.91) becomes:

$$D_t = \lambda + \int_0^t \frac{1}{k} Z_s^P\left[dB_s - \frac{1}{k}Z_s^A ds\right];$$
$$W_t^A = U_A(I_P(D_T)) - \int_t^T \frac{1}{2k}|Z_s^A|^2 ds - \int_t^T Z_s^A\left[dB_s - \frac{1}{k}Z_s^A ds\right]; \quad (6.46)$$
$$\tilde{W}_t^P = U_P(I_P(D_T)) - \int_t^T Z_s^P\left[dB_s - \frac{1}{k}Z_s^A ds\right].$$

Moreover, the principal's optimal utility and the optimal contract are

$$V_P(\lambda) = W_0^P = \tilde{W}_0^P + \lambda W_0^A, \qquad C_T = I_P(D_T). \tag{6.47}$$

Comparing the equations for D_t and $\tilde{W}_t^P$ in (6.46), we see that

$$D_t = \frac{1}{k}\tilde{W}_t^P + \tilde{\lambda}$$

for some constant $\tilde{\lambda}$. In particular,

$$D_T = \frac{1}{k}U_P(C_T) + \tilde{\lambda}.$$

This means, using (6.47), that the optimal C_T can be obtained from the following generalization of Borch's rule to the hidden action case:

$$\frac{U_P'(C_T)}{U_A'(C_T)} = -\frac{1}{k}U_P(C_T) - \tilde{\lambda}. \tag{6.48}$$

Assume the above equation determines uniquely

$$C_T = \xi(\tilde{\lambda}) \quad \text{for some random variable } \xi(\tilde{\lambda}). \tag{6.49}$$

We then have the BSDE system

$$W_t^A = U_A(\xi(\tilde{\lambda})) - \int_t^T \frac{1}{2k}|Z_s^A|^2 ds - \int_t^T Z_s^A \left[dB_s - \frac{1}{k} Z_s^A ds \right];$$

$$\tilde{W}_t^P = U_P(\xi(\tilde{\lambda})) - \int_t^T Z_s^P \left[dB_s - \frac{1}{k} Z_s^A ds \right].$$

We have

$$\lambda = D_0 = \frac{1}{k}\tilde{W}_0^P + \tilde{\lambda} \quad \text{and thus } V_P(\lambda) = \tilde{W}_0^P + \left[\frac{1}{k}\tilde{W}_0^P + \tilde{\lambda}\right] W_0^A.$$

Finally, if we can find $\tilde{\lambda}^*$ such that the corresponding agent's initial wealth satisfies $W_0^{A,*} = R_0$, then we have

$$V_P = \tilde{W}_0^{P,*} + \left[\frac{1}{k}\tilde{W}_0^{P,*} + \tilde{\lambda}^*\right] R_0.$$

Remark 6.2.1 In this remark we assume that U_A is a deterministic function and

$$U_P(C_T) = \tilde{U}_P(X_T - C_T)$$

for some deterministic function $\tilde{U}_P$. Then, (6.48) is a nonlinear equation:

$$\frac{\tilde{U}_P'(X_T - C_T)}{U_A'(C_T)} = \frac{1}{k} U_P(X_T - C_T) + \tilde{\lambda} \tag{6.50}$$

and thus the optimal contract C_T is a function of the terminal value X_T only:

$$C_T = \Phi(X_T) \quad \text{for some deterministic function } \Phi. \tag{6.51}$$

For an economic discussion of this nonlinear equation see Remark 6.2.4 below.

We next study the principal's problem following the approach in Sect. 5.2.4. Let u be the principal's target action, $w^A \geq R_0$ be the agent's initial utility. Denote

$$J_A := (U_A')^{-1} \quad \text{and} \quad \hat{U}_P := U_P(J_A). \tag{6.52}$$

Then, by (6.38) and (6.36),

$$Z^A = ku, \quad C_T = J_A(W_T^A), \tag{6.53}$$

and thus

$$W_t^A = w^A + \int_0^t \frac{ku_s^2}{2} ds + \int_0^t ku_s dB_s^u;$$

$$W_t^P = \hat{U}_P(W_T^A) - \int_t^T Z_s^P dB_s^u.$$

6.2 General Risk Preferences, Quadratic Cost, and Lump-Sum Payment

We assume the standard condition

U_A is increasing and concave, U_P is decreasing and concave.   (6.54)

Then, clearly W^A is increasing in w^A and W^P is decreasing in w^A. Thus, the principal would chooses $w^A = R_0$. Therefore, the principal's problem becomes

$$V_P := \sup_u W_0^P, \qquad (6.55)$$

where

$$\begin{aligned} W_t^A &= R_0 + \int_0^t \frac{ku_s^2}{2} ds + \int_0^t ku_s dB_s^u; \\ W_t^P &= \hat{U}_P(W_T^A) - \int_t^T Z_s^P dB_s^u. \end{aligned} \qquad (6.56)$$

In this case, (5.46) becomes

$$\begin{aligned} \Gamma_t &= 1 + \int_0^t \Gamma_s u_s dB_s; \\ \bar{Y}_t &= \hat{U}'_P(W_T^A)\Gamma_T - \int_t^T \bar{Z}_s dB_s, \end{aligned} \qquad (6.57)$$

and the optimization condition (5.47) (see also (5.92)) becomes

$$\Gamma_t Z_t^P - [\bar{Y}_t u_t - \bar{Z}_t]k = 0. \qquad (6.58)$$

Applying Itô's rule, we have

$$\begin{aligned} &d\left(\Gamma_t W_t^P + k\bar{Y}_t\right) \\ &= -\Gamma_t Z_t^P u_t dt + \Gamma_t Z_t^P dB_t + W_t^P \Gamma_t u_t dB_t + Z^P \Gamma_t u_t dt + k\bar{Z}_t dB_t \\ &= \left[\Gamma_t Z_t^P + W_t^P \Gamma_t u_t + k\bar{Z}_t\right] dB_t = \left[\Gamma_t W_t^P + k\bar{Y}_t\right] u_t dB_t, \end{aligned}$$

thanks to (6.58). This implies that

$$\Gamma_t W_t^P + \bar{Y}_t = \hat{\lambda} \Gamma_t$$

for some constant $\hat{\lambda}$. In particular,

$$\hat{\lambda}\Gamma_T = \Gamma_T W_T^P + k\bar{Y}_T = \Gamma_T \hat{U}_P(W_T^A) + k\hat{U}'_P(W_T^A)\Gamma_T.$$

Then,

$$\hat{U}_P(W_T^A) + k\hat{U}'_P(W_T^A) = \hat{\lambda}.$$

This, together with (6.52) and (6.53), leads to (6.48) again, for an appropriately chosen constant $\tilde{\lambda}$.

6.2.3 A Direct Approach

In this section we provide a direct approach for solving the problem, without using the results of Chap. 5, except for the general model described in Sect. 5.1.

We start with the agent's problem. Note that the agent's utility process satisfies

$$W_t^A = U_A(C_T) - \int_t^T \left[\frac{ku_s^2}{2} - u_s Z_s^A\right] ds - \int_t^T Z_s^A dB_s. \qquad (6.59)$$

We have then immediately the following result.

Proposition 6.2.2 *Assume, for a given C_T, that the admissible set of u is given such that the BSDE (6.59) is well-posed and satisfies the comparison principle. Then, the necessary and sufficient condition for the agent's optimal effort is*

$$u_t = \frac{1}{k} Z_t^A. \qquad (6.60)$$

Proof By the comparison principle for BSDEs the optimal u is obtained by minimizing the integrand $\frac{ku^2}{2} - uZ^A$ in (6.59), which implies (6.60). □

By (6.60), the agent's optimal utility W^A satisfies

$$dW_t^A = -\frac{k}{2}u_t^2 + ku_t dB_t.$$

Then,

$$de^{W_t^A/k} = e^{W_t^A/k} u_t dB_t. \qquad (6.61)$$

This implies that, recalling (5.3),

$$e^{W_t^A/k} = e^{W_0^A/k} M_t^u.$$

Noting that $W_T^A = U_A(C_T)$, we get

$$M_T^u = \exp\left(\frac{1}{k}[U_A(C_T) - W_0^A]\right).$$

Moreover, under condition (6.54), as analyzed in the paragraph right after (6.54), it is optimal for the principal to offer contract C_T so that $W_0^A = R_0$. Therefore, for such contract and for the agent's optimal action, we have

$$M_T^u = e^{-R_0/k} e^{U_A(C_T)/k}. \qquad (6.62)$$

This turns out to be exactly the reason why this problem is tractable: the fact that the choice of the probability measure corresponding to the optimal action u has an explicit functional relation with the promised payoff C_T.

We now turn to the principal's problem. Recall that

$$W_0^P = E^u[\tilde{U}_P(X_T - C_T)] = E[M_T^u \tilde{U}_P(X_T - C_T)].$$

6.2 General Risk Preferences, Quadratic Cost, and Lump-Sum Payment

Then, the principal's problem is

$$V_P := e^{-R_0/k} \sup_{C_T} E\left[e^{U_A(C_T)/k} \tilde{U}_P(X_T - C_T)\right]$$

subject to $E\left[e^{U_A(C_T)/k}\right] = e^{R_0/k}$. (6.63)

As usual, we consider the following relaxed problem with a Lagrange multiplier λ:

$$V_P(\lambda) := e^{-R_0/k} \sup_{C_T} E\left[e^{U_A(C_T)/k}\left[\tilde{U}_P(X_T - C_T) + \lambda\right]\right]. \quad (6.64)$$

The following result is then obvious:

Proposition 6.2.3 *Assume that the contract C_T is required to satisfy*

$$L \leq C_T \leq H$$

for some $\mathcal{F}_T$-measurable random variables L, H, which may take infinite values. If, with probability one, there exists a finite value $C_T^\lambda(\omega) \in [L(\omega), H(\omega)]$ that maximizes

$$e^{U_A(C_T)/k}\left[\tilde{U}_P(X_T - C_T) + \lambda\right] \quad (6.65)$$

and λ can be found so that

$$E\left[e^{U_A(C_T^\lambda)/k}\right] = e^{R_0/k},$$

then C_T^λ is the optimal contract.

Remark 6.2.4 Since (6.65) is considered ω by ω, we have reduced the problem to a one-variable deterministic optimization problem. In particular, if C_T is not constrained, the first order condition for optimal C_T is of the form

$$\frac{\tilde{U}_P'(X_T - C_T)}{U_A'(C_T)} = \frac{1}{k}\tilde{U}_P(X_T - C_T) + \lambda \quad (6.66)$$

and thus the optimal contract C_T is a function of the terminal value X_T only.

(i) The difference between Borch's rule (2.3) or (4.5) and condition (6.66) is the term with $\tilde{U}_P$: the ratio of marginal utilities of the agent and the principal is no longer constant, but a linear function of the utility of the principal. Increase in global utility of the principal also makes him happier at the margin, relative to the agent, and decrease in global utility makes him less happy at the margin. This will tend to make the contract "more nonlinear" than in the first best case. For example, if both utility functions are exponential, and we require $C_T \geq L > -\infty$ (for technical reasons), it is easy to check from Borch's rule that the first best contract C_T will be linear in X_T for $C_T > L$. On the other hand, as can be seen from (6.66), the second best contract will be nonlinear. Finally, we see that if cost k tends to infinity, the second best contract will tend to the first best contract.

(ii) By (6.66), omitting the functions arguments, we can find that

$$\frac{\partial}{\partial X_T} C_T = 1 - \frac{\tilde{U}_P' U_A''}{\tilde{U}_P'' U_A' + \tilde{U}_P' U_A'' - \frac{1}{k}\tilde{U}_P'(U_A')^2}.$$

Thus, under the standard conditions that

$$U_A \text{ and } \tilde{U}_P \text{ are increasing and concave,} \qquad (6.67)$$

the contract is a non-decreasing function of X_T, and its slope with respect to X_T is not higher than one. In the first best case, Borch's rule gives us

$$\frac{\partial}{\partial X_T} C_T = 1 - \frac{\tilde{U}'_P U''_A}{\tilde{U}''_P U'_A + \tilde{U}'_P U''_A}.$$

We see that the sensitivity of the contract is higher in the second best case, partly because more incentives are needed to induce the agent to provide optimal effort when the effort is hidden. The term which causes the increase in the slope of the contract is $\frac{1}{k}\tilde{U}'_P(U'_A)^2$ in the denominator. We see that this term is dominated by the agent's marginal utility, but it also depends on the principal's marginal utility. Higher marginal utility for either party causes the slope of the contract to increase relative to the first best case. As already mentioned above, higher cost k makes it closer to the first best case.

6.2.4 Example: Risk-Neutral Principal and Log-Utility Agent

Example 6.2.5 Suppose $k = 1$ and the principal is risk-neutral while the agent is risk-averse with

$$\tilde{U}_P(C_T) = X_T - C_T, \qquad U_A(C_T) = \log C_T.$$

Since the log utility does not allow nonpositive output values, let us change the model to, with $\sigma_t > 0$ being a given process,

$$dX_t = \sigma_t X_t dB_t = \sigma_t u_t X_t dt + \sigma_t X_t dB_t^u.$$

Then, $X_t > 0$ for all t. Moreover, assume that

$$\lambda_0 := 2e^{R_0} - X_0 > 0.$$

In this case, the first order condition of (6.66) becomes

$$C_T = X_T - C_T + \lambda.$$

This gives a linear contract

$$C_T = \frac{1}{2}(X_T + \lambda),$$

and in order to satisfy the IR constraint in (6.63)

$$e^{R_0} = E[C_T] = \frac{1}{2}(X_0 + \lambda),$$

we need to take

6.2 General Risk Preferences, Quadratic Cost, and Lump-Sum Payment

$$\lambda = \lambda_0.$$

By assumption $\lambda_0 > 0$, we have $C_T > 0$, and C_T is then the optimal contract.

By (6.61), agent's optimal effort u is obtained by solving the BSDE

$$\tilde{W}_t^A = E_t[C_T] = C_T + \int_0^t \tilde{W}_s^A u_s dB_s.$$

Noting that

$$E_t[C_T] = \frac{1}{2}(X_t + \lambda_0) = e^{R_0} + \int_0^t \sigma_t X_t dB_t,$$

we get

$$\tilde{W}_t^A = \frac{1}{2}(X_t + \lambda_0), \qquad \tilde{W}_t^A u_t = \sigma_t X_t,$$

and thus

$$u_t = 2\sigma_t \frac{X_t}{X_t + \lambda_0}.$$

Since $\lambda_0 > 0$, we see that the effort goes down as the output decreases, and goes up when the output goes up. Thus, the incentive effect coming from the fact that the agent is paid an increasing function of the output at the end, translates into earlier times, so when the promise of the future payment gets higher, the agent works harder. Also notice that the effort is bounded in this example by $2\sigma_t$.

Assume now that σ is deterministic. The principal's optimal utility can be computed to be equal to

$$V_P = e^{-R_0} E\left[e^{U_A(C_T)} \tilde{U}_P(X_T - C_T)\right] = e^{-R_0} E\left[C_T[X_T - C_T]\right]$$

$$= e^{-R_0} E\left[\frac{1}{4}[X_T + \lambda_0][X_T - \lambda_0]\right]$$

$$= \frac{1}{4} e^{-R_0} E\left[X_0^2 \exp\left(2\int_0^T \sigma_t dB_t - \int_0^T \sigma_t^2 dt\right) - \left[2e^{R_0} - X_0\right]^2\right]$$

$$= \frac{1}{4} e^{-R_0} \left[X_0^2 \exp\left(\int_0^T \sigma_t^2 dt\right) - 4e^{2R_0} + 4e^{R_0} X_0 - X_0^2\right]$$

$$= X_0 - e^{R_0} + \frac{1}{4} e^{-R_0} X_0^2 \left[\exp\left(\int_0^T \sigma_t^2 dt\right) - 1\right].$$

The first term, $X_0 - e^{R_0}$, is what the principal can get if he pays a constant payoff C_T, in which case the agent would choose $u \equiv 0$. The second term is the extra benefit of inducing the agent to apply non-zero effort. The extra benefit increases quadratically with the initial output X_0, increases exponentially with the volatility squared, and decreases exponentially with the agent's reservation utility. While the principal would like best to have the agent with the lowest R_0, the cost of hiring expensive agents is somewhat offset when the volatility is high (which is not surprising, given that the principal is risk-neutral).

For comparison, we look now at the first best case in this example. Interestingly, we have

Proposition 6.2.6 *Assume that $\sigma_t > 0$ is deterministic and bounded. Then, the principal's first best optimal utility is infinite.*

Proof We see from Borch's rule (2.3) that, whenever the principal is risk-neutral, a candidate for an optimal contract is a constant contract C_T. With log-utility for the agent, we set
$$C_T = \lambda$$
where λ is obtained from the IR constraint, and the optimal utility of the principal is obtained from
$$\sup_u E[X_T - \lambda] = \sup_u \left[E\left\{ X_0 e^{\int_0^T [u_t \sigma_t - \frac{1}{2}\sigma_t^2] dt + \int_0^T \sigma_t dB_t} \right\} - e^R e^{E\{\int_0^T \frac{1}{2} u_t^2 dt\}} \right]. \quad (6.68)$$

Under the assumption that σ is deterministic and bounded, we show now that the right-hand side of (6.68) is infinite. In fact, for any n, set
$$A_n := \left\{ \int_0^{\frac{T}{2}} \sigma_t dB_t > n \right\} \in \mathcal{F}_{\frac{T}{2}}; \qquad \alpha_n := P(A_n) \to 0;$$
and
$$u_t^n(\omega) := \begin{cases} \alpha_n^{-\frac{1}{2}}, & \frac{T}{2} \le t \le T, \omega \in A_n; \\ 0, & \text{otherwise.} \end{cases} \quad (6.69)$$

Then, the cost is finite:
$$E\left\{ \int_0^T \frac{1}{2}(u_t^n)^2 dt \right\} = \frac{T}{4}.$$

However, for a generic constant $c > 0$,
$$E\left\{ x \exp\left(\int_0^T \left[u_t^n \sigma_t - \frac{1}{2}\sigma_t^2 \right] dt + \int_0^T \sigma_t dB_t \right) \right\}$$
$$= E\left\{ x \exp\left(\alpha_n^{-\frac{1}{2}} \int_{\frac{T}{2}}^T \sigma_t dt \mathbf{1}_{A_n} - \int_0^T \frac{1}{2}\sigma_t^2 dt + \int_0^T \sigma_t dB_t \right) \right\}$$
$$\ge E\left\{ x \exp\left(\alpha_n^{-\frac{1}{2}} \int_{\frac{T}{2}}^T \sigma_t dt - \int_0^T \frac{1}{2}\sigma_t^2 dt + \int_0^T \sigma_t dB_t \right) \mathbf{1}_{A_n} \right\}$$
$$= E\left\{ x \exp\left(\alpha_n^{-\frac{1}{2}} \int_{\frac{T}{2}}^T \sigma_t dt - \int_0^{\frac{T}{2}} \frac{1}{2}\sigma_t^2 dt + \int_0^{\frac{T}{2}} \sigma_t dB_t \right) \mathbf{1}_{A_n} \right\}$$
$$\ge cE\left\{ x \exp\left(\alpha_n^{-\frac{1}{2}} \int_{\frac{T}{2}}^T \sigma_t dt + n \right) \mathbf{1}_{A_n} \right\}$$
$$= cx \exp\left(\alpha_n^{-\frac{1}{2}} \int_{\frac{T}{2}}^T \sigma_t dt + n \right) P(A_n)$$
$$= cx \exp\left(\alpha_n^{-\frac{1}{2}} \int_{\frac{T}{2}}^T \sigma_t dt + n \right) \alpha_n \ge cx \alpha_n e^{c\alpha_n^{-\frac{1}{2}}},$$

which diverges to infinity as $\alpha_n \to 0$. □

6.3 Risk-Neutral Principal and Infinite Horizon

We note that another completely solvable example in this special framework is the case of both the principal and the agent having linear utilities. However, in that case it is easily shown that the first best and the second best are the same, so there is no need to consider the second best.

6.3 Risk-Neutral Principal and Infinite Horizon

We now present a model which is a variation on the one in Sannikov (2008).

6.3.1 The Model

We consider a Markov model on infinite horizon with constant volatility, with risk-neutral principal and risk-averse agent with separable utility paid at rate c, both having the same discount rate r. More precisely, we have

$$T = \infty, \quad U_A = 0, \quad U_P = 0, \quad v(t,x) = v,$$

and

$$g(t,u) = re^{-rt}g(u), \quad u_A(t,c) = re^{-rt}u_A(c),$$
$$u_P(t, X_t, c) = re^{-rt}[X_t - X_0 - c]. \tag{6.70}$$

Moreover, the IR constraint is only at initial time:

$$W_0^A \geq R_0. \tag{6.71}$$

Then,

$$X_t = x + vB_t;$$
$$W_t^A = \int_t^\infty re^{-rs}[u_A(c_s) - g(u_s)]ds - \int_t^\infty Z_s^A dB_s^u; \tag{6.72}$$
$$W_t^P = \int_t^\infty re^{-rs}[vB_s - c_s]ds - \int_t^\infty Z_s^P dB_s^u.$$

The goal of this section is to show that in this case the solution boils down to solving a differential equation in one variable, the agent's promised (remaining) utility process. Once that equation is obtained, it is possible to get economic conclusions by solving it numerically.

6.3.2 Necessary Conditions Derived from the General Theory

Note to the Reader The reader not interested in the use of general theory of Chap. 5 can skip this section and go to the following section that provides a more direct approach for dealing with the above model.

As usual we start with the agent's problem. In this case, as in (5.24) we have that the agent's optimal control satisfies
$$-re^{-rt}g'(u_t) + Z_t^A = 0. \tag{6.73}$$
Denote
$$\hat{W}_t^A := e^{rt} W_t^A, \quad \hat{Z}_t^A := r^{-1} e^{rt} Z_t^A. \tag{6.74}$$
Then,
$$u_t = I_A(\hat{Z}_t^A), \quad \text{where } I_A := (g')^{-1}, \tag{6.75}$$
and, given c, the agent's optimal utility satisfies
$$\hat{W}_t^A = \int_t^\infty re^{-r(s-t)} \big[u_A(c_s) - g(I_A(\hat{Z}_s^A)) + \hat{Z}_s^A I_A(\hat{Z}_s^A)\big] ds$$
$$- \int_t^\infty re^{-r(s-t)} \hat{Z}_s^A dB_s. \tag{6.76}$$
For the principal's problem, the optimization condition (5.89) becomes
$$D_t r e^{-rt} u_A'(c) - re^{-rt} = 0, \tag{6.77}$$
and thus, the principal's optimal control is
$$c_t = I_P(D_t) \quad \text{where } I_P := (1/u_A')^{-1}. \tag{6.78}$$
Then, the FBSDE (5.99) becomes
$$D_t = \lambda + \int_0^t Z_s^P r^{-1} e^{rs} I_A'(\hat{Z}_s^A) \big[dB_s - I_A(\hat{Z}_s^A) ds\big];$$
$$W_t^A = \int_t^\infty re^{-rs} \big[u_A(I_P(D_s)) - g(I_A(\hat{Z}_s^A))\big] ds$$
$$- \int_t^\infty re^{-rs} \hat{Z}_s^A \big[dB_s - I_A(\hat{Z}_s^A) ds\big]; \tag{6.79}$$
$$W_t^P = \int_t^\infty re^{-rs} \big[vB_s - I_P(D_s)\big] ds - \int_t^\infty \tilde{Z}_s^P \big[dB_s - I_A(\hat{Z}_s^A) ds\big].$$
Note that
$$\int_t^\infty re^{-rs} B_s ds = e^{-rt} B_t + \int_t^\infty e^{-rs} dB_s$$
$$= e^{-rt} B_t + \int_t^\infty e^{-rs} u_s ds + \int_t^\infty e^{-rs} dB_s^u. \tag{6.80}$$
Then,
$$W_t^P = ve^{-rt} B_t + \int_t^\infty e^{-rs} \big[vI_A(\hat{Z}_s^A) - rI_P(D_s)\big] ds$$
$$- \int_t^\infty [Z_s^P - ve^{-rs}] \big[dB_s - I_A(\hat{Z}_s^A) ds\big].$$

6.3 Risk-Neutral Principal and Infinite Horizon

Denote
$$\hat{W}_t^P := e^{rt} W_t^P - vB_t, \qquad \hat{Z}_t^P := r^{-1}\left[e^{rt} Z_t^P - v\right]. \tag{6.81}$$

Then, (6.79) becomes

$$D_t = \lambda + \int_0^t [\hat{Z}_s^P + v/r] I_A'(\hat{Z}_s^A)[dB_s - I_A(\hat{Z}_s^A) ds];$$

$$\hat{W}_t^A = \int_t^\infty r e^{-r(s-t)} \left[u_A(I_P(D_s)) - g(I_A(\hat{Z}_s^A))\right] ds$$

$$\quad - \int_t^\infty r e^{-r(s-t)} \hat{Z}_s^A [dB_s - I_A(\hat{Z}_s^A) ds]; \tag{6.82}$$

$$\hat{W}_t^P = \int_t^\infty e^{-r(s-t)} \left[v I_A(\hat{Z}_s^A) - r I_P(D_s)\right] ds$$

$$\quad - \int_t^\infty r e^{-r(s-t)} \hat{Z}_s^P [dB_s - I_A(\hat{Z}_s^A) ds].$$

The above FBSDE (6.82) is Markovian and time homogeneous. Thus, we expect to have

$$\hat{W}_t^A = \varphi_A(D_t), \qquad \hat{W}_t^P = \varphi_P(D_t), \quad \text{for some deterministic functions } \varphi_A, \varphi_P.$$

Assume φ_A has an inverse, and denote

$$\psi := (\varphi_A)^{-1}, \qquad \hat{F}(x) := \varphi_P(\psi(x)). \tag{6.83}$$

Then, ψ and $\hat{F}$ are independent of λ, and we have

$$\hat{W}_t^P = \hat{F}(\hat{W}_t^A), \qquad D_t = \psi(\hat{W}_t^A). \tag{6.84}$$

This implies that

$$V_P = \sup_{w_A \geq R_0} \hat{F}(w_A). \tag{6.85}$$

In particular, when the function $\hat{F}$ is decreasing, then

$$V_P = \hat{F}(R_0). \tag{6.86}$$

We now formally derive the equation which the function $\hat{F}$ should satisfy. Let $(D, \hat{W}^A, \hat{Z}^A, \hat{W}^P, \hat{Z}^P)$ solve (6.82). Denote

$$u_t := I_A(\hat{Z}_t^A), \qquad c_t := I_P(D_t). \tag{6.87}$$

Then,

$$dD_t = \frac{\hat{Z}_t^P + v}{g''(u_t)} dB_t^u;$$

$$d\hat{W}_t^A = d(e^{rt} W_t^A) = r\hat{W}_t^A dt - r[u_A(c_t) - g(u_t)] dt + r g'(u_t) dB_t^u;$$

$$d\hat{W}_t^P = d(e^{rt}[W_t^P - vre^{-rt} B_t]) = r\hat{W}_t^P dt - r[vu_t - c_t] dt + r[\hat{Z}_t^P + v] dB_t^u.$$

On the other hand, applying Itô's rule, we have

$$dD_t = d(\psi(\hat{W}_t^A)) = \psi'(\hat{W}_t^A)r[\hat{W}_t^A - u_A(c_t) + g(u_t)]dt + \psi'(\hat{W}_t^A)r\hat{Z}_t^A dB_t^u$$
$$+ \frac{1}{2}\psi''(\hat{W}_t^A)|rg'(u_t)|^2 dt;$$
$$d\hat{W}_t^P = d(\hat{F}(\hat{W}_t^A)) = \hat{F}'(\hat{W}_t^A)r[\hat{W}_t^A - u_A(c_t) + g(u_t)]dt + \hat{F}'(\hat{W}_t^A)r\hat{Z}_t^A dB_t^u$$
$$+ \frac{1}{2}\hat{F}''(\hat{W}_t^A)|rg'(u_t)|^2 dt.$$

Comparing the above expressions, we get

$$\psi'(\hat{W}_t^A)[\hat{W}_t^A - u_A(c_t) + g(u_t)] + \frac{r}{2}\psi''(\hat{W}_t^A)|g'(u_t)|^2 = 0;$$

$$\psi'(\hat{W}_t^A)g'(u_t) = \frac{\hat{Z}_t^P + v}{rg''(u_t)};$$

$$\hat{F}'(\hat{W}_t^A)[\hat{W}_t^A - u_A(c_t) + g(u_t)] + \frac{r}{2}\hat{F}''(\hat{W}_t^A)|g'(u_t)|^2 = \hat{F}(\hat{W}_t^A) - vu_t + c_t;$$

$$\hat{F}'(\hat{W}_t^A)g'(u_t) = \hat{Z}_t^P + v.$$

Using these, we obtain

$$vu_t - c_t - \hat{F}(\hat{W}_t^A) + \hat{F}'(\hat{W}_t^A)[\hat{W}_t^A - u_A(c_t) + g(u_t)] + \frac{r}{2}\hat{F}''(\hat{W}_t^A)|g'(u_t)|^2 = 0,$$
(6.88)

and the optimal c, u, together with the function ψ, satisfy:

$$\psi(\hat{W}_t^A)u'_A(c_t) = 1;$$
$$r\psi'(\hat{W}_t^A)g''(u_t) = \hat{F}'(\hat{W}_t^A); \quad (6.89)$$
$$\psi'(\hat{W}_t^A)[\hat{W}_t^A - u_A(c_t) + g(u_t)] + \frac{r}{2}\psi''(\hat{W}_t^A)|g'(u_t)|^2 = 0.$$

This gives us a differential equation for function $\hat{F}$ (and ψ), and shows how optimal u and c depend in a deterministic way on the agent utility process $\hat{W}_t^A$.

6.3.3 A Direct Approach

In this subsection, we solve the problem directly by using the standard approach in Stochastic Control Theory, the Hamilton–Jacobi–Bellman (HJB) equation. This approach is briefly reviewed in Sect. 5.4.4.

The usual argument implies that the agent's optimal effort satisfies

$$re^{-rt}g'(u_t) = Z_t^A.$$

(This is (6.73) in the previous section.)

6.3 Risk-Neutral Principal and Infinite Horizon

For any given w_A, we restrict our control (c, u) to be an element of the set $\mathcal{A}(w_A)$ of all controls that, besides the standard measurability and integrability conditions, satisfy

$$\lim_{t \to \infty} W_t^A = 0,$$

where $W_t^A = w_A - \int_0^t re^{-rs}[u_A(c_s) - g(u_s)]ds + \int_0^t re^{-rs} g'(u_s) dB_s^u.$ (6.90)

Here, the limit is in L^2 sense.

Denote

$$\tilde{W}_t^P := W_t^P - vre^{-rt} B_t, \qquad \tilde{Z}_t^P := Z_t^P - vre^{-rt}. \qquad (6.91)$$

Using

$$\int_t^\infty re^{-rs} B_s ds = e^{-rt} B_t + \int_t^\infty e^{-rs} dB_s = e^{-rt} B_t + \int_t^\infty e^{-rs} u_s ds$$

$$+ \int_t^\infty e^{-rs} dB_s^u, \qquad (6.92)$$

we get

$$\tilde{W}_t^P = \int_t^\infty re^{-rs}[vu_s - c_s]ds - \int_t^\infty \tilde{Z}_s^P dB_s^u. \qquad (6.93)$$

Note that $W_0^P = \tilde{W}_0^P$. Assume, for each $(c, u) \in \mathcal{A}(w_A)$, that c is implementable using effort u that is optimal for the agent. Then,

$$V_P = \sup_{w_A \geq R_0} F(w_A) \quad \text{where} \quad F(w_A) := \sup_{(c,u) \in \mathcal{A}(w_A)} \tilde{W}_0^P. \qquad (6.94)$$

We now derive formally the HJB equation that F should satisfy. For any $(c, u) \in \mathcal{A}(w_A)$ and $t \geq \delta > 0$, note that

$$e^{r\delta} W_t^A = e^{r\delta} W_\delta^A - \int_\delta^t re^{-r(s-\delta)}[u_A(c_s) - g(u_s)]ds + \int_\delta^t re^{-r(s-\delta)} g'(u_s) dB_s^u,$$

$$e^{r\delta} \tilde{W}_t^P = \int_t^\infty re^{-r(s-\delta)}[vu_s - c_s]ds - \int_t^\infty e^{r\delta} \tilde{Z}_s^P dB_s^u. \qquad (6.95)$$

Following the standard arguments in Stochastic Control Theory, we have the following Dynamic Programming Principle:

Proposition 6.3.1 *Assume that function F above is continuous. Then, for any $\delta > 0$,*

$$F(w_A) = \sup_{c, u \in \mathcal{A}(w_A)} E^u \left[\int_0^\delta re^{-rs}[vu_s - c_s]ds + e^{-r\delta} F(\hat{W}_\delta^A) \right] \qquad (6.96)$$

where

$$\hat{W}_t^A := e^{rt} W_t^A \quad \text{and}$$

$$W_\delta^A = w_A - \int_0^\delta re^{-rs}[u_A(c_s) - g(u_s)]ds + \int_0^\delta re^{-rs} g'(u_s) dB_s^u. \qquad (6.97)$$

Assume now that F is sufficiently smooth. Applying Itô's rule, we have

$$d(\hat{W}_t^A) = d(e^{rt}W_t^A) = r\hat{W}_t^A dt - r[u_A(c_t) - g(u_t)]dt + rg'(u_t)dB_t^u,$$

and thus

$$\begin{aligned}d(e^{-rt}F(\hat{W}_t^A)) &= -re^{-rt}F(\hat{W}_t^A)dt + \frac{1}{2}e^{-rt}F''(\hat{W}_t^A)|rg'(u_t)|^2 dt \\ &+ e^{-rt}F'(\hat{W}_t^A)\big[r\hat{W}_t^A dt - r[u_A(c_t) - g(u_t)]dt \\ &+ rg'(u_t)dB_t^u\big].\end{aligned} \quad (6.98)$$

Plugging this into (6.96), dividing both sides by δ, and then sending $\delta \to 0$, we get

$$\sup_{c,u}\left[vu - c - F(w_A) + \frac{r}{2}F''(w_A)|g'(u)|^2 + F'(w_A)[w_A - u_A(c) + g(u)]\right] = 0. \quad (6.99)$$

Furthermore, if the supremum is attained by a control couple $(c, u) \in \mathcal{A}(w_A)$ such that

$$\begin{aligned}c_t &= \operatorname*{argmin}_{c}\big[c + F'(\hat{W}_t^A)u_A(c)\big], \\ u_t &= \operatorname*{argmax}_{u}\left[vu + F'(\hat{W}_t^A)g(u) + \frac{r}{2}F''(\hat{W}_t^A)[g'(u)]^2\right],\end{aligned} \quad (6.100)$$

then they are optimal.

We now prove a verification result, under quite strong conditions. More general results can be obtained following the viscosity solution approach, as in Fleming and Soner (2006) and Yong and Zhou (1999).

Proposition 6.3.2 *Assume that the HJB equation (6.99) has a classical solution $\tilde{F}$ that has linear growth. Then, $F \leq \tilde{F}$. Moreover, if there exists a pair $(c, u) \in \mathcal{A}(w_A)$ such that (6.100) holds, then $F = \tilde{F}$, and c and u are optimal.*

Proof For any $(c, u) \in \mathcal{A}(w_A)$, by (6.98) and (6.99) we have

$$\begin{aligned}d(e^{-rt}\tilde{F}(\hat{W}_t^A)) &= re^{-rt}\bigg[-\tilde{F}(\hat{W}_t^A) + \frac{r}{2}\tilde{F}''(\hat{W}_t^A)|g'(u_t)|^2 \\ &\quad + \tilde{F}'(\hat{W}_t^A)[\hat{W}_t^A - u_A(c_t) + g(u_t)]\bigg]dt \\ &\quad + re^{-rt}\tilde{F}'(\hat{W}_t^A)g'(u_t)dB_t^u \\ &\leq re^{-rt}[c_t - vu_t]dt + re^{-rt}\tilde{F}'(\hat{W}_t^A)g'(u_t)dB_t^u.\end{aligned} \quad (6.101)$$

Then, (6.93) leads to

$$d(e^{-rt}\tilde{F}(\hat{W}_t^A) - \tilde{W}_t^P) \leq [re^{-rt}\tilde{F}'(\hat{W}_t^A)g'(u_t) - \tilde{Z}_t^P]dB_t^u. \quad (6.102)$$

Note that, by the linear growth of F and (6.90),

$$|e^{-rt}\tilde{F}(\hat{W}_t^A)| \leq Ce^{-rt}[1 + |\hat{W}_t^A|] \leq C[e^{-rt} + |W_t^A|] \to 0, \quad \text{as } t \to \infty.$$

6.3 Risk-Neutral Principal and Infinite Horizon

Since we are assuming enough integrability, we obtain from (6.102) that

$$\tilde{F}(w^A) - \tilde{W}_0^P = \tilde{F}(\hat{W}_0^A) - \tilde{W}_0^P \geq 0.$$

This, together with the arbitrariness of (c, u), implies $\tilde{F}(w^A) \geq F(w_A)$.

On the other hand, if $(c, u) \in \mathcal{A}(w_A)$ satisfies (6.100), then the inequality in (6.101) becomes an equality. Consequently, (6.102) becomes an equation, and thus $\tilde{F}(w^A) = \tilde{W}_0^P$ for this (c, u). Then clearly $\tilde{F}(w^A) = F(w_A)$ and (c, u) is an optimal control. □

Remark 6.3.3 For the readers who are familiar with the previous section, we remark that function F here is in general different from function $\hat{F}$ of that section. Indeed, let

$$\mathcal{A} := \bigcup_{w_A \geq R_0} \mathcal{A}(w_A). \tag{6.103}$$

Then, the system (6.79) is obtained by solving the optimization problem

$$V_P(\lambda) := \sup_{(c,u) \in \mathcal{A}} \left[W_0^{P,c,u} + \lambda W_0^{A,c,u} \right], \tag{6.104}$$

and

$$W_0^P = \hat{W}_0^P = \varphi_P(\lambda) = V_P(\lambda) - \lambda \varphi_A(\lambda). \tag{6.105}$$

For given $w_A \geq R_0$, if we choose $\lambda := \psi(w_A)$, then

$$\begin{aligned}
\hat{F}(w_A) = \hat{W}_0^P &= V_P(\psi(w_A)) - w_A \psi(w_A) \\
&= \sup_{(c,u) \in \mathcal{A}} \left[W_0^{P,c,u} + \psi(w_A) W_0^{A,c,u} \right] - w_A \psi(w_A) \\
&\geq \sup_{(c,u) \in \mathcal{A}(w_A)} \left[W_0^{P,c,u} + \psi(w_A) W_0^{A,c,u} \right] - w_A \psi(w_A) \\
&= \sup_{(c,u) \in \mathcal{A}(w_A)} W_0^{P,c,u} = F(w_A).
\end{aligned} \tag{6.106}$$

However, we note that

$$\sup_{w_A \geq R_0} \hat{F}(w_A) = V_P = \sup_{w_A \geq R_0} F(w_A). \tag{6.107}$$

6.3.4 Interpretation and Discussion

(i) From (6.100) we see that the principal faces a tradeoff between minimizing the payment c and maximizing the agent's utility $u_A(c)$, but weighted by the marginal change $F'(\hat{W}_t^A)$ in the principal's utility relative to the agent's utility.

(ii) From (6.100), we see that the optimally induced effort faces a tradeoff between maximizing the drift of the output, minimizing the cost of the effort, and minimizing the risk to which the agent is exposed. The latter risk is represented by the term

$$-F''(\hat{W}_t^A)\left[re^{-rt}g'(u_t)\right]^2$$

thus, equal to a (minus) product of the marginal change in sensitivity of the principal's utility with respect to the agent's promised utility, with the squared volatility of the agent's promised utility.

(iii) If c_t is unconstrained, it follows from (6.100) that the first order condition for optimality in c_t is

$$c_t = I_P(F'(\hat{W}_t^A))$$

in the notation of (6.78). On the other hand, by that equation the Stochastic Maximum Principle approach gives us $c_t = I_P(D_t)$ which implies

$$D_t = F'(\hat{W}_t^A)$$

and provides an economic meaning to the process D as the marginal change in the principal's utility with respect to the agent's utility.

(iv) In order to solve the problem using HJB equation (6.99), we need boundary conditions. These will depend on the specifics of the model. For example, suppose that we allow only non-negative effort, $u_t \geq 0$. Then, at the minimum possible value for the agent's utility, denoted w_L, the agent will apply minimal effort zero, and also be paid minimal possible consumption, denoted c_L, which will make the principal's utility equal to $-r\int_0^\infty e^{-rt}c_L dt = -c_L$. Thus, the boundary condition at the bottom range is $F(w_L) = -c_L$. If there is no upper bound on W^A and no lower bound on $W^{P,\lambda}$, then we also have $F(+\infty) = -\infty$. Sannikov (2008) works with more realistic assumptions, discussed next.

6.3.5 Further Economic Conclusions and Extensions

Having obtained the HJB differential equation, Sannikov (2008) is able to do numerical computations and discuss economic consequences of the model. We list some of them here. First, there are some additional conditions in his model: The consumption c_t and effort u_t are restricted to be non-negative, as is the utility process W_t^A. It is assumed that $g(u) \geq \varepsilon u$ for all $u \geq 0$, and $g(0) = 0$. The utility function u_A is bounded below, with $u_A(0) = 0$. Thus, the boundary conditions for the PDE are, first:

$$F(0) = 0.$$

That is, once the agent's remaining utility hits zero, the principal retires him with zero payment. The second boundary condition for the solution is

$$F(w_{gp}) = -u_A^{-1}(w_{gp}), \qquad F'(w_{gp}) = -\left[u_A^{-1}\right]'(w_{gp})$$

6.3 Risk-Neutral Principal and Infinite Horizon

where w_{gp} is an unknown point ("gp" stands for "golden parachute", an expression for the retirement payment). The interpretation of this condition is that when the agent's promised utility reaches too high a point w_{gp}, the principal retires the agent and continues paying him constant consumption, which then has to be equal to $u_A^{-1}(w_{gp})$. Hence the expected value of output is zero after that time, and the principal's remaining utility is equal to $[-u_A^{-1}(w_{gp})]$. For the use below, denote the principal's retirement profit

$$F_0(w) = -u_A^{-1}(w).$$

Another condition on F is

$$F(w) \geq F_0(w) \quad \text{for all } w \geq 0.$$

That is, the principal's profit is no less than the value obtained by retiring the agent.

In this model where the consumption c is constrained from below, it may happen that the agent is paid more than his reservation value $R_0 \geq 0$. This is because the function F, under the above conditions and restrictions, is not necessarily decreasing in the area where $c \equiv 0$. The principal gives the agent the value $W_0^A = w_0$ which maximizes $F(w)$ on $[R_0, w_{gp}]$, if $F(w_0) > 0$. Otherwise, if $F(w) \leq 0$ for all w on $[R_0, w_{gp}]$ the principal does not hire the agent.

Using the above machinery, it is possible to show the following general principle for this model: a change in boundary conditions that makes the principal's utility $F(w)$ uniformly higher, increases the agent's optimal effort $u = u(w)$ for all wage levels w.

Changing boundary conditions allows us to consider the following extensions:

1. **The agent can quit at any time and take an outside job** ("outside option") with expected utility $\tilde{R}_0 < R_0$. $\tilde{R}_0$ is interpreted as the value of new employment minus the search costs. In this case, the solution $\tilde{F}$ is obtained by solving the same HJB equation, except that the boundary conditions change: now $\tilde{F}(\tilde{R}_0) = 0$, since $w^A = \tilde{R}_0$ is the low retirement point, not $w^A = 0$. The boundary conditions for the new high retirement point $\tilde{w}_{gp}$ are the same as before, except we have a constraint $\tilde{w}_{gp} > \tilde{R}_0$. Numerical computations and/or analytical results in Sannikov (2008) show that what happens is:
   (i) $\tilde{F} \leq F$: the principal's profit is lower;
   (ii) $\tilde{w}_{gp} < w_{gp}$: the high retirement point occurs sooner, as the principal's profit is lower;
   (iii) the agent works less hard;
   (iv) the consumption payment c is lower: the payments are "backloaded" when the principal is trying to tie the agent more closely to the firm;
   (v) the agent's promised utility W_0^A is at least as large as without the outside option.

2. **The principal can replace the agent with another agent** of the same reservation value R_0, at a fixed cost C. In this case the principal's retirement profit will be higher, $\tilde{F}_0(w) = F_0(w) + D$, for D of the form $D = F(w_0) - C$, where w_0 has to be determined. The boundary conditions are $F(0) = \tilde{F}_0(0) = D$,

$F(w_{gp}) = \tilde{F}_0(w_{gp})$ and $F'(w_{gp}) = \tilde{F}'_0(w_{gp})$. Then, $w_0 = w_0(D)$ has to be chosen so that F is maximized on the interval $[R_0, w_{gp}]$. Since we don't know D in advance, numerically we start with an arbitrary value of D. If, doing the above procedure, we get $D = F(w_0) - C$, we are done. Otherwise, we have to adjust the value of D up or down, and repeat the procedure.

What happens is:
 (i) the principal's profit is higher;
 (ii) the agent works harder;
 (iii) w_{gp} is increasing in C.
 (iv) The principal's utility may be higher than the first best utility with only one agent.

3. **The principal can train and promote the agent** at a cost $K \geq 0$, instead of retiring him. When promotion happens, it increases the drift from u to θu for $\theta > 1$, but also increases the agent's outside option from zero to $\tilde{R} > 0$. If we denote by F_1 the principal's profit without promotion and with F_2 her profit with promotion, we will have a boundary condition $F_1(w_p) = F_2(w_p) - K$, $F'_1(w_p) = F'_2(w_p)$, where w_p is the point of promotion. What happens is:
 (i) the principal's profit is higher;
 (ii) the agent works harder until promotion;
 (iii) the consumption payment c is lower;
 (iv) the agent's promised utility W_0^A is at least as large as without promotion.
 (v) The promotion is not offered right away for the following reasons: (a) the agent first has to show good performance; (b) the agent's outside option may increase with promotion, making him more likely to leave; (c) training for promotion is costly.

4. **The principal cannot commit to the payments and to not replacing the agent**. What happens is:
 (i) the principal's profit is lower;
 (ii) the agent works less hard;
 (iii) the consumption payment c is higher;
 (iv) the agent's promised utility W_0^A is smaller;
 (v) with lack of commitment, the principal's profit with replacement or promotion options can actually be lower than without those options.

6.4 Further Reading

Section 6.1 summarizes some of the main results from Hölmstrom and Milgrom (1987). The example of Sect. 6.2 is a generalization of a case studied in Cvitanić et al. (2006). The setting and the results of Sect. 6.3 are taken from Sannikov (2008). An interesting application of Sannikov (2008) can be found in Fong (2009).

There is a growing literature extending methods of this chapter to various new applications involving moral hazard in continuous-time. These include: (i) processes driven by jumps and not by Brownian motion, see Zhang (2009) and Biais et al. (2010); (ii) imperfect information and learning, see Adrian and Westerfield (2009), DeMarzo and Sannikov (2011), Giat and Subramanian (2009), Prat and Jovanovic (2010), He et al. (2010), and Giat et al. (2011); (iii) asset pricing, see Ou-Yang (2005); (iv) executive compensation, see He (2009); (v) stochastic interest rates and mortgage contracts, see Piskorski and Tchistyi (2010). Additional references can be found in a nice survey paper, Sannikov (2012).

References

Adrian, T., Westerfield, M.: Disagreement and learning in a dynamic contracting model. Rev. Financ. Stud. **22**, 3839–3871 (2009)

Biais, B., Mariotti, T., Rochet, J.-C., Villeneuve, S.: Large risks, limited liability, and dynamic moral hazard. Econometrica **78**, 73–118 (2010)

Cvitanić, J., Wan, X., Zhang, J.: Optimal contracts in continuous-time models. J. Appl. Math. Stoch. Anal. **2006**, 1–27 (2006)

DeMarzo, P.M., Sannikov, Y.: Learning, termination and payout policy in dynamic incentive contracts. Working paper, Princeton University (2011)

Fong, K.G.: Evaluating skilled experts: optimal scoring rules for surgeons. Working paper, Stanford University (2009)

Giat, Y., Subramanian, A.: Dynamic contracting under imperfect public information and asymmetric beliefs. Working paper, Georgia State University (2009)

Giat, Y., Hackman, S.T., Subramanian, A.: Investment under uncertainty, heterogeneous beliefs and agency conflicts. Rev. Financ. Stud. **23**(4), 1360–1404 (2011)

He, Z.: Optimal executive compensation when firm size follows geometric Brownian motion. Rev. Financ. Stud. **22**, 859–892 (2009)

He, Z., Wei, B., Yu, J.: Permanent risk and dynamic incentives. Working paper, Baruch College (2010)

Holmström, B., Milgrom, P.: Aggregation and linearity in the provision of intertemporal incentives. Econometrica **55**, 303–328 (1987)

Ou-Yang, H.: An equilibrium model of asset pricing and moral hazard. Rev. Financ. Stud. **18**, 1219–1251 (2005)

Piskorski, T., Tchistyi, A.: Optimal mortgage design. Rev. Financ. Stud. **23**, 3098–3140 (2010)

Prat, J., Jovanovic, B.: Dynamic incentive contracts under parameter uncertainty. Working paper, NYU (2010)

Sannikov, Y.: A continuous-time version of the principal-agent problem. Rev. Econ. Stud. **75**, 957–984 (2008)

Sannikov, Y.: Contracts: the theory of dynamic principal-agent relationships and the continuous-time approach. Working paper, Princeton University (2012)

Zhang, Y.: Dynamic contracting with persistent shocks. J. Econ. Theory **144**, 635–675 (2009)

Chapter 7
An Application to Capital Structure Problems: Optimal Financing of a Company

Abstract In this chapter we present an application to corporate finance: how to optimally structure financing of a company (a project), in the presence of moral hazard. In the model the agent can misreport the firm earnings and transfer money to his own savings account, but there is an optimal contract under which the agent will report truthfully, and will not save, but consume everything he is paid. The model leads to a relatively simple and realistic financing structure, consisting of equity (dividends), long-term debt and a credit line. These instruments are used for financing the initial capital needed, as well as for the agent's salary and covering possible operating losses. The agent is paid by a fixed fraction of dividends, which he has control over. At the optimum, the dividends are paid locally, when the agent's expected utility process hits a certain boundary point, or equivalently, after the credit line balance has been paid off. Because the agent receives enough in dividends, he has no incentives to misreport. When having a larger credit line is optimal, in order to be allowed such credit, it may happen that the debt is negative, meaning that the firm needs to maintain a margin account on which it receives less interest than it pays on the credit line, as also sometimes happens in the real world. The continuous-time framework and the associated mathematical tools enable one to compute many comparative statics, by solving appropriate differential equations, and/or by computing appropriate expected values.

7.1 The Model

We follow mostly the approach and the model from DeMarzo and Sannikov (2006), while similar results have been obtained also by Biais et al. (2007).

Instead of assuming that the agent applies unobservable effort, we assume that the principal cannot observe the output process, and the agent may misreport its value, and keep the difference for himself. We assume that the real output process is

$$dX_t^u := \mu dt + v dB_t^{-u} = (\mu + u_t v)dt + v dB_t, \qquad (7.1)$$

and the agent reports the process X:

$$dX_t := (\mu - u_t v)dt + v dB_t^{-u} = \mu dt + v dB_t. \qquad (7.2)$$

We assume μ, v are constants and, as before, u is $\mathbb{F}^B$-adapted.

The agent receives a proportion $0 < \lambda \leq 1$ of the difference in the case of reporting a lower profit than is true, that is, in case $u_t \geq 0$. In principle, we could also allow him to over-report the change in cash flows, $u_t < 0$, but it can be shown that this will not happen at the optimum. Thus, we assume

$$u \geq 0. \tag{7.3}$$

It is clear that this model is equivalent to the usual moral hazard model with unobservable effort, when the benefit of exhorting effort $-u_t v$ is equal to additional $\lambda u_t v dt$ in consumption.

The agent maintains a savings account S, unobserved by the principal, with the dynamics

$$dS_t = \rho S_t dt + \lambda u_t v dt + di_t - dc_t, \qquad S_{0-} = 0. \tag{7.4}$$

Here, ρ is the agent's savings rate, $di_t \geq 0$ is his increase in income paid by the principal, and $dc_t \geq 0$ is his increase in consumption. We assume c and i are right continuous with left limits, hence so is S. It is required that

$$S_t \geq 0, \tag{7.5}$$

which is a constraint on the agent's consumption c.

The agent is hired until (a possibly random) time τ, which is specified in the contract. It is assumed that he receives a payoff $R \geq 0$ at the time of termination τ, and that he is risk-neutral, maximizing

$$W_{0-}^{A,c,u} := W_0^{A,c,u} + \Delta c_0$$

$$\text{where } W_0^{A,c,u} := W_0^{A,\tau,c,u} := E^{-u}\left[\int_0^\tau e^{-\gamma s} dc_s + e^{-\gamma \tau} R\right] \tag{7.6}$$

where Δc_0 is a possible initial jump in consumption. Here, we use the convention that

$$\int_{t_1}^{t_2} \cdot dc_t := \int_{(t_1,t_2]} \cdot dc_t.$$

Thus, the agent's problem is

$$W_{0-}^A := \sup_{(c,u)\in\mathcal{A}(\tau,i)} W_{0-}^{A,c,u} = \sup_{(c,u)\in\mathcal{A}(\tau,i)} E^{-u}\left[\int_{[0,\tau]} e^{-\gamma s} dc_s + e^{-\gamma \tau} R\right], \tag{7.7}$$

where the agent's admissible control set $\mathcal{A}(\tau, i)$ will be specified later.

The principal's controls are τ and i, which are $\mathbb{F}^X$-adapted and thus, equivalently, $\mathbb{F}^B$-adapted. We only consider implementable contracts (τ, i), that is, such that there exists $(c, u) \in \mathcal{A}(\tau, i)$ with $W_{0-}^A = W_{0-}^{A,c,u}$. Similarly, the principal is also risk-neutral, with expected utility

$$W_{0-}^{P,\tau,i,c,u} := W_0^{P,\tau,i,c,u} - \Delta i_0$$

$$:= E^{-u}\left[\int_0^\tau e^{-rs} d(X_s - i_s) + e^{-r\tau} L\right] - \Delta i_0. \tag{7.8}$$

7.2 Agent's Problem

We will specify the principal's admissible set $\mathcal{A}$ later. The principal's problem is

$$W_{0-}^P := \sup\{W_{0-}^{P,\tau,i,c,u} : (\tau, i) \in \mathcal{A} \text{ and } (c, u) \in \mathcal{A}(\tau, i) \text{ is the agent's}$$
$$\text{optimal control}\}. \tag{7.9}$$

As usual in the literature, when the agent is indifferent between different actions, we assume he will choose one among those which are best for the principal. We assume that the principal has to pay an amount $K \geq 0$ in order to get the project started. However, this does not change the principal's problem (7.9).

We fix $W_{0-}^A = w_A$ and we want to find a contract which maximizes principal's expected utility, while delivering w_A to the agent if he applies the strategy which is optimal for the given contract.

We now specify some initial assumptions, with more being specified later.

Assumption 7.1.1

(i) $\gamma > r \geq \rho$ and $rL + \gamma R \leq \mu$.
(ii) Given (τ, i), each (c, u) in the agent's admissible set $\mathcal{A}(\tau, i)$ consists of a pair of $\mathbb{F}^B$-adapted processes, c is non-decreasing and right continuous with left limit, (7.3) and (7.5) hold, and Girsanov's theorem holds for $-u 1_{[0,\tau]}$.
(iii) Each (τ, i) in the principal's admissible set $\mathcal{A}$ consists of $\mathbb{F}^B$-adapted processes, i is non-decreasing and right continuous with left limit, $E[(\int_0^\tau e^{-\gamma t} di_t)^2] < \infty$, and (τ, i) is implementable.

7.2 Agent's Problem

We first study the agent's problem (7.7). We start by showing that the agent will consume everything he gets, and will keep his savings at zero. This is because the rate at which the principal can save is higher or equal to the agent's rate.

Proposition 7.2.1 *Assume Assumption 7.1.1 holds. Given a contract $(\tau, i) \in \mathcal{A}$, for any u, the agent's optimal consumption is*

$$dc_t = \lambda u_t v dt + di_t \quad \text{and thus} \quad S = 0. \tag{7.10}$$

Proof Note that (7.4) leads to

$$d(e^{-\rho t} S_t) = \lambda e^{-\rho t} u_t v dt + e^{-\rho t} di_t - e^{-\rho t} dc_t.$$

For any c such that (7.5) holds, applying the integration by parts formula and thanks to the assumption that $\gamma \geq \rho$, we have

$$W_{0-}^{A,c,u} = E^{-u}\left[\int_{[0,\tau]} e^{-(\gamma-\rho)s}e^{-\rho s}dc_s + e^{-\gamma\tau}R\right]$$

$$= E^{-u}\left[\int_{[0,\tau]} e^{-(\gamma-\rho)s}\left[\lambda e^{-\rho s}u_s v\,ds + e^{-\rho s}di_s - d\left(e^{-\rho s}S_s\right)\right] + e^{-\gamma\tau}R\right]$$

$$= E^{-u}\left[\int_{[0,\tau]} e^{-(\gamma-\rho)s}\left[\lambda e^{-\rho s}u_s v\,ds + e^{-\rho s}di_s\right] + e^{-\gamma\tau}R\right.$$
$$\left. - e^{-(\gamma-\rho)\tau}e^{-\rho\tau}S_\tau - (\gamma-\rho)\int_0^\tau e^{-(\gamma-\rho)s}e^{-\rho s}S_s\,ds\right]$$

$$\le E^{-u}\left[\int_{[0,\tau]} e^{-(\gamma-\rho)s}\left[\lambda e^{-\rho s}u_s v\,ds + e^{-\rho s}di_s\right] + e^{-\gamma\tau}R\right],$$

and equality holds if and only if (7.10) holds. □

Thus, the agent's problem can be rewritten as

$$W_{0-}^A = \sup_{u\in\mathcal{A}(\tau,i)} W_{0-}^{A,u} := \sup_{u\in\mathcal{A}(\tau,i)} E^{-u}\left[\int_{[0,\tau]} e^{-\gamma s}[\lambda u_s v\,ds + di_s] + e^{-\gamma\tau}R\right]. \quad (7.11)$$

We next specify a technical condition on the agent's admissible set $\mathcal{A}(\tau,i)$.

Assumption 7.2.2 Given (τ,i), $\mathcal{A}(\tau,i)$ is the set of $\mathbb{F}^B$-adapted process $u \ge 0$ such that

$$E^{-u}\left[\left(\int_0^\tau e^{-\gamma t}|u_t|dt + \int_0^\tau e^{-\gamma t}di_t\right)^2\right] + E\left[|M_\tau^{-u}|^3 + [M_\tau^{-u}]^{-3}\right] < \infty. \quad (7.12)$$

We now characterize the optimal control u. Denote by $W_t^{A,u}$ and $\hat{W}_t^{A,u}$ the agent's remaining utility and the discounted remaining utility, respectively:

$$W_t^{A,u} = E_t^{-u}\left[\int_t^\tau e^{-\gamma(s-t)}[\lambda u_s v\,ds + di_s] + e^{-\gamma(\tau-t)}R\right], \qquad \hat{W}_t^{A,u} = e^{-\gamma t}W_t^{A,u}. \tag{7.13}$$

By (7.12) and the Martingale Representation Theorem of Lemma 10.4.6, there exists $Z^{A,u}$ such that

$$\hat{W}_t^{A,u} = e^{-\gamma\tau}R + \int_t^\tau e^{-\gamma s}[\lambda u_s v\,ds + di_s] - \int_t^\tau e^{-\gamma s}v Z_s^{A,u}dB_s^{-u}.$$

This leads to

$$d\hat{W}_t^{A,u} = e^{-\gamma t}u_t v[Z_t^{A,u} - \lambda]dt - e^{-\gamma t}di_t + e^{-\gamma t}v Z_t^{A,u}dB_t \quad \text{and}$$
$$\hat{W}_\tau^{A,u} = e^{-\gamma\tau}R. \tag{7.14}$$

Equivalently,

$$dW_t^{A,u} = \gamma W_t^{A,u}dt + u_t v[Z_t^{A,u} - \lambda]dt - di_t + v Z_t^{A,u}dB_t \quad \text{and} \quad W_\tau^{A,u} = R. \tag{7.15}$$

7.2 Agent's Problem

Theorem 7.2.3 *Assume Assumptions 7.1.1 and 7.2.2 hold. For any $(\tau, i) \in \mathcal{A}$, $u \in \mathcal{A}(\tau, i)$ is optimal if and only if*

$$Z_t^{A,u} = \lambda \quad \text{when } u_t > 0 \quad \text{and} \quad Z_t^{A,u} \geq \lambda \quad \text{when } u_t = 0. \quad (7.16)$$

Proof (i) We first prove the sufficiency. Assume $u \in \mathcal{A}(\tau, i)$ satisfies (7.16). For any $\tilde{u} \in \mathcal{A}(\tau, i)$. Denote

$$\Delta u := u - \tilde{u}, \quad \Delta \hat{W}_t^A := \hat{W}_t^{A,u} - \hat{W}_t^{A,\tilde{u}}, \quad \Delta Z_t^A := Z_t^{A,u} - Z_t^{A,\tilde{u}}.$$

Then,

$$\Delta \hat{W}_{0-}^A = \Delta \hat{W}_0^A = \int_0^\tau e^{-\gamma t} v \big[\lambda - Z_t^{A,u}\big] \Delta u_t dt - \int_0^\tau e^{-\gamma t} v \Delta Z_t^A dB_t^{-\tilde{u}}. \quad (7.17)$$

By (7.16) we have

$$\big[\lambda - Z_t^{A,u}\big] \Delta u_t \geq 0,$$

and thus

$$\Delta \hat{W}_{0-}^A \geq -\int_0^\tau e^{-\gamma t} v \Delta Z_t^A dB_t^{-\tilde{u}}.$$

Note that

$$E^{-\tilde{u}} \bigg[\bigg(\int_0^\tau |e^{-\gamma t} Z_t^{A,u}|^2 dt\bigg)^{\frac{1}{2}}\bigg]$$

$$= E^{-u} \bigg[(M_\tau^{-u})^{-1} M^{-\tilde{u}} \bigg(\int_0^\tau |e^{-\gamma t} Z_t^{A,u}|^2 dt\bigg)^{\frac{1}{2}}\bigg]$$

$$\leq \big(E^{-u}\big[(M_\tau^{-u})^{-2}(M^{-\tilde{u}})^2\big]\big)^{\frac{1}{2}} \bigg(E^{-u}\bigg[\int_0^\tau |e^{-\gamma t} Z_t^{A,u}|^2 dt\bigg]\bigg)^{\frac{1}{2}}$$

$$= \big(E\big[(M_\tau^{-u})^{-1}(M^{-\tilde{u}})^2\big]\big)^{\frac{1}{2}} \bigg(E^{-u}\bigg[\int_0^\tau |e^{-\gamma t} Z_t^{A,u}|^2 dt\bigg]\bigg)^{\frac{1}{2}}$$

$$\leq \big(E\big[(M_\tau^{-u})^{-3}\big]\big)^{\frac{1}{6}} \big(E\big[(M_\tau^{-\tilde{u}})^3\big]\big)^{\frac{1}{3}} \bigg(E^{-u}\bigg[\int_0^\tau |e^{-\gamma t} Z_t^{A,u}|^2 dt\bigg]\bigg)^{\frac{1}{2}}$$

$$< \infty, \quad (7.18)$$

thanks to (7.12). Then,

$$E^{-\tilde{u}} \bigg[\int_0^\tau e^{-\gamma t} v \Delta Z_t^A dB_t^{-\tilde{u}}\bigg] = 0. \quad (7.19)$$

Therefore, $\Delta \hat{W}_{0-}^A \geq 0$ for any $\tilde{u} \in \mathcal{A}(\tau, i)$, that is, $\hat{u}$ is optimal.

(ii) We next prove necessity. Assume $u \in \mathcal{A}(\tau, i)$ is optimal. Set

$$\theta_t := (u_t \wedge 1)\text{sgn}\big(\lambda - Z_t^{A,u}\big) + 1_{\{u_t = 0\}} 1_{\{\lambda > Z_t^{A,u}\}}.$$

Then,

$$u_t + \theta_t \geq 0.$$

For each n, denote

$$\tau_n := \inf\left\{t \geq 0 : M_t^{-(u+\theta)} \geq nM_t^{-u} \text{ or } M_t^{-(u+\theta)} \leq \frac{1}{n}M_t^{-u}\right\} \wedge \tau.$$

Then, $\tau_n \uparrow \tau$. Set

$$u_t^n = u_t + \theta_t 1_{[0,\tau_n]}(t).$$

One can easily check that u^n satisfies Assumption 7.2.2 and thus $u^n \in \mathcal{A}(\tau, i)$. By (7.17) we have

$$\hat{W}_{0-}^{A,u} - \hat{W}_{0-}^{A,u^n} = -\int_0^{\tau_n} e^{-\gamma t} v[\lambda - Z_t^{A,u}]\theta_t dt - \int_0^\tau e^{-\gamma t} v \Delta Z_t^A dB_t^{-\tilde{u}}$$

$$= -\int_0^{\tau_n} e^{-\gamma t} v\big[(u_t \wedge 1)|\lambda - Z_t^{A,u}| + 1_{\{u_t = 0\}}(\lambda - Z_t^{A,u})^+\big]dt$$

$$- \int_0^\tau e^{-\gamma t} v \Delta Z_t^A dB_t^{-\tilde{u}}.$$

Since u is optimal, then (7.19) implies that

$$0 \leq \hat{W}_{0-}^{A,u} - \hat{W}_{0-}^{A,u^n}$$

$$= -E^{-\tilde{u}}\left[\int_0^{\tau_n} e^{-\gamma t} v\big[(u_t \wedge 1)|\lambda - Z_t^{A,u}| + 1_{\{u_t = 0\}}(\lambda - Z_t^{A,u})^+\big]dt\right].$$

This proves (7.16) on $[0, \tau_n]$. Sending $n \to \infty$, we obtain the result. $\square$

Remark 7.2.4 If we assume $u \leq C_0$ for all $u \in \mathcal{A}(\tau, i)$, then one has $\Delta u \geq 0$ when $u = C_0$. Following similar arguments, one can easily see that $u \in \mathcal{A}(\tau, i)$ is optimal if and only if

$$\begin{aligned} Z_t^{A,u} &= \lambda & \text{when } 0 < u_t < C_0; \\ Z_t^{A,u} &\geq \lambda & \text{when } u_t = 0; \quad \text{and} \\ Z_t^{A,u} &\leq \lambda & \text{when } u_t = C_0. \end{aligned} \quad (7.20)$$

The next result shows that truth-telling is always one possible optimal strategy for the agent.

Corollary 7.2.5 *Assume Assumptions 7.1.1 and 7.2.2 hold. A contract $(\tau, i) \in \mathcal{A}$ is implementable if and only if $Z^{A,0} \geq \lambda$. In particular, for any implementable $(\tau, i) \in \mathcal{A}$, $u = 0$ is optimal.*

Proof If $Z^{A,0} \geq \lambda$, by Theorem 7.2.3 we know that $u = 0$ is optimal, and in particular, (τ, i) is implementable. On the other hand, assume $(\tau, i) \in \mathcal{A}$ is implementable and $u \in \mathcal{A}(\tau, i)$ is an arbitrary optimal control. Then, (7.16) holds. This implies that

$$u_t\big[Z_t^{A,u} - \lambda\big] = 0,$$

and thus, by (7.14),

$$\hat{W}_t^{A,u} = e^{-\gamma\tau}R + \int_t^\tau e^{-\gamma s}di_s - \int_t^\tau e^{-\gamma s}vZ_s^{A,u}dB_s.$$

This is the same BSDE as for $\hat{W}^{A,0}$. By (7.18) with $\tilde{u} = 0$, we see that $\hat{W}_{0-}^{A,u} = \hat{W}_{0-}^{A,0}$. That is, $u = 0$ is also optimal. □

Considering now $W^{A,u}$ instead of $\hat{W}^{A,u}$, we state

Corollary 7.2.6 *Assume Assumptions 7.1.1 and 7.2.2 hold. Suppose that*

$$Z^A \geq \lambda, \qquad E\left[\int_0^\tau |e^{-\gamma t}Z_t^A|^2 dt\right] < \infty, \qquad W_\tau^A = R,$$

where

$$dW_t^A = \gamma W_t^A dt - di_t + vZ_t^A dB_t, \qquad W_{0-}^A = w_A. \tag{7.21}$$

Then, $u \in \mathcal{A}(\tau, i)$ is optimal if and only if $u_t = 0$ when $Z_t^A > \lambda$, and in this case we have $W_{0-}^{A,u} = W_{0-}^A = w_A$. In particular, if $Z^A \equiv \lambda$, then any $u \in \mathcal{A}(\tau, i)$ is optimal. That is, the agent is indifferent with respect to the choice of u.

7.3 Principal's Problem

We now investigate the principal's problem (7.9). The principal's utility (7.8) can be written as

$$W_{0-}^{P,\tau,i,u} := W_0^{P,\tau,i,u} - \Delta i_0$$

$$:= E^{-u}\left[\int_0^\tau e^{-rs}\left[(\mu - u_t v)dt - di_t\right] + e^{-r\tau}L\right] - \Delta i_0. \tag{7.22}$$

7.3.1 Principal's Problem Under Participation Constraint

Fix the agent's optimal utility at time zero at value w_A. We modify the principal's problem by assuming that the agent's promised utility has to be larger than R at all times (otherwise, he would quit and take his "outside option"):

$$b(w_A) := \sup_{(\tau,i)\in\mathcal{A}(w_A)} \sup_{u\in\mathcal{A}_0(\tau,i)} W_{0-}^{P,\tau,i,u} \tag{7.23}$$

where, for $w_A \geq R$,

$$\mathcal{A}(w_A) := \{(\tau, i) \in \mathcal{A} : W_{0-}^A = w_A \text{ and } W_t^A \geq R, 0 \leq t \leq \tau\};$$

$$\mathcal{A}_0(\tau, i) := \text{the set of the agent's optimal controls } u \in \mathcal{A}(\tau, i). \tag{7.24}$$

We need the latter notation because the agent's optimal u may not be unique. In particular, by Corollary 7.2.5, $u \equiv 0 \in \mathcal{A}_0(\tau, i)$ for any $(\tau, i) \in \mathcal{A}(w_A)$. We note that, by Corollary 7.2.6, process W^A depends only on (τ, i)—it stays the same for different $u \in \mathcal{A}_0(\tau, i)$.

We first have:

Lemma 7.3.1 *For any $w_A \geq R$, $\mathcal{A}(w_A)$ is not empty.*

Proof Fix any $\bar{R} \geq w_A$. Let $Z_t^A := \lambda$ and i be the reflection process which keeps W_t^A within $[R, \bar{R}]$, and the contract terminates once W_t^A hits R. That is, i is the smallest increasing process such that

$$W_t^A = w_A + \int_0^t \gamma W_s^A ds - i_t + \lambda v B_t$$

stays within $[R, \bar{R}]$, and

$$\tau := \inf\{t : W_t^A = R\}.$$

We will show that $(\tau, i) \in \mathcal{A}(w_A)$. We first show that $\tau < \infty$, a.s. Denote $p(w_A) := P(\tau = \infty | W_0^A = w_A)$. Clearly, p is increasing in w_A. Then,

$$p(\bar{R}) = P\left(W_t^A > R \text{ for all } t \geq 0 | W_0^A = \bar{R}\right) = E\left[1_{\{\inf_{0 \leq t \leq 1} W_t^A > R\}} p(W_1^A) | W_0^A = \bar{R}\right]$$

$$\leq E\left[1_{\{\inf_{0 \leq t \leq 1} W_t^A > R\}} p(\bar{R}) | W_0^A = \bar{R}\right] = P\left(\inf_{0 \leq t \leq 1} W_t^A > R\right) p(\bar{R}).$$

It is obvious that $P(\inf_{0 \leq t \leq 1} W_t^A > R) < 1$. Then, $p(\bar{R}) = 0$ and thus $p(w_A) = 0$.

We next show that

$$E\left[\left(\int_0^\tau e^{-\gamma t} di_t\right)^2\right] < \infty.$$

In fact,

$$e^{-\gamma t} W_t^A = w_A - \int_0^t e^{-\gamma s} di_s + \lambda v \int_0^t e^{-\gamma s} dB_s.$$

Then,

$$\int_0^\tau e^{-\gamma s} di_s = w_A - e^{-\gamma \tau} R + \lambda v \int_0^\tau e^{-\gamma s} dB_s,$$

and thus,

$$E\left[\left(\int_0^\tau e^{-\gamma t} di_t\right)^2\right] = E\left[|w_A - e^{-\gamma \tau} R|^2 + |\lambda v|^2 \int_0^\tau e^{-2\gamma s} ds\right]$$

$$\leq |w_A - R|^2 + \frac{|\lambda v|^2}{2\gamma} < \infty. \tag{7.25}$$

Finally, by definition $W_t^A \geq R$ for $t \leq \tau$. Therefore, $(\tau, i) \in \mathcal{A}(w_A)$. □

7.3 Principal's Problem

Lemma 7.3.2 *For any $w_A \geq R$ and $(\tau, i) \in \mathcal{A}(w_A)$, it must hold that*
$$\tau = \inf\{t \geq 0 : W_t^A \leq R\}. \tag{7.26}$$

Consequently,
$$b(R) = L. \tag{7.27}$$

Proof By Corollary 7.2.5, for Z^A such that (7.21) holds, we have $Z^A \geq \lambda$ and $W_\tau^A = R$. Denote
$$\hat{\tau} := \inf\{t \geq 0 : W_t^A \leq R\}.$$
Then, $\hat{\tau} \leq \tau$. Since $W^A \geq R$ and W^A is right continuous, we have $W_{\hat{\tau}}^A = R$. Note that
$$W_{\hat{\tau}+\delta}^A - W_{\hat{\tau}}^A = \int_{\hat{\tau}}^{\hat{\tau}+\delta} \gamma W_s^A ds - \int_{\hat{\tau}}^{\hat{\tau}+\delta} di_s + \int_{\hat{\tau}}^{\hat{\tau}+\delta} v Z_s^A dB_s$$
$$\leq \int_{\hat{\tau}}^{\hat{\tau}+\delta} v Z_s^A [dB_s - \gamma W_s^A / v Z_s^A ds],$$
and that $\gamma W_s^A / v Z_s^A$ is locally bounded. Then, by applying Girsanov's theorem we see that W^A can not satisfy the IR constraint $W_t^A \geq R$ for $t > \hat{\tau}$, and therefore, $\hat{\tau} = \tau$. $\square$

By the above results, we see that (τ, i) is in one-to-one correspondence with (Z^A, i). Let $\hat{\mathcal{A}}(w_A)$ denote the set of $\mathbb{F}^B$-adapted processes (Z^A, i) such that

(i) i is non-decreasing and right continuous with left limits;
(ii) $Z^A \geq \lambda$ and $\int_0^t |Z_s^A|^2 ds < \infty$, P-a.s. for any t;
(iii) for W^A and τ defined by (7.21) and (7.26), it holds that $\tau < \infty$, a.s. and $W_\tau^A = R$;
(iv) $E[\int_0^\tau |e^{-\gamma t} Z_t^A|^2 dt] < \infty$.

Moreover, for each $(Z^A, i) \in \hat{\mathcal{A}}(w_A)$, let $\hat{\mathcal{A}}_0(Z^A, i)$ denote the set of $u \in \mathcal{A}(\tau, i)$ for the corresponding τ such that $u_t = 0$ whenever $Z_t^A > \lambda$. Then, it is clear that, for $w_A \geq R$,
$$b(w_A) = \sup_{(Z^A,i) \in \hat{\mathcal{A}}(w_A)} \sup_{u \in \hat{\mathcal{A}}_0(Z^A,i)} E^{-u}\left[-\Delta i_0 + \int_0^\tau e^{-rs}[(\mu - u_t v)dt - di_t]\right.$$
$$\left. + e^{-r\tau} L \right]. \tag{7.28}$$

Lemma 7.3.3 *Let $\tilde{Z}^A, \tilde{i}, \tilde{\tau}$ satisfy $E[\int_0^{\tilde{\tau}} |e^{-\gamma t} \tilde{Z}_t^A|^2 dt + |\int_0^{\tilde{\tau}} e^{-\gamma t} d\tilde{i}_t|^2] < \infty$. For any $w_A \geq R$, denote*
$$d\tilde{W}_t^A = \gamma \tilde{W}_t^A dt - d\tilde{i}_t + v \tilde{Z}_t^A dB_t, \qquad \tilde{W}_{0-}^A = w_A.$$
If $\tilde{W}_t^A \geq R$, $0 \leq t \leq \tilde{\tau}$, then there exists $(Z^A, i) \in \hat{\mathcal{A}}(w_A)$ such that $Z_t^A = \tilde{Z}_t^A$, $i_t = \tilde{i}_t$, $0 \leq t \leq \tilde{\tau}$.

Proof On $[0, \tilde{\tau}]$ set $Z_t^A := \tilde{Z}_t^A$, $i_t := \tilde{i}_t$, and thus $W_t^A = \tilde{W}_t^A$. For $t > \tilde{\tau}$, let $Z_t^A := \lambda$ and i be the reflection process (which is continuous) such that the following process W_t^A stays within $[R, \tilde{W}_{\tilde{\tau}}^A]$:

$$dW_t^A = \gamma W_t^A dt - di_t + \lambda v dB_t, \qquad W_{\tilde{\tau}}^A = \tilde{W}_{\tilde{\tau}}^A.$$

By Lemma 7.3.1, we know $\tau := \inf\{t : W_t^A = R\} < \infty$ a.s. Moreover, by (7.25) we see that

$$E_{\tilde{\tau}}\left[\left(\int_{\tilde{\tau}}^{\tau} e^{-\gamma(s-\tilde{\tau})} di_s\right)^2\right] \le |\tilde{W}_{\tilde{\tau}}^A - R|^2 + \frac{|\lambda v|^2}{2\gamma} \le |\tilde{W}_{\tilde{\tau}}^A|^2 + \frac{|\lambda v|^2}{2\gamma}.$$

By our conditions it is clear that

$$E\left[|e^{-\gamma \tilde{\tau}} \tilde{W}_{\tilde{\tau}}^A|^2\right] < \infty.$$

Then,

$$E\left[\left(\int_0^{\tau} e^{-\gamma s} di_s\right)^2\right]$$
$$= E\left[\left(\int_0^{\tilde{\tau}} e^{-\gamma s} d\tilde{i}_s + e^{-\gamma \tilde{\tau}} \int_{\tilde{\tau}}^{\tau} e^{-\gamma(s-\tilde{\tau})} di_s\right)^2\right]$$
$$\le 2E\left[\left(\int_0^{\tilde{\tau}} e^{-\gamma s} d\tilde{i}_s\right)^2 + e^{-2\gamma \tilde{\tau}}\left[|\tilde{W}_{\tilde{\tau}}^A|^2 + \frac{|\lambda v|^2}{2\gamma}\right]\right] < \infty.$$

Then, it follows that $(Z^A, i) \in \hat{\mathcal{A}}(w_A)$. □

Next, following standard arguments, one has the following Dynamic Programming Principle for the problem (7.28), for any stopping time $\tilde{\tau}$:

$$b(w_A) = \sup_{(Z^A, i) \in \hat{\mathcal{A}}(w_A)} \sup_{u \in \hat{\mathcal{A}}_0(Z^A, i)} E^{-u}[G_{\tau \wedge \tilde{\tau}}] \qquad (7.29)$$

where

$$G_t := -\Delta i_0 + \int_0^t e^{-rs}\left[(\mu - u_s v)ds - di_s\right] + e^{-rt} b(W_t^A). \qquad (7.30)$$

If b is sufficiently smooth, by Itô's formula we have

$$e^{rt} dG_t = \left[\mu - u_t v[1 + Z_t^A b'(W_t^A)] + \gamma W_t^A b'(W_t^A) + \frac{1}{2}(vZ_t^A)^2 b''(W_t^A)\right.$$
$$\left. - rb(W_t^A)\right] dt - [1 + b'(W_{t-}^A)] di_t + v Z_t^A b'(W_t^A) dB_t^{-u}$$
$$= \left[\mu - u_t v[1 + \lambda b'(W_t^A)] + \gamma W_t^A b'(W_t^A) + \frac{1}{2}(vZ_t^A)^2 b''(W_t^A)\right.$$
$$\left. - rb(W_t^A)\right] dt - [1 + b'(W_{t-}^A)] di_t + v Z_t^A b'(W_t^A) dB_t^{-u}, \qquad (7.31)$$

7.3 Principal's Problem

thanks to the fact that $u_t[Z_t^A - \lambda] = 0$. Thus, noting that $W_0^A = w_A - \Delta i_0$, we get

$$b(w_A) = \sup_{(Z^A,i)\in\hat{A}} \sup_{u\in\hat{A}_0(Z^A,i)} E^{-u}\Bigg[b(w_A - \Delta i_0) - \Delta i_0$$
$$- \int_0^{\tau\wedge\tilde{\tau}} e^{-rt}\big[1 + b'(W_{t-}^A)\big]di_t$$
$$+ \int_0^{\tau\wedge\tilde{\tau}} e^{-rt}\Big[\mu - u_t v\big[1 + \lambda b'(W_t^A)\big] + \gamma W_t^A b'(W_t^A)$$
$$+ \frac{1}{2}(vZ_t^A)^2 b''(W_t^A) - rb(W_t^A)\Big]dt\Bigg]. \qquad (7.32)$$

7.3.2 Properties of the Principal's Value Function

In this subsection we derive heuristically the form of the principal's value function b. First, the principal can always pay a lump-sum of $di > 0$ to the agent, which means that we have $b(w) \geq b(w - di) - di$. This would imply $b'(w) \geq -1$. Moreover, as long as we have strict inequality, there will be no payments. More precisely, for any $w_A \geq R$ and any $w > w_A$, applying (7.29) to $b(w)$ by setting $\tilde{\tau} := 0$ and $\Delta i_0 := w - w_A$, we get

$$b(w) \geq -\Delta i_0 + b(W_0^A) = b(w - \Delta i_0) - \Delta i_0 = b(w_A) - w + w_A. \qquad (7.33)$$

Sending $w \downarrow w_A$, this implies that, assuming b' exists,

$$b'(w_A) \geq -1 \quad \text{for all } w_A \geq R. \qquad (7.34)$$

Next, we guess that

$$b \in C^2 \quad \text{and is concave.} \qquad (7.35)$$

Denote

$$R^* := \inf\{w \geq R : b'(w) \leq -1\}. \qquad (7.36)$$

Since b' is decreasing, then $b'(w) \leq -1$ for all $w \geq R^*$. This, together with (7.34), implies that

on $[R^*, \infty)$, $b'(w_A) = -1$ and thus $b(w_A) = b(R^*) - (w_A - R^*)$. (7.37)

We now consider $w \in [R, R^*)$. Setting

$$Z^A := \lambda, \quad i := 0, \quad u := 0$$

and plugging $\tilde{\tau} := \delta > 0$ into (7.32), we have

$$E\Bigg[\int_0^{\tau\wedge\delta} e^{-rt}\Big[\mu + \gamma W_t^A b'(W_t^A) + \frac{1}{2}(v\lambda)^2 b''(W_t^A) - rb(W_t^A)\Big]dt\Bigg] \leq 0.$$

Dividing by δ and sending $\delta \to 0$, we get

$$\mu + \gamma w_A b'(w_A) + \frac{1}{2}v^2\lambda^2 b''(w_A) - rb(w_A) \leq 0. \quad (7.38)$$

On the other hand, for any (Z^A, i, u), plugging $\tilde{\tau} := \delta$ into (7.32), we get

$$\sup_{(Z^A,i)\in \hat{\mathcal{A}}, u\in \hat{\mathcal{A}}_0(Z^A,i)} E^{-u}\left[-\int_0^{\tau\wedge\delta} e^{-rt}[1+b'(W_t^A)]di_t \right.$$

$$+ \int_0^{\tau\wedge\delta} e^{-rt}\left[\mu - u_t v[1+\lambda b'(W_t^A)] + \gamma W_t^A b'(W_t^A)\right.$$

$$\left.\left. + \frac{1}{2}(vZ_t^A)^2 b''(W_t^A) - rb(W_t^A)\right]dt \right] = 0.$$

By (7.34), $\lambda \leq 1$, and $b'' \leq 0$, we have

$$\sup_{(Z^A,i)\in \hat{\mathcal{A}}, u\in \hat{\mathcal{A}}_0(Z^A,i)} E\left[M_{\tau\wedge\delta}^{-u} \int_0^{\tau\wedge\delta} e^{-rt}\left[\mu + \gamma W_t^A b'(W_t^A)\right.\right.$$

$$\left.\left. + \frac{1}{2}(v\lambda)^2 b''(W_t^A) - rb(W_t^A)\right]dt \right] \geq 0.$$

Dividing by δ and sending $\delta \to 0$, formally we obtain

$$\mu + \gamma w_A b'(w_A) + \frac{1}{2}(v\lambda)^2 b''(w_A) - rb(w_A) \geq 0.$$

This, together with (7.38), leads to

$$\mu + \gamma w b'(w) + \frac{1}{2}\lambda^2 v^2 b''(w) - rb(w) = 0, \quad w\in[R, R^*]. \quad (7.39)$$

Finally, by (7.37) we have $b'(R^*) = -1$ and $b''(R^*) = 0$. The condition $b'(R^*) = -1$ is the "smooth-pasting" or "smooth-fit" condition, guaranteeing that the derivative from the left and the right agree at $w = R^*$. Moreover, in order to have smooth fit for the second derivative, $b''(R^*) = 0$, the differential equation for b implies also the condition

$$rb(R^*) + \gamma R^* = \mu. \quad (7.40)$$

Intuitively, the payments are postponed until the expected return of the project μ is completely used up by the principal's and the agent's expected returns.

In conclusion, given (7.35), function b should be determined by (7.37) and (7.39), together with the boundary condition (7.27) and the free boundary conditions (7.40). We state the precise result next.

7.3.3 Optimal Contract

The main result of this chapter is:

7.3 Principal's Problem

Theorem 7.3.4 *Assume Assumptions* 7.1.1 *and* 7.2.2 *hold. Consider the ODE system* (7.37), (7.39), *and* (7.40):

$$\mu + \gamma w b'(w) + \frac{1}{2}\lambda^2 v^2 b''(w) - rb(w) = 0, \quad w \in [R, R^*);$$
$$\text{on } [R^*, \infty), \quad b'(w_A) = -1 \quad \text{and thus} \quad b(w_A) = b(R^*) - (w_A - R^*);$$
$$b(R) = L, \quad rb(R^*) + \gamma R^* = \mu;$$

and assume it has a concave solution $b \in C^2$. *Then,*

(i) b *is the value function defined by* (7.23) (*or equivalently by* (7.28)).
(ii) *When* $w_A \in [R, R^*]$, *truth-telling is optimal,* $u \equiv 0$. *Moreover, it is optimal to set*

$$Z_t^A \equiv \lambda,$$

and the payments i *to be the reflection process which keeps* W_t^A *within* $[R, R^*]$. *The optimal contract terminates once* W_t^A *hits* R. *That is,* i *is the smallest increasing process such that*

$$W_t^A = w_A + \int_0^t rW_s^A ds - i_t + \lambda B_t$$

stays within $[R, R^*]$. *In particular, when* $W_t^A \in (R, R^*)$, $di_t = 0$.
(iii) *When* $w_A > R^*$, *then the optimal contract pays an immediate payment of* $w_A - R^*$ *to the agent, and the contract continues with the agent's new initial utility* R^*.

Proof Let b denote the solution to the ODE system and $\hat{b}$ denote the value function defined by (7.28). We first show that

$$\hat{b} \leq b. \tag{7.41}$$

To see that, given $(Z^A, i) \in \hat{\mathcal{A}}(w_A)$ and $u \in \hat{\mathcal{A}}_0(\tau, i)$, recall the process G in (7.29) and its dynamics (7.31). By (7.37) and the assumption that b is concave, we see that $b'(w_A) \geq -1$ for $w_A \in [R, R^*]$ and thus

$$1 + \lambda b'(W_t^A) \geq 1 + b'(W_t^A) \geq 0.$$

This, together with the assumptions that b is concave and $Z^A \geq \lambda$, implies that

$$e^{rt} dG_t \leq \left[\mu + \gamma W_t^A b'(W_t^A) + \frac{1}{2}\lambda^2 v^2 b''(W_t^A) - rb(W_t^A)\right] dt$$
$$+ Z_t^A b'(W_t^A) dB_t^{-u}.$$

When $W_t^A \in [R, R^*]$, by (7.39) we have

$$\mu + \gamma W_t^A b'(W_t^A) + \frac{1}{2}\lambda^2 v^2 b''(W_t^A) - rb(W_t^A) = 0.$$

When $W_t^A > R^*$, by (7.37), (7.40), and the assumption that $\gamma > r$, we have

$$\mu + \gamma W_t^A b'(W_t^A) + \frac{1}{2}\lambda^2 v^2 b''(W_t^A) - rb(W_t^A)$$
$$= rb(R^*) + \gamma R^* - \gamma W_t^A - r[b(R^*) - W_t^A + R^*] = [r - \gamma][W_t^A - R^*] < 0.$$

Therefore, in all the cases we have
$$e^{rt} dG_t \leq Z_t^A b'(W_t^A) dB_t^{-u}.$$

Since b' is bounded, G is a P^{-u}-super-martingale. Notice that $b(w_A) = G_0$ and $b(W_\tau^A) = b(R) = L$. We have then
$$b(w_A) = G_0 \geq E^{-u}[G_\tau]. \tag{7.42}$$

This, together with (7.29), setting $\tilde{\tau} := \tau$, proves (7.41).

We next prove $\hat{b}(w_A) = b(w_A)$ for $w_A \in [R, R^*]$. Let (Z^A, i, u) be specified as in (ii). By Lemma 7.3.1 we see that $(Z^A, i) \in \hat{\mathcal{A}}(w_A)$.

Since $di_t > 0$ implies that $W_t^A = R^*$ and thus $b'(W_t^A) = -1$, we have $(1 + b'(W_t^A))di_t = 0$. Moreover, since $W_t^A \in [R, R^*]$, by (7.39) and (7.31) we see that
$$e^{rt} dG_t = Z_t^A b'(W_t^A) dB_t.$$

This, together with the fact that b' is bounded, implies that G is a P-martingale. Then, we have
$$b(w_A) = G_0 = E[G_\tau].$$

Thus, $\hat{b}(w_A) = b(w_A)$ for $w_A \in [R, R^*]$ and (Z^A, i, u) given in (ii) is an optimal contract.

Finally, for $w_A > R^*$, note that the value function should satisfy (7.33), and then by (7.40) we have $\hat{b}(w_A) \geq b(w_A)$. This, together with (7.41), implies that $\hat{b}(w_A) = b(w_A)$. It is obvious that the contract described in (iii) is optimal. □

Remark 7.3.5 In this remark we discuss how to find the above function b. Consider the following elliptic ODE with parameter $\theta \geq -1$:
$$\begin{cases} \mu + \gamma w b'_\theta(w) + \frac{1}{2}\lambda^2 v^2 b''_\theta(w) - rb_\theta(w) = 0, & w \in [R, \infty); \\ b_\theta(R) = L, \quad b'_\theta(R) = \theta. \end{cases} \tag{7.43}$$

By standard results in the ODE literature, the above ODE has a unique smooth solution b_θ. By Assumption 7.1.1(i),
$$\frac{1}{2}\lambda^2 v^2 b''_\theta(R) = -[\mu + \gamma R b'_\theta(R) - rb_\theta(R)]$$
$$= -[\mu + \gamma R\theta - rL] \leq \gamma R + rL - \mu \leq 0.$$

Denote
$$R^*_\theta := \inf\{w \geq R : b''_\theta(w) = 0\}.$$

If we can find $\theta \geq -1$ such that
$$R^*_\theta < \infty \quad \text{and} \quad b'_\theta(R^*_\theta) = -1, \tag{7.44}$$

then one can check straightforwardly that

$$b(w) := b_\theta(w) 1_{[R, R_\theta^*]}(w) + \left[b_\theta\left(R_\theta^*\right) + R_\theta^* - w\right] 1_{(R_\theta^*, \infty)} \quad (7.45)$$

satisfies all our requirements.

7.4 Implementation Using Standard Securities

We now want to show that the above contract can be implemented using real-world securities, namely, equity, long-term debt and credit line. The implementation will be accomplished using the following:

- The firm starts with initial capital K and possibly an additional amount needed for initial dividends or cash reserves.
- The firm has access to a credit line up to a limit of C^L. The interest rate on the credit line balance is r^C. The agent decides on borrowing money from the credit line and on repayments to the credit line. If the limit C^L is reached, the firm/project is terminated.
- Shareholders receive dividends which are paid from cash reserves or the credit line, at the discretion of the agent.
- The firm issues a long (infinite) term debt with continuous coupons paying at rate x, with face value of the debt equal to $D = x/r$. If the firm cannot pay a coupon payment, the project is terminated.

The agent will be paid by a fraction of dividends. We assume that once the project is terminated the agent does not receive anything from his holdings of equity. Here is the result that shows precisely how the optimal contract is implemented.

Theorem 7.4.1 *Suppose that the credit line has interest rate $r^C = \gamma$, and that the long-term debt satisfies*

$$x = rD = \mu - \gamma R/\lambda - \gamma C^L. \quad (7.46)$$

Assume the dividends $\delta_t dt$ are paid only at the times the credit line balance hits zero, so that $\lambda \delta_t$ is the reflection process that keeps the credit line above zero. If the agent is paid by a proportion $\lambda \delta_t dt$ of the firm's dividends, he will not misreport the cash flows, and will use them to pay the debt coupons and the credit line before issuing dividends. Denoting the current balance of the credit line by M_t, the agent's expected utility process satisfies

$$W_t^A = R + \lambda(C^L - M_t). \quad (7.47)$$

If in addition

$$C^L = (R^* - R)/\lambda \quad (7.48)$$

then the above capital structure of the firm implements the optimal contract.

Proof Denote by δ_t the cumulative dividends process. By that, we mean that $\delta_t dt$ is equal to whatever money is left after paying the interest $\gamma M_t dt$ on the credit line and the debt coupons xdt. Since the total amount of available funds is equal to the balance of the credit line M plus the reported profit X, and since $M + X$ is divided between the credit line interest payments, debt coupon payments and dividends, we have

$$dM_t = \gamma M_t dt + x dt + d\delta_t - dX_t.$$

With W_t^A as in (7.47), and from $x = rD$ and (7.46), we have

$$dW_t^A = -\lambda dM_t = \gamma W_t^A dt - \lambda d\delta_t + \lambda dB_t.$$

If we set $di_t = \lambda d\delta_t$, then this corresponds to the agent's utility with zero savings and $Z^A \equiv \lambda$, which implies that the agent will not have incentives to misreport. Moreover, since the dividends are paid when $M_t = 0$, which, by (7.48) is equivalent to $W_t^A = R^*$, we see that the optimal strategy is implemented by this capital structure. □

Remark 7.4.2 (i) Truth-telling is a consequence of providing the agent with a fraction λ of dividends, and giving him control over the timing of dividends. The agent will not pay the dividends too soon because of (7.47)—if he did empty the credit line instantaneously at time t to pay the dividends and then immediately default, he would be getting W_t^A in expected utility, which is also what he is getting by waiting until the credit line balance is zero. He does pay the dividends once that happens because $\gamma > r$.

(ii) The choice of credit line limit C^L resolves the trade-off between delaying the agent's payments and delaying termination. The level of debt payments $x = rD$ cannot be too high in order to ensure that the agent does not use the credit line too soon; it cannot be too low either, otherwise the agent would save excess cash even after the credit line is paid off, in order to delay termination.

(iii) It is possible to have $D < 0$, which is to be interpreted as a margin account the firm may have to keep in order to have access to the credit line. This account earns interest r, cannot be withdrawn from, and is exhausted by creditors in case of termination.

7.5 Comparative Statics

The above framework enables us to get many economics conclusions, either by analytic derivations, or by numerical computations. We list here some of those reported in DeMarzo and Sannikov (2006).

7.5 Comparative Statics

7.5.1 Example: Agent Owns the Firm

Suppose $\lambda = 1$. In this case the agent gets all the dividends, and the firm is financed by debt. It can be verified that $D = b(R^*)$. Suppose also that W_0^A is chosen so that $b(W_0^A) = K$, the lowest payoff the investors require. With extremely high volatility v, it may happen that such W_0^A does not exist, and no contract is offered. In the cases this is not a problem, numerical examples show that with higher volatility the principal's expected utility $b(w)$ gets smaller, the credit line limit C^L gets larger, the debt level $D = b(R^*)$ gets smaller. In fact, D becomes negative for very high levels of volatility. The required capital K and the margin balance $(-D)$ are financed by a large initial draw of $R^* - W_0^A$ on the credit line (recall that $dW^A = -\lambda dM$). The margin balance pays interest to the project, and this provides incentives for the agent to keep running the firm. This interest is received even after the credit line is paid off, and thus, the upfront financing by investors of the margin balance is a way to guarantee a long-term commitment by the investors to the firm.

With medium volatility we may have $0 < D < K$, so that part of the initial capital K is raised from debt, and part by a draw $R^* - W_0^A$ on the credit line. With very low volatility, we may have $W_0^A > R^*$, so that an immediate dividend of $W_0^A - R^*$ is paid.

7.5.2 Computing Parameter Sensitivities

We now show how to compute partial derivatives of the values determining the optimal contract with respect to a given parameter θ. We first have the following Feynman–Kac type result:

Lemma 7.5.1 *Suppose the process W has the dynamics*

$$dW_t = \gamma W_t dt + \lambda v dB_t - di_t$$

where i is a local-time process that makes W reflect at R^*. The process W is stopped at time $\tau = \min\{t : W_t = R\}$. Given a function g bounded on interval $[R, R^*]$ and constants r, k, L, suppose a function G defined on $[R, R^*]$ solves the ordinary differential equation

$$rG(w) = g(w) + \gamma w G'(w) + \frac{1}{2}\lambda^2 v^2 G''(w) \tag{7.49}$$

with boundary conditions

$$G(R) = L, \qquad G'(R^*) = -k.$$

Then, G can be written as

$$G(w) = E_{W_0 = w}\left[\int_0^\tau e^{-rt} g(W_t) dt - k \int_0^\tau e^{-rt} di_t + e^{-r\tau} L\right]. \tag{7.50}$$

Proof Define the process

$$H_t = \int_0^t e^{-rs} g(W_s) ds - k \int_0^t e^{-rs} di_s + e^{-rt} G(W_t).$$

Using Itô's rule, we get

$$e^{rt} dH_t = \left[g(W_t) + \gamma W_t G'(W_t) + \frac{1}{2} \lambda^2 v^2 G''(W_t) - r G(W_t) \right] dt$$
$$- (k + G'(W_t)) di_t + G'(W_t) \lambda v d B_t.$$

Since G satisfies the ODE, the dt term is zero. So is the di_t term, because $G'(R^*) = -k$ and the process i changes only when $W_t = R^*$. Moreover, since also $G'(w)$ is bounded on $[R, R^*]$, the process H_t is a martingale, and $E[H_t] = H_0 = G(W_0)$. Moreover, as G is bounded on $[R, R^*]$, then also $E[H_\tau] = G(W_0)$. □

Let θ be one of the parameters L, μ, γ, v^2 or λ, and denote by $b_\theta(w)$ the function representing the optimal principal's utility given that parameter. We can compute then its derivative using

Proposition 7.5.2 *We have*

$$\partial_\theta b_\theta(w)$$
$$= E_{W_0^A = w} \left[\int_0^\tau e^{-rt} \left(\partial_\theta \mu + (\partial_\theta \gamma) W_t^A b_\theta'(W_t^A) + \frac{1}{2} [\partial_\theta (\lambda^2 v^2)] b_\theta''(W_t^A) \right) dt \right.$$
$$\left. + e^{-r\tau} \partial_\theta L \right].$$

Proof Denote by $b(w) = b_{\theta, R^*}(w)$ the principal's expected utility function given θ and the reflecting point R^*. This function satisfies

$$rb(w) = \mu + \gamma w b'(w) + \frac{1}{2} \lambda^2 v^2 b''(w) \tag{7.51}$$

with boundary conditions $b(R) = L$, $b'(R^*) = -1$. Denote by $R^*(\theta)$ the value of R^* which maximizes $b_{\theta, R^*}(W_0^A)$, so that $b_\theta = b_{\theta, R^*(\theta)}$. By the Envelope Theorem we have

$$\partial_\theta b_\theta(w) = \partial_\theta b_{\theta, R^*(\theta)}(w) = \partial_\theta b_{\theta, R^*}(w) \big|_{R^* = R^*(\theta)}.$$

Using this and differentiating (7.51) with respect to θ at $R^* = R^*(\theta)$, we get

$$r \partial_\theta b_\theta(w) = \partial_\theta \mu + (\partial_\theta \gamma) w b_\theta'(w) + \gamma w \partial_w \partial_\theta b_\theta(w) + \frac{1}{2} [\partial_\theta (\lambda^2 v^2)] b_\theta''(w)$$
$$+ \frac{1}{2} \lambda^2 v^2 \partial_{w^2}^2 \partial_\theta b_\theta(w) \tag{7.52}$$

with boundary conditions $\partial_\theta b_\theta(R) = \partial_\theta L$, $\partial_w \partial_\theta b_\theta(R^*) = 0$. The statement follows from Lemma 7.5.1. □

7.5 Comparative Statics

Using the knowledge of $\partial_\theta b(w)$, we can find the effect on θ on debt and credit line by differentiating the boundary condition $rb(R^*) + \gamma R^* = \mu$, as well as the definition of the agent's starting value, for example $b(W_0^A) = K$.

As an example, we get that

$$\partial_L b(w) = E_{W_0=w}\left[e^{-r\tau}\right].$$

Moreover, differentiating the boundary condition (considering b as a function of two variables, L and w), we get

$$r\left[\partial_L b(R^*) + b'(R^*)\partial_L R^*\right] + \gamma \partial_L R^* = 0.$$

Since $b'(R^*) = -1$, we get

$$\partial_L R^* = -\frac{r}{\gamma - r} E_{W_0^A=R^*}\left[e^{-r\tau}\right].$$

Thus, larger L means shorter time until paying dividends, and hence shorter credit line, because liquidation is less inefficient. One can similarly compute $\partial_\theta b(w)$ for other θ, and also the sensitivities of debt and credit line to θ, and we report some of the conclusions below. Since $\theta = R$ was not included in the cases above, let us also mention that we have

$$\partial_R b(w) = -b'(R) E_{W_0^A=w}\left[e^{-r\tau}\right].$$

This is because when changing the agent's liquidation value R by dR while simultaneously changing the principal's liquidation value L by $b'(R)dR$, the principal's expected utility will not change.

7.5.3 Some Comparative Statics

Using the above method one can compute many comparative statics, reported in DeMarzo and Sannikov (2006). We mention some conclusions next:

- As L increases, C^L decreases, since termination is less undesirable.
- As R increases, C^L and debt payments decrease to reduce the agent's desire to default sooner.
- As μ increases, C^L increases to delay termination, and debt increases since the cash flows are higher.
- As the agent's discount rate γ increases, C^L decreases as the agent is more impatient to start consuming. The debt value can either increase or decrease.
- As volatility v increases, C^L increases and debt decreases, as discussed in the example above.
- If we choose the highest possible W_0^A, that is, such that $b(W_0^A) = K$, W_0^A increases with L and μ and decreases with v^2, γ, λ and R. Similarly, if we choose the highest possible amount for $b(W_0^A)$, it behaves in the same way.

7.6 Further Reading

This chapter is based on DeMarzo and Sannikov (2006). Biais et al. (2007) obtain equivalent results in the limit of a discrete model. Earlier discrete-time dynamic agency models of the firm include Spear and Srivastava (1987), Leland (1998), Quadrini (2004), DeMarzo and Fishman (2007a, 2007b). Survey paper Sannikov (2012) provides additional references to recent papers.

References

Biais, B., Mariotti, T., Plantin, G., Rochet, J.-C.: Dynamic security design: convergence to continuous time and asset pricing implications. Rev. Econ. Stud. **74**, 345–390 (2007)
DeMarzo, P.M., Fishman, M.: Optimal long-term financial contracting. Rev. Financ. Stud. **20**, 2079–2128 (2007a)
DeMarzo, P.M., Fishman, M.: Agency and optimal investment dynamics. Rev. Financ. Stud. **20**, 151–188 (2007b)
DeMarzo, P.M., Sannikov, Y.: Optimal security design and dynamic capital structure in a continuous-time agency model. J. Finance **61**, 2681–2724 (2006)
Leland, H.E.: Agency costs, risk management, and capital structure. J. Finance **53**, 1213–1243 (1998)
Quadrini, V.: Investment and liquidation in renegotiation-proof contracts with moral hazard. J. Monet. Econ. **51**, 713–751 (2004)
Sannikov, Y.: Contracts: the theory of dynamic principal-agent relationships and the continuous-time approach. Working paper, Princeton University (2012)
Spear, S., Srivastava, S.: On repeated moral hazard with discounting. Rev. Econ. Stud. **53**, 599–617 (1987)

Part IV
Third Best: Contracting Under Hidden Action and Hidden Type—The Case of Moral Hazard and Adverse Selection

Chapter 8
Adverse Selection

Abstract The continuous-time adverse selection problems we consider can be transformed into calculus of variations problems on choosing the optimal expected utility for the agent. When the cost is quadratic, the optimal contract is typically a nonlinear function of the final output value and it may also depend on the underlying source of risk. With risk-neutral agent and principal, a range of lower type agents gets non-incentive cash contracts. As the cost of the effort gets higher, the non-incentive range gets wider, and only the highest type agents get informational rent. The rent gets smaller with higher values of cost, as do the incentives.

8.1 The Model and the PA Problem

We adopt the following variation of the hidden action model (5.4):

$$dX_t = (u_t v_t + \theta)dt + v_t dB_t^{u+\bar{\theta}} \quad \text{where } \bar{\theta}_t := \frac{\theta}{v_t}. \tag{8.1}$$

Here, θ is the skill parameter of the agent. For example, it can be interpreted as the return that the agent can achieve with zero effort. We assume here that u is the effort chosen by the agent and that the process v is fixed. Even if v was an action to be chosen, since the output process X is observed continuously, v is also observed as its quadratic variation process, and thus the principal can tell the agent which v to use. We discuss this case in a latter section.

The agent to be hired by the principal is of type $\theta \in [\theta_L, \theta_H]$, where θ_L, θ_H are known to the principal. The principal does not know θ, but has a prior distribution F on $[\theta_L, \theta_H]$, while the agent knows the value of θ. The principal offers a **menu** of lump-sum contract payoffs $C_T(\theta)$, to be delivered at time T, and agent θ can choose payoff $C_T(\theta')$, where θ' may or may not be equal to his true type θ. The agent's problem is defined to be

$$R(\theta) := \sup_{\theta' \in [\theta_L, \theta_H]} R(\theta, \theta') := \sup_{\theta' \in [\theta_L, \theta_H]} \sup_u E^{u+\bar{\theta}}[U_A(C_T(\theta')) - G_T(u;\theta)], \tag{8.2}$$

where U_A is the agent's utility function and $G_T(u, \theta)$ is the cost of effort. There is no continuous payment c to the agent.

8.1.1 Constraints Faced by the Principal

First, we assume that the IR constraint for the minimal agent's utility is

$$R(\theta) \geq r(\theta) \tag{8.3}$$

where $r(\theta)$ is a given function representing the reservation utility of the type θ agent. In other words, agent θ will not work for the principal unless he can attain expected utility of at least $r(\theta)$. For example, it might be natural that $r(\theta)$ is increasing in θ, so that higher type agents require higher minimal utility. The principal knows the function $r(\theta)$.

Second, by the standard **revelation principle** of the Principal–Agent theory, we may restrict ourselves to the truth-telling contracts, that is, to such contracts for which the agent θ will choose optimally the contract $C_T(\theta)$. In other words, we will have

$$R(\theta) = R(\theta, \theta), \quad \forall \theta \in [\theta_L, \theta_H]. \tag{8.4}$$

This is because if this was not satisfied by the optimal menu of contracts, the principal could relabel the contracts in the optimal menu so that the contract meant for agent θ is, indeed, chosen by him.

Third, as usual, we consider only the implementable contracts. That is, such contracts for which for any θ, there exists a unique optimal effort of the agent, denoted $\hat{u}(\theta)$, such that

$$R(\theta) = E^{\hat{u}(\theta)+\bar{\theta}}\big[U_A\big(C_T(\theta)\big) - G_T\big(\hat{u}(\theta), \theta\big)\big].$$

Under these constraints, the principal's problem is to maximize, over $C_T(\theta)$ in a suitable admissible set to be defined below, the expression

$$\int_{\theta_L}^{\theta_H} E^{\hat{u}(\theta)+\bar{\theta}}\big[U_P\big(X_T - C_T(\theta)\big)\big] dF(\theta). \tag{8.5}$$

8.2 Quadratic Cost and Lump-Sum Payment

It is hard to get a handle on the constraint (8.4), and we do not have a comprehensive general theory. Instead, we restrict ourselves to the setting of Sect. 6.2: there is no continuous payment c, and we assume quadratic cost of effort

$$G_T(u, \theta) = \frac{k}{2} \int_0^T (u_s v_s)^2 ds. \tag{8.6}$$

We could use the FBSDE approach, as we did in Sect. 6.2.2, and get necessary conditions for the agent's problem for a fixed choice of θ'. However, we opt here to apply the approach of Sect. 6.2.3, and identify alternative sufficient and necessary conditions for solving the agent's problem.

8.2.1 Technical Assumptions

The approach of Sect. 6.2.3 can be applied under the assumptions that we list next.

Assumption 8.2.1 Assume U_A, U_P are twice differentiable and
$$v_t \equiv v$$
is a constant. Consequently, $\bar{\theta} := \frac{\theta}{v}$ is also a constant.

Assumption 8.2.2 For each θ, the set $\mathcal{U}(\theta)$ of admissible effort processes u of the agent of type θ is the space of $\mathbf{F}^B$-adapted processes u such that

(i) $M^{u+\bar{\theta}}$ is a P-martingale, or equivalently, $M^{\bar{\theta},u}$ is a $P^{\bar{\theta}}$-martingale, where
$$M_t^{\bar{\theta},u} := \exp\left(\int_0^t u_s dB_s^{\bar{\theta}} - \frac{1}{2}\int_0^t |u_s|^2 ds\right) = M_t^{u+\bar{\theta}}\left(M_t^{\bar{\theta}}\right)^{-1}.$$

(ii) It holds that
$$E^{u+\bar{\theta}}\left[\left(\int_0^T |u_t|^2 dt\right)^2 + M_T^{u+\bar{\theta}}\right] < \infty \quad \text{and} \quad E^{\bar{\theta}}\left[|M_T^{\bar{\theta},u}|^2\right] < \infty. \quad (8.7)$$

By Remark 10.4.9, we know the above conditions imply that
$$E^{\bar{\theta}}\left[\sup_{0 \le t \le T} |M_t^{\bar{\theta},u}|^2\right] < \infty.$$

Given a contract C_T and θ, θ', consider the following BSDE
$$\tilde{W}_t^{A,\theta,\theta'} = e^{\kappa U_A(C_T(\theta'))} - \int_t^T \tilde{Z}_s^{A,\theta,\theta'} dB_s^{\bar{\theta}}, \quad (8.8)$$

where
$$\kappa := \frac{1}{kv^2}. \quad (8.9)$$

Assumption 8.2.3 Let $\mathcal{A}_0$ denote the set of contracts C_T which satisfy:

(i) For any $\theta \in [\theta_L, \theta_H]$, $C_T(\theta)$ is $\mathcal{F}_T$-measurable and
$$E\left[|U_A(C_T(\theta))|^4 + M_T^{\bar{\theta}} e^{2\kappa U_A(C_T(\theta))}\right] < \infty. \quad (8.10)$$

(ii) Denoting
$$\hat{u}_t(\theta, \theta') := \frac{\tilde{Z}_t^{A,\theta,\theta'}}{\tilde{W}_t^{A,\theta,\theta'}}, \quad (8.11)$$
we have $\hat{u}(\theta, \theta') \in \mathcal{U}(\theta)$ and $E^{\bar{\theta}}[\int_0^T |\hat{u}_t(\theta, \theta')|^2 dt] < \infty$.

(iii) For dF-a.s. θ, $C_T(\theta)$ is differentiable in θ and $\{e^{\kappa U_A(C_T(\theta'))} U_A'(C_T(\theta')) \times |\partial_\theta C_T(\theta')|\}$ is uniformly integrable under $P^{\bar{\theta}}$, uniformly in θ'.

(iv) $\sup_{\theta \in [\theta_L, \theta_H]} E^{\bar{\theta}}[e^{\kappa U_A(C_T(\theta))}|U_P(X_T - C_T(\theta))|] < \infty.$

Assumption 8.2.4 The admissible set $\mathcal{A}$ of the principal consists of contracts $C_T \in \mathcal{A}_0$ which satisfy the IR constraint (8.3) and the revelation principle (8.4).

Under (8.10), clearly BSDE (8.8) is well-posed and $\tilde{W}^{A,\theta,\theta'} > 0$, so that $\hat{u}(\theta, \theta')$ is well defined. We note that a direct corollary of Theorem 8.2.5 below is that any $C_T \in \mathcal{A}$ is implementable.

We assume

$$\mathcal{A} \neq \phi.$$

8.2.2 Solution to the Agent's Problem

We have the following results, analogous to those in Sect. 6.2.3:

Theorem 8.2.5 *Assume Assumptions 8.2.1 and 8.2.2 hold, and C_T satisfies Assumption 8.2.3(i) and (ii).*

(i) *For any $\theta, \theta' \in [\theta_L, \theta_H]$, the optimal effort $\hat{u}(\theta, \theta') \in \mathcal{U}(\theta)$ of the agent of type θ, faced with the contract $C(\theta')$, is defined as in (8.11), after solving BSDE (8.8), and we have*

$$\kappa R(\theta, \theta') = \log\big(E^{\bar{\theta}}\big[e^{\kappa U_A(C_T(\theta'))}\big]\big) = \log \tilde{W}_0^{A,\theta,\theta'}. \tag{8.12}$$

(ii) *In particular, for a truth-revealing contract $C_T \in \mathcal{A}$, the optimal effort $\hat{u}(\theta) \in \mathcal{U}(\theta)$ for the agent is obtained by solving the BSDE*

$$\tilde{W}_t^{A,\theta} = e^{\kappa U_A(C_T(\theta))} - \int_t^T \hat{u}_s(\theta) \tilde{W}_s^{A,\theta} dB_s^{\bar{\theta}}, \tag{8.13}$$

and the agent's optimal expected utility is given by

$$\kappa R(\theta) = \log\big(E^{\bar{\theta}}\big[e^{\kappa U_A(C_T(\theta))}\big]\big) = \log \tilde{W}_0^{A,\theta}. \tag{8.14}$$

(iii) *For optimal $\hat{u}(\theta, \theta')$, the change of measure process $M^{\hat{u}(\theta,\theta')+\bar{\theta}}$ satisfies*

$$M_T^{\hat{u}(\theta,\theta')+\bar{\theta}} = e^{-\kappa R(\theta,\theta')} M_T^{\bar{\theta}} e^{\kappa U_A(C_T(\theta'))}. \tag{8.15}$$

Proof (i) For any $u \in \mathcal{U}(\theta)$, denote the agent's remaining utility:

$$W_t^{A,u} := W_t^{A,u,\theta,\theta'} := E_t^{u+\bar{\theta}}\bigg[U_A\big(C_T(\theta')\big) - \frac{k}{2}\int_t^T (u_s v)^2 ds\bigg].$$

By (8.10) and (8.7) we have

8.2 Quadratic Cost and Lump-Sum Payment

$$E^{u+\bar{\theta}}[|U_A(C_T(\theta'))|^2] = E[M_T^{u+\bar{\theta}}|U_A(C_T(\theta'))|^2]$$
$$\leq \frac{1}{2} E[|M_T^{u+\bar{\theta}}|^2 + |U_A(C_T(\theta'))|^4]$$
$$= \frac{1}{2} E^{u+\bar{\theta}}[M_T^{u+\bar{\theta}}] + \frac{1}{2} E[|U_A(C_T(\theta'))|^4] < \infty;$$

$$E^{u+\bar{\theta}}\left[\left(\int_0^T |u_t|^2\right)^2\right] < \infty.$$

Applying Lemma 10.4.6, there exists $Z^{A,u} := Z^{A,u,\theta,\theta'}$ such that

$$W_t^{A,u} = U_A(C_T(\theta')) - \frac{1}{2\kappa} \int_t^T |u_s|^2 ds - \int_t^T Z_s^{A,u} dB_s^{u+\bar{\theta}}. \qquad (8.16)$$

Next, denote

$$W_t^{A,\theta,\theta'} := \frac{1}{\kappa} \ln(\tilde{W}_t^{A,\theta,\theta'}), \qquad Z_t^{A,\theta,\theta'} := \frac{1}{\kappa} \frac{\tilde{Z}_t^{A,\theta,\theta'}}{\tilde{W}_t^{A,\theta,\theta'}} = \frac{1}{\kappa} \hat{u}(\theta, \theta').$$

Applying Itô's formula we have

$$W_t^{A,\theta,\theta'} = U_A(C_T(\theta')) - \int_t^T \frac{\kappa}{2} |Z_s^{A,\theta,\theta'}|^2 ds - \int_t^T Z_s^{A,\theta,\theta'} dB_s^{\bar{\theta}}$$
$$= U_A(C_T(\theta')) - \int_t^T \left[\frac{\kappa}{2} |Z_s^{A,\theta,\theta'}|^2 + u_s Z_s^{A,\theta,\theta'}\right] ds$$
$$- \int_t^T Z_s^{A,\theta,\theta'} dB_s^{u+\bar{\theta}}. \qquad (8.17)$$

Then,

$$W_0^{A,u} - W_0^{A,\theta,\theta'} = -\int_0^T \left[\frac{1}{2\kappa}|u_s|^2 + \frac{\kappa}{2}|Z_s^{A,\theta,\theta'}|^2 - u_s Z_s^{A,\theta,\theta'}\right] ds$$
$$- \int_0^T [Z_s^{A,u} - Z_s^{A,\theta,\theta'}] dB_s^{u+\bar{\theta}}$$
$$= -\int_0^T \frac{1}{2\kappa}[|u_s|^2 + |\hat{u}_s(\theta,\theta')|^2 - u_s \hat{u}_s(\theta,\theta')] ds$$
$$- \int_0^T [Z_s^{A,u} - Z_s^{A,\theta,\theta'}] dB_s^{u+\bar{\theta}}$$
$$= -\int_0^T \frac{1}{2\kappa}|u_s - \hat{u}_s(0,0')|^2 - \int_0^T [Z_s^{A,u} - Z_s^{A,\theta,\theta'}] dB_s^{u+\bar{\theta}}$$
$$\leq -\int_0^T [Z_s^{A,u} - Z_s^{A,\theta,\theta'}] dB_s^{u+\bar{\theta}},$$

with the equality holding if and only if $u = \hat{u}(\theta, \theta')$. Note that

$$E^{u+\bar{\theta}}\left[\left(\int_0^T |Z_t^{A,\theta,\theta'}|^2 dt\right)^{\frac{1}{2}}\right] = E^{\bar{\theta}}\left[M_T^{\bar{\theta},u}\left(\int_0^T |\hat{u}_t(\theta,\theta')|^2 dt\right)^{\frac{1}{2}}\right]$$
$$\leq \frac{1}{2\kappa} E^{\bar{\theta}}\left[|M_T^{\bar{\theta},u}|^2 + \int_0^T |\hat{u}_t(\theta,\theta')|^2 dt\right] < \infty.$$

Then,
$$E^{u+\bar{\theta}}\left[\int_0^T [Z_s^{A,u} - Z_s^{A,\theta,\theta'}] dB_s^{u+\bar{\theta}}\right] = 0,$$

and thus
$$W_0^{A,u} \leq W_0^{A,\theta,\theta'}.$$

This implies that $\hat{u}(\theta,\theta')$ is the agent's optimal control. Therefore,
$$R(\theta,\theta') = W_0^{A,\theta,\theta'} = \frac{1}{\kappa} \ln \tilde{W}_0^{A,\theta,\theta'} = \frac{1}{\kappa} \ln\left(E^{\bar{\theta}}\left[e^{\kappa U_A(C_T(\theta'))}\right]\right).$$

(ii) This is a direct consequence of (i), by setting $\theta' = \theta$.
(iii) Note that
$$d\tilde{W}_t^{A,\theta,\theta'} = \tilde{W}_t^{A,\theta,\theta'} \hat{u}_t(\theta,\theta') dB_t^{\bar{\theta}}.$$

Then, $\tilde{W}_t^{A,\theta,\theta'} = \tilde{W}_0^{A,\theta,\theta'} M_t^{\bar{\theta},\tilde{u}(\theta,\theta')}$, and thus
$$M_T^{\hat{u}(\theta,\theta')+\bar{\theta}} = M_T^{\bar{\theta},\tilde{u}(\theta,\theta')} M_T^{\bar{\theta}} = (\tilde{W}_0^{A,\theta,\theta'})^{-1} \tilde{W}_T^{A,\theta,\theta'} M_T^{\bar{\theta}}$$
$$= e^{-\kappa R(\theta,\theta')} e^{\kappa U_A(C_T(\theta'))} M_T^{\bar{\theta}}.$$

This completes the proof. $\square$

Clearly Assumption 8.2.3(ii) is important for the agent's problem. We next provide a sufficient condition for it to hold:

Lemma 8.2.6 *If $C_T(\theta')$ is bounded, then Assumption 8.2.3(ii) holds.*

Proof Since U_A is continuous, then $U_A(C_T(\theta'))$ is also bounded. Let $K > 0$ denote a generic constant which may depend on the bound of $U_A(C_T(\theta'))$ and κ and may vary from line to line. Then, by BSDE (8.8),
$$e^{-K} \leq \tilde{W}_t^{A,\theta,\theta'} \leq e^K \quad \text{and} \quad E_t\left[\int_t^T |\tilde{Z}_s^{A,\theta,\theta'}|^2 ds\right] \leq K.$$

This implies that
$$E_t\left[\int_t^T |\hat{u}_s(\theta,\theta')|^2 ds\right] \leq K.$$

Applying Lemma 9.6.5 and (9.53) we know that $M^{\bar{\theta},\hat{u}(\theta,\theta')}$ is a $P^{\bar{\theta}}$-martingale and

8.2 Quadratic Cost and Lump-Sum Payment

$$E^{\bar{\theta}}\left[\left(\int_0^T |\hat{u}_t(\theta,\theta')|^2 dt\right)^4\right] < \infty.$$

Moreover, by the arguments in Theorem 8.2.5(iii), it is clear that

$$M_T^{\hat{u}(\theta,\theta')+\bar{\theta}} = \left(\tilde{W}_0^{A,\theta,\theta'}\right)^{-1}\tilde{W}_T^{A,\theta,\theta'} M_T^{\bar{\theta}} \leq K M_T^{\bar{\theta}}.$$

Then, it is straightforward to check (8.7). □

8.2.3 Principal's Relaxed Problem

We now have workable expressions (8.12) and (8.14) for the expected utility of the type θ agent, when declaring type θ' and when declaring the true type θ, respectively. The approach we will take is standard in PA literature: we will find the first order condition for truth-telling and use it as the additional constraint on the menu of contracts to be offered. Eventually, once the problem is solved under such a constraint, one has to verify that the first order condition is sufficient, that is, one has to verify that the obtained contract is, in fact, truth-telling.

Note that the first order condition for truth-telling is

$$\partial_{\theta'} R(\theta,\theta')\big|_{\theta'=\theta} = 0.$$

Under this condition we get

$$\kappa e^{\kappa R(\theta)} R'(\theta) = \frac{d}{d\theta} e^{\kappa R(\theta,\theta)} = \kappa e^{\kappa R(\theta,\theta)}\left[\partial_\theta R(\theta,\theta) + \partial_{\theta'} R(\theta,\theta)\right]$$
$$= \kappa e^{\kappa R(\theta)} \partial_\theta R(\theta,\theta).$$

From this, recalling the definition of $M^{\bar{\theta}} = M^{\frac{\theta}{v}}$, and differentiating the exponential version of the first equality of (8.12) with respect to θ, we get the first order condition for truth-telling:

$$\kappa e^{\kappa R(\theta)} R'(\theta) = \frac{1}{v} E^{\bar{\theta}}\left[e^{\kappa U_A(C_T(\theta))} B_T^{\bar{\theta}}\right]. \tag{8.18}$$

In accordance with the above, and recalling (8.15), the principal's problem of maximizing (8.5) is replaced by a new, relaxed principal's problem, given by the following

Definition 8.2.7 The relaxed principal's problem is

$$\sup_{R(\cdot)} \int_{\theta_L}^{\theta_H} \sup_{C_T(\cdot) \in \mathcal{A}_0} e^{-\kappa R(\theta)} E^{\bar{\theta}}\left[e^{\kappa U_A(C_T(\theta))} U_P(X_T - C_T(\theta))\right] dF(\theta) \tag{8.19}$$

under the constraints

$$R(\theta) \geq r(\theta), \qquad E^{\bar{\theta}}\left[e^{\kappa U_A(C_T(\theta))}\right] = e^{\kappa R(\theta)},$$
$$E^{\bar{\theta}}\left[e^{\kappa U_A(C_T(\theta))} B_T^{\bar{\theta}}\right] = v\kappa e^{\kappa R(\theta)} R'(\theta). \tag{8.20}$$

Introducing Lagrange multipliers λ and μ, the Lagrangian of the constrained optimization problem inside the integral above becomes

$$V_P(\theta, R, \lambda, \mu) := e^{-\kappa R(\theta)} \sup_{C_T(\cdot) \in \mathcal{A}_0} E^{\bar{\theta}} \{ e^{\kappa U_A(C_T(\theta))} [U_P(X_T - C_T(\theta)) - \lambda(\theta) - \mu(\theta) B_T^{\bar{\theta}}] \}. \tag{8.21}$$

Remark 8.2.8 If we can solve the latter problem over $C_T(\theta)$ and then find the Lagrangian multipliers $\lambda(\theta), \mu(\theta)$ so that the constraints are satisfied, then the principal's relaxed problem (8.19) reduces to a deterministic calculus of variation problem over the function $R(\theta)$, the agent's expected utility. In the classical, single-period adverse selection problem with a risk-neutral principal, a continuum of types, but no moral hazard, it is also possible to reduce the problem to a calculus of variations problem, typically over the payment $C_T(\theta)$. Under the so-called Spence-Mirrlees condition on the agent's utility function and with a risk-neutral principal, a contract $C_T(\theta)$ is truth-telling if and only if it is a non-decreasing function of θ and the first order truth-telling condition is satisfied. In our model, where we also have moral hazard and risk-averse principal, the calculus of variation problem cannot be reduced to the problem over $C_T(\theta)$, but remains to be a problem over the agent's utility $R(\theta)$. Unfortunately, for a general utility function U_A of the agent, we have not been able to formulate a condition on U_A under which we could find necessary and sufficient conditions on $R(\theta)$ to induce truth-telling. Later below, we are able to show that the first order approach works for a risk-neutral principal and agent, when the hazard function of θ is increasing, in agreement with the classical theory.

8.2.4 Properties of the Candidate Optimal Contract

The above problem is very difficult in general. We focus on the special case of the risk-neutral principal agent in a later section. Here, we get some qualitative conclusions, assuming that the solution to the relaxed problem exists, and that it is equal to the solution of the original problem.

The first order condition for the problem (8.21) can be written as

$$\frac{U_P'(X_T - C_T)}{U_A'(C_T)} = \kappa [U_P(X_T - C_T(\theta)) - \lambda(\theta) - \mu(\theta) B_T^{\bar{\theta}}]. \tag{8.22}$$

We see that, compared to the moral hazard case (6.50) (or (6.66)), there is an extra term $\kappa \mu(\theta) B_T^{\bar{\theta}}$. The optimal contract is a function of the output value X_T and of the "benchmark" random risk level $B_T^{\bar{\theta}}$. Here, with constant volatility v, we can write $B_T^{\bar{\theta}} = \frac{1}{v}[X_T - x - \theta T]$, and the contract is still a function only of the final output value X_T. If, on the other hand, the volatility were a time-varying process, then the optimal contract would depend on X_T and the underlying risk level

$B_T^{\bar\theta} = \int_0^T \frac{1}{v_t}[dX_t - \theta dt]$, and would thus depend on the history of the output X. Random variable $B_T^{\bar\theta}$ can be interpreted as a benchmark value that the principal needs to use to distinguish between different agent types.

8.3 Risk-Neutral Agent and Principal

Because the first order condition (8.22) generally leads to a nonlinear equation for C_T, it is hard or impossible to solve our adverse selection problem for most utility functions. We here discuss the case of linear utility functions and uniform prior on θ. The main results and economic conclusions thereof are contained in Theorem 8.3.1 and the remarks thereafter.

Suppose that

$$U_A(x) = x, \qquad U_P(x-c) = x - c, \qquad X_t = x + vB_t, \qquad G_T(\theta) = \frac{k}{2}\int_0^T u_t^2 dt \tag{8.23}$$

for some positive constants k, x, v. From (8.22) we get a linear relationship between the payoff C_T and B_T (equivalently, X_T)

$$1 = \kappa\big[x + vB_T - C_T(\theta) - \lambda(\theta) - \mu(\theta)(B_T - \bar\theta T)\big].$$

From this we can write

$$C_T(\theta) = a(\theta) + b(\theta)B_T \tag{8.24}$$

for some deterministic functions a, b. Note that, under $P^{\bar\theta}$, B_T has normal distribution with mean $\bar\theta T$ and variance T. Then, for any constant α,

$$E^{\bar\theta}\big[e^{\alpha B_T}\big] = e^{\alpha\bar\theta T + \frac{1}{2}\alpha^2 T},$$
$$E^{\bar\theta}\big[B_T^{\bar\theta} e^{\alpha B_T}\big] = \alpha T e^{\alpha\bar\theta T + \frac{1}{2}\alpha^2 T}, \tag{8.25}$$
$$E^{\bar\theta}\big[B_T e^{\alpha B_T}\big] = [\bar\theta + \alpha]T e^{\alpha\bar\theta T + \frac{1}{2}\alpha^2 T}.$$

From this, the last two equations in (8.20) imply

$$e^{\kappa a(\theta) + \kappa b(\theta)\bar\theta T + \frac{1}{2}\kappa^2 b(\theta)^2 T} = e^{\kappa R(\theta)},$$
$$\kappa bT e^{\kappa a(\theta) + \kappa b(\theta)\bar\theta T + \frac{1}{2}\kappa^2 b(\theta)^2 T} = v\kappa e^{\kappa R(\theta)} R'(\theta). \tag{8.26}$$

We can solve this system, and we get, recalling (8.9) and omitting the argument θ,

$$b = \frac{v}{T}R', \qquad a = R - \theta R' - \frac{(R')^2}{2kT}. \tag{8.27}$$

Substituting into the principal's relaxed problem (8.19), we see that she needs to maximize

$$\int_{\theta_L}^{\theta_H} e^{\kappa a - \kappa R} E^{\bar\theta}\big[e^{\kappa b B_T}[x - a + (v - b)B_T]\big] dF(\theta)$$

which is, using (8.25), (8.26) and (8.27), equal to

$$\int_{\theta_L}^{\theta_H} \left\{ x - R(\theta) + \theta R'(\theta) + \frac{(R'(\theta))^2}{2kT} \right. \\ \left. + \left(v - v\frac{R'(\theta)}{T}\right)\left(\frac{\theta}{v} + \frac{1}{kv^2}\frac{vR'(\theta)}{T}\right)T \right\} dF(\theta). \quad (8.28)$$

Maximizing this is equivalent to minimizing

$$\int_{\theta_L}^{\theta_H} \left\{ R(\theta) + \frac{1}{2kT}(R'(\theta))^2 - \frac{1}{k}R'(\theta) \right\} dF(\theta) \quad (8.29)$$

and it has to be done under the constraint

$$R(\theta) \geq r(\theta)$$

for some given function $r(\theta)$. If this function is constant, and the distribution F is uniform, we have the following result:

Theorem 8.3.1 *Assume Assumptions 8.2.1, 8.2.2, 8.2.3, and 8.2.4 hold. Assume further that (8.23) holds, θ is uniform on $[\theta_L, \theta_H]$, and the IR lower bound is $r(\theta) \equiv r_0$. Then, the principal's problem (8.19) under the first two constraints in (8.20) and the revelation principle (4.4) has a unique solution as follows. Denote $\theta^* := \max\{\theta_H - 1/k, \theta_L\}$. The optimal choice of agent's utility R by the principal is given by*

$$R(\theta) = \begin{cases} r_0, & \theta_L \leq \theta < \theta^*; \\ r_0 + kT\theta^2/2 + T(1 - k\theta_H)\theta - kT(\theta^*)^2/2 - T(1 - k\theta_H)\theta^*, & (8.30) \\ \theta^* \leq \theta \leq \theta_H \end{cases}$$

and consequently,

$$b(\theta) = \begin{cases} 0, & \theta_L \leq \theta < \theta^*; \\ v[1 + k(\theta - \theta_H)], & \theta^* \leq \theta \leq \theta_H; \end{cases} \quad (8.31)$$

$$a(\theta) = \begin{cases} r_0, & \theta_L \leq \theta < \theta^*; \\ r_0 - kT\theta^2 - T(1 - k\theta_H)(\theta - \theta^*) - \frac{T}{2k}(1 - k\theta_H)^2 - \frac{kT}{2}(\theta^*)^2, & (8.32) \\ \theta^* \leq \theta \leq \theta_H. \end{cases}$$

The optimal agent's effort is given by

$$u(\theta) = \begin{cases} 0, & \theta_L \leq \theta < \theta^*; \\ \frac{1}{v}[1/k + \theta - \theta_H], & \theta^* \leq \theta \leq \theta_H. \end{cases} \quad (8.33)$$

The optimal contract is linear, of the form

$$C_T(\theta) = \begin{cases} a(\theta), & \theta_L \leq \theta < \theta^*; \\ a(\theta) + (1 + k\theta - k\theta_H)(X_T - x), & \theta^* \leq \theta \leq \theta_H. \end{cases} \quad (8.34)$$

8.3 Risk-Neutral Agent and Principal

Note that when the agent is risk-neutral this is in agreement with the single-period case (2.23).

Remark 8.3.2 (i) If $1/k < \theta_H - \theta_L$, a range of lower type agents gets no "rent" above the reservation value r_0, the corresponding contract is not incentive as it does not depend on X, and the effort u is zero. The higher type agents get utility $R(\theta)$ which is quadratically increasing in their type θ. It can also be computed that the principal's utility is linear in θ. As the cost k gets higher, the non-incentive range gets wider, and only the highest type agents get informational rent. The rent gets smaller with higher values of cost k, as do the incentives (the slope of C_T with respect to X_T).

(ii) Some analogous results can be obtained for the general distribution F of θ, that has a density $f(\theta)$, using the fact that the solution y to the Euler equation (8.37) in the proof below satisfies:

$$y(\theta) = \beta + T\theta + \alpha \int_{\theta_L}^{\theta} \frac{dx}{f(x)} + kT \int_{\theta_L}^{\theta} \frac{F(x)}{f(x)} dx \tag{8.35}$$

for some constants α and β.

Proof of Theorem 8.3.1 We first show that (8.30)–(8.34) solve the relaxed principal's problem (8.19)–(8.20). Then, in Lemma 8.3.3 below, we check that the truth-telling constraint is indeed satisfied.

First, one can prove straightforwardly that $u \in \mathcal{U}$ and $C_T \in \mathcal{A}$. Next, if F has density f, in light of the integrand in (8.29), denote

$$\varphi(y, y') := \left[y + \frac{1}{2kT}(y')^2 - \frac{1}{k}y' \right] f. \tag{8.36}$$

Here, y is a function on $[\theta_L, \theta_H]$ and y' is its derivative. Then, the Euler ODE for the calculus of variations problem (8.29), denoting by y the candidate solution, is (see, for example, Kamien and Schwartz 1991)

$$\varphi_y = \frac{d}{d\theta} \varphi_{y'}$$

or, in our example,

$$y'' = kT + (T - y')\frac{f'}{f}. \tag{8.37}$$

Since θ is uniformly distributed on $[\theta_L, \theta_H]$, this gives

$$y(\theta) = kT\theta^2/2 + \alpha\theta + \beta$$

for some constants α, β. According to the calculus of variations, on every interval R is either of the same quadratic form as y, or is equal to r_0. One possibility is that, for some $\theta_L \leq \theta^* \leq \theta_H$,

$$R(\theta) = \begin{cases} r_0, & \theta_L \leq \theta < \theta^*; \\ kT\theta^2/2 + \alpha\theta + \beta, & \theta^* \leq \theta \leq \theta_H. \end{cases} \tag{8.38}$$

In this case, $R(\theta)$ is not constrained at $\theta = \theta_H$. By standard results of calculus of variations, the free boundary condition is, recalling notation (8.36), $0 = \varphi_{y'}(\theta_H)$, which implies

$$T = y'(\theta_H) \tag{8.39}$$

from which we get

$$\alpha = T(1 - k\theta_H).$$

Moreover, by the principle of smooth fit, if $\theta_L < \theta^* < \theta_H$, we need to have

$$0 = R'(\theta^*) = kT\theta^* + \alpha$$

which gives

$$\theta^* = \theta_H - \frac{1}{k}$$

if $1/k < \theta_H - \theta_L$. If $1/k \geq \theta_H - \theta_L$ then we can take

$$\theta^* = \theta_L.$$

In either case the candidate for the optimal solution is given by (8.30).

Another possibility would be

$$R(\theta) = \begin{cases} kT\theta^2/2 + \alpha\theta + \beta, & \theta_L \leq \theta < \theta^*; \\ r_0, & \theta^* \leq \theta \leq \theta_H. \end{cases} \tag{8.40}$$

In this case the free boundary condition at $\theta = \theta_L$ would give $\alpha = T(1 - k\theta_L)$, but this is incompatible with the smooth fit condition $kT\theta^* + \alpha = 0$.

The last possibility is that $R(\theta) = T\theta^2/2 + \alpha\theta + \beta$, everywhere. We would get again that at the optimum $\alpha = T(1 - k\theta_H)$, and β would be chosen so that $R(\theta^*) = r_0$ at its minimum point θ^*. Doing computations and comparing to the case (8.30), it is readily checked that (8.30) is still optimal.

Now (8.31) and (8.32) follow directly from (8.27), and combining with (8.24) we get (8.34) immediately.

To obtain the agent's optimal action $\hat{u}(\theta)$, we note that the BSDE (8.13) leads to

$$\tilde{W}_t^{A,\theta} = E_t^{\bar{\theta}}\left[e^{\kappa C_T(\theta)}\right] = E_t^{\bar{\theta}}\left[e^{\kappa[a(\theta)+b(\theta)B_T]}\right] = e^{\kappa a(\theta)+\kappa b(\theta)B_t} E_t^{\bar{\theta}}\left[e^{\kappa b(\theta)(B_T - B_t)}\right].$$

By the first equation in (8.25) we have

$$\tilde{W}_t^{A,\theta} = E_t^{\bar{\theta}}\left[e^{\kappa C_T(\theta)}\right] = E_t^{\bar{\theta}}\left[e^{\kappa[a(\theta)+b(\theta)B_T]}\right]$$
$$= e^{\kappa a(\theta)+\kappa b(\theta)B_t + \kappa b(\theta)\bar{\theta}(T-t) + \frac{1}{2}|\kappa b(\theta)|^2(T-t)}.$$

This leads to

$$d\tilde{W}_t^{A,\theta} = \tilde{W}_t^{A,\theta} \kappa b(\theta) dB_t^{\bar{\theta}},$$

and thus $\hat{u}(\theta) = \kappa b(\theta)$, which implies (8.33). □

It remains to check that the contract is truth-telling. This follows from the following lemma, which is stated for general density f.

8.4 Controlling Volatility

Lemma 8.3.3 *Let f be the density of F. Consider the hazard function $h = f/(1-F)$, and assume that $h' > 0$. Then, the contract $C_T = a(\theta) + b(\theta) B_T$, where a and b are chosen as in (8.27), is truth-telling.*

Proof From (8.12), (8.25), and (8.27), it is straightforward to compute

$$R(\theta, \theta') = \log E^{\bar{\theta}}\left[e^{\kappa a(\theta') + \kappa b(\theta') B_T}\right] = a(\theta') + b(\theta') \bar{\theta} T + \frac{\kappa}{2} |b(\theta')|^2 T$$
$$= R(\theta') + R'(\theta')(\theta - \theta').$$

We have then

$$\partial_{\theta'} R(\theta, \theta') = R''(\theta')(\theta - \theta'). \tag{8.41}$$

Here, either $R(\theta') = r_0$ or $R(\theta') = y(\theta')$ where y is the solution (8.35) to the Euler ODE. If $\theta' < \theta^*$ so that $R(\theta') = r_0$, then we see that $R(\theta, \theta') = r_0$, which is the lowest the agent can get, so he has no reason to pretend to be of type θ'. Otherwise, with $\theta' \geq \theta^*$ and $R = y$ and omitting the argument θ, note that

$$R' = T + \alpha/f + kTF/f;$$
$$R'' = kT - (\alpha + kTF) f'/f^2.$$

The free boundary condition (8.39) for $y = R$ is still the same, and gives

$$\alpha = -kTF(\theta_H) = -kT.$$

Notice that this implies

$$R'' = kT + kT \frac{f'}{f^2}(1 - F).$$

Thus, $R'' > 0$ if and only if

$$f'(1 - F) > -f^2. \tag{8.42}$$

This is equivalent to $h' > 0$, which is assumed. From (8.41), we see, that under condition (8.42), $R(\theta, \theta')$ is increasing for $\theta' < \theta$ and decreasing for $\theta' > \theta$, so $\theta' = \theta$ is the maximum. □

8.4 Controlling Volatility

In this section we allow for the control of the risk, that is, of the diffusion coefficient (volatility) of the output process.

8.4.1 The Model

We now study the model

$$dX_t = \theta v_t dt + v_t dB_t^{\theta} = v_t dB_t, \qquad X_0 = x \tag{8.43}$$

where we change the drift to θv_t. Here, volatility v_t is to be chosen, at no cost. Since v is the quadratic variation process of process X, and since we assume that X is continuously observed, v can be observed. Thus, we assume that v is in fact dictated by the principal. Actually, the optimal contract below will be such that the agent is indifferent between various choices of v.

The main application of this model would be to portfolio management. In this case θ is the return rate the manager can attain by his skills (say, by his choice of risky assets in which to invest), while v is, up to a linear transformation depending on the standard deviations and correlations of the underlying risky assets, the amount invested in the corresponding portfolio. In other words, the investors (which here constitute the principal) only have a prior distribution on the return of the portfolio of the assets the manager will pick to invest into, but they do observe his trading strategy and can estimate exactly the variance-covariance structure. This is consistent with real-world applications, as it is well known that it is much harder for the principal to estimate expected return of a portfolio, than to estimate its volatility.

We assume that v is an $\mathbf{F}^B$-adapted process such that $E \int_0^T v_t^2 dt < \infty$, so that X is a martingale process under P. We derive some qualitative conclusions in this section, without providing technical analysis.

Note that there is a "budget constraint" on the output X_T, which is the martingale property

$$E[X_T] = x. \tag{8.44}$$

We already used the Martingale Representation Theorem in the chapter on risk-sharing, that says that for any $\mathcal{F}_T$-measurable random variable Y_T that satisfies $E[Y_T] = x$, there exists an admissible volatility process v such that $X_T = X_T^v = Y_T$. This is what makes the budget constraint (8.44) a constraint on the possible choices of v.

8.4.2 Main Result: Solving the Relaxed Problem

The agent's utility, when declaring the true type θ, is denoted

$$R(\theta) := E\big[M_T^\theta U_A\big(C_T(\theta)\big)\big] \tag{8.45}$$

and the IR constraint is $R(\theta) \geq r(\theta)$. Note that, unlike in Sect. 8.3 where we used $M_T^{\bar{\theta}}$, here we use M_T^θ. Denote

$$I_A(x) = \big(U_A'\big)^{-1}(x), \qquad I_P(x) = \big(U_P'\big)^{-1}(x).$$

Proposition 8.4.1 *Suppose that the contract payoff has to satisfy the **limited liability** constraint:*

$$C_T(\theta) \geq L$$

8.4 Controlling Volatility

for some constant L, and that

$$E^\theta[U_A(L)] \geq r(\theta), \quad \theta \in [\theta_L, \theta_H].$$

Then, the optimal payoff $C_T(\theta)$ for the principal's relaxed problem, defined in (8.49) below, is given by

$$C_T(\theta) = L \vee I_A\left(\frac{-\nu(\theta)}{M_T^\theta(\lambda(\theta) + \mu(\theta) B_T)}\right) \mathbf{1}_{\{\lambda(\theta) + \mu(\theta) B_T < 0\}} + L \mathbf{1}_{\{\lambda(\theta) + \mu(\theta) B_T \geq 0\}}, \tag{8.46}$$

where λ, μ, ν are Lagrange multipliers for the IR constraint, the truth-telling first order condition and the budget constraint, respectively. Moreover, volatility v will be chosen so that

$$X_T = C_T(\theta) + I_P\left(\frac{\nu(\theta)}{M_T^\theta}\right). \tag{8.47}$$

Proof From (8.45), similarly as before, we get that the first order truth-telling constraint is

$$E\left[M_T^\theta U_A'(C_T(\theta)) \partial_\theta C_T(\theta)\right] = 0.$$

Differentiating (8.45) with respect to θ, we have

$$E\left[M_T^\theta U_A(C_T(\theta))[B_T - \theta T] + M_T^\theta U_A'(C_T(\theta)) \partial_\theta C_T(\theta)\right] = R'(\theta),$$

which implies that

$$E\left[B_T M_T^\theta U_A(C_T(\theta))\right] = \left[R'(\theta) + T\theta R(\theta)\right]. \tag{8.48}$$

If we denote by ν the Lagrange multiplier corresponding to the budget constraint (8.44), the Lagrangian relaxed problem for the principal is then to maximize, over X_T, C_T,

$$E\left[\int_{\theta_L}^{\theta_H} \left(M_T^\theta U_P(X_T - C_T(\theta)) - \nu(\theta) X_T - M_T^\theta U_A(C_T(\theta))\right.\right.$$
$$\left.\left. \times \left[\lambda(\theta) + \mu(\theta) B_T\right]\right) dF(\theta)\right]. \tag{8.49}$$

If we take derivatives with respect to X_T inside the expectation and the integral, we get that the optimal X_T is obtained from

$$M_T^\theta U_P'(X_T - C_T(\theta)) = \nu(\theta) \tag{8.50}$$

which implies (8.47). Substituting this back into the principal's problem, and noticing that

$$Y_T(\theta) := X_T - C_T(\theta)$$

is fixed by (8.50), we see that we need to maximize over $C_T(\theta)$ the expression

$$E\left[\int_{\theta_L}^{\theta_H} \left(-\nu(\theta)[Y_T(\theta) + C_T(\theta)] - M_T^\theta U_A(C_T(\theta))[\lambda(\theta) + \mu(\theta) B_T]\right) dF(\theta)\right].$$

If $\lambda(\theta) + \mu(\theta) B_T < 0$, the integrand is maximized at $C_T = I_A(\frac{-\nu}{M_T^\theta(\lambda+\mu B_T)})$. However, if $\lambda(\theta) + \mu(\theta) B_T \geq 0$, the maximum is attained at the smallest possible value of C_T, namely L. Thus, the optimal $C_T(\theta)$ is given by (8.46). □

Remark 8.4.2 (i) Note that the optimal contract does not depend on the agent's action process v_t or the output X, but only on his type θ and the underlying (fixed) noise B_T. Thus, the agent is indifferent between different choices of action process v given this contract, and he has to be told by the principal which process v to use.

(ii) Notice that we can write

$$B_T = \theta T + B_T^\theta = \int_0^T \frac{dX_t}{v_t}$$

so that the optimal payoff C_T is a function of the volatility weighted average of the accumulated output value, which is a sufficient statistic for the unknown parameter θ. We can think of the optimal contract as a function of the underlying benchmark risk B_T.

8.4.3 Comparison with the First Best

Consider the model

$$dX_t = \theta v_t dt + v_t dB_t$$

where everything is observable. We now recall results from Chap. 3 adapted to the constraint $C_T(\theta) \geq L$. Denoting

$$Z_t^\theta = e^{-t\theta^2/2 - \theta B_t}$$

we have, at the optimum,

$$X_T - C_T(\theta) = I_P(\nu(\theta) Z_T^\theta)$$
$$C_T(\theta) = L \vee I_A(\lambda(\theta) Z_T^\theta)$$

where $\nu(\theta)$ and $\lambda(\theta)$ are determined so that $E[Z_T^\theta X_T] = x$ and the IR constraint is satisfied.

Thus, the optimal contract is of a similar form to the one we obtain for the relaxed problem in the adverse selection case, except that in the latter case there is an additional randomness in determining when the contract is above its lowest possible level L; see (8.46).

In the first best case the ratio of marginal utilities U_P'/U_A' is constant, if $C_T > L$. In the adverse selection relaxed problem, we have, omitting dependence on θ,

$$\frac{U_P'(X_T - C_T)}{U_A'(C_T)} = -\mathbf{1}_{\{C_T > L\}}[\lambda + \mu B_T] + \mathbf{1}_{\{C_T = L\}} \frac{\nu}{U_A'(L) M_T^\theta}$$

where C_T is given in (8.46). Similarly as in the case of controlling the drift, this ratio is random.

In the first best case it is also optimal to offer the contract

$$C_T = X_T - I_P\left(vZ_T^\theta\right) \tag{8.51}$$

and this contract is incentive compatible in the sense that it will induce the agent to implement the first best action process v, without the principal telling him what to do. This is not the case with adverse selection, in which the agent is given the contract payoff (8.46) and has to be told which v to use.

8.5 Further Reading

Classical adverse selection models are covered in books Laffont and Martimort (2001), Salanie (2005), and Bolton and Dewatripont (2005). Two papers in continuous-time that have both adverse selection and moral hazard are Sung (2005), analyzing a model in which the principal observes only the initial and the final value of the underlying process, and Sannikov (2007). Our approach expands slightly on the models from Cvitanić and Zhang (2007). An extension of the Sannikov (2008) model to adverse selection is analyzed in Cvitanić et al. (2012).

References

Bolton, P., Dewatripont, M.: Contract Theory. MIT Press, Cambridge (2005)
Cvitanić, J., Zhang, J.: Optimal compensation with adverse selection and dynamic actions. Math. Financ. Econ. **1**, 21–55 (2007)
Cvitanić, J., Wan, X., Yang, H.: Dynamics of contract design with screening. Manag. Sci. (2012, forthcoming)
Laffont, J.J., Martimort, D.: The Theory of Incentives: The Principal–Agent Model. Princeton University Press, Princeton (2001)
Salanie, B.: The Economics of Contracts: A Primer, 2nd edn. MIT Press, Cambridge (2005)
Sannikov, Y.: Agency problems, screening and increasing credit lines. Working paper, Princeton University (2007)
Sannikov, Y.: A continuous-time version of the principal agent problem. Rev. Econ. Stud. **75**, 957–984 (2008)
Sung, J.: Optimal contracts under adverse selection and moral hazard: a continuous-time approach. Rev. Financ. Stud. **18**, 1121–1173 (2005)

Part V
Backward SDEs and Forward-Backward SDEs

Chapter 9
Backward SDEs

Abstract In this chapter we first introduce Backward SDEs by means of a popular application, option pricing and hedging. We show how these problems lead naturally to BSDEs, and then, we provide the basic theory. We present the important Comparison Theorem for BSDEs. Existence and uniqueness are first shown under Lipschitz and square-integrability conditions. Then, the case of quadratic growth is studied, often encountered in applications. In Markovian models a connection to PDEs is established, which can be useful for numerical solutions.

9.1 Introduction

Note to the Reader For simplicity, in most of this part of the book we use only the one-dimensional notation. However, unless we note otherwise, all the results can be extended straightforwardly to the high-dimensional case.

We first introduce the standard notation in the BSDE literature. Let $(\Omega, \mathcal{F}, P)$ be a probability space, B a Brownian motion, $T > 0$ a fixed terminal time, and $\mathbb{F} := \{\mathcal{F}_t\}_{0 \le t \le T}$ the augmented filtration generated by B. For any $1 \le p, q \le \infty$, let

$L^p(\mathcal{F}_T) :=$ the space of $\mathcal{F}_T$-measurable random variables ξ such that

$$\|\xi\|_p := \left(E\{|\xi|^p\}\right)^{\frac{1}{p}} < \infty,$$

$L^{p,q}(\mathbb{F}) :=$ the space of $\mathbb{F}$-adapted processes η such that

$$\|\eta\|_{p,q} := \left(E\left\{\left(\int_0^T |\eta_t|^p dt\right)^{\frac{q}{p}}\right\}\right)^{\frac{1}{q}} < \infty;$$

$L^p(\mathbb{F}) := L^{p,p}(\mathbb{F})$ and $\|\eta\|_p := \|\eta\|_{p,p}$,

where, when $p = \infty$, as usual $\|\xi\|_\infty$ and $\|\eta\|_\infty$ denote the L^∞-norm of ξ and η, respectively.

In this chapter we study the following Backward SDE:

$$dY_t = -f(t, \omega, Y_t, Z_t)dt + Z_t dB_t, \qquad Y_T = \xi, \tag{9.1}$$

where ξ, the *terminal condition*, is an $\mathcal{F}_T$-measurable random variable, and f is progressively measurable in all the variables and $f(\cdot, \cdot, y, z)$ is $\mathbb{F}$-adapted for any fixed (y, z). A solution to the BSDE consists of a pair (Y, Z) of $\mathbb{F}$-*adapted* processes satisfying the above equation.

J. Cvitanić, J. Zhang, *Contract Theory in Continuous-Time Models*, Springer Finance, DOI 10.1007/978-3-642-14200-0_9, © Springer-Verlag Berlin Heidelberg 2013

9.1.1 Example: Option Pricing and Hedging

An option pricing application is presented here for the benefit of readers who are sufficiently familiar therewith, so they could use that familiarity to develop intuition for BSDEs and FBSDEs.

Consider a financial market consisting of a riskless asset R, called the bank account, and a risky asset S, called the stock, whose dynamics are as follows:

$$dR_t = r_t R_t dt, \qquad dS_t = S_t[\mu_t dt + \sigma_t dB_t]. \tag{9.2}$$

Let π_t be a portfolio process, that is, an adapted process representing the amount of money held in stock at time t. We say that the portfolio is *self-financing* if the corresponding wealth process Y, consisting of the total money in bank and stock, satisfies

$$dY_t = (Y_t - \pi_t) dR_t/R_t + \pi_t dS_t/S_t. \tag{9.3}$$

Using (9.2), we obtain

$$dY_t = \big[r_t(Y_t - \pi_t) + \mu_t \pi_t\big] dt + \sigma_t \pi_t dB_t. \tag{9.4}$$

Now, let $\xi = g(S.)$ be a contingent claim with maturity time T, that is, an $\mathcal{F}_T$-measurable random variable. For example, $\xi = (S_T - K)^+$ is the payoff of a European call option. We say that a self-financing portfolio (perfectly) *hedges* ξ if $Y_T = \xi$. This is equivalent to finding (Y, π) such that

$$dY_t = \big[r_t(Y_t - \pi_t) + \mu_t \pi_t\big] dt + \sigma_t \pi_t dB_t \quad \text{and} \quad Y_T = \xi. \tag{9.5}$$

This is called a Backward SDE because the terminal condition Y_T is given. We emphasize again that the solution to the BSDE is a *pair* of $\mathbb{F}$-*adapted* processes (Y, π). In particular, if a unique solution exists,

> Y_0 is the unique arbitrage-free price of payoff ξ, and process π is the hedging portfolio.

We remark that BSDE (9.5) is linear, which can be solved explicitly by using the Martingale Representation Theorem, as we will see in the next section. In particular, when r, μ and σ are all constants and ξ is the payoff of the European call option, Y_0 can be computed via the famous Black–Scholes formula.

In (9.2) we assume the interest rate is the same for both borrowers and lenders, which is usually not the case in practice. Let us assume instead that the borrowing rate r_t is greater than the lending rate $\tilde{r}_t$. Then, the BSDE (9.5) becomes nonlinear:

$$dY_t = \big[\tilde{r}_t(Y_t - \pi_t)^+ - r_t(Y_t - \pi_t)^- + \mu_t \pi_t\big] dt + \sigma_t \pi_t dB_t \quad \text{and} \quad Y_T = \xi. \tag{9.6}$$

The well-posedness of such a nonlinear BSDE will be studied later in the chapter. We note that in general there are no closed-form expressions for solutions to nonlinear BSDEs.

If we want to model an investor who may have a price impact on the risky asset price dynamics, we may want to use a model like this:

9.2 Linear Backward SDEs

$$\begin{cases} dS_t = \mu(t, S_t, Y_t, \pi_t)dt + \sigma(t, S_t, Y_t, \pi_t)dB_t; \\ dY_t = \left[\tilde{r}_t(Y_t - \pi_t)^+ - r_t(Y_t - \pi_t)^- + \mu_t\pi_t\right]dt + \sigma_t\pi_t dB_t; \\ S_0 = s, \qquad Y_T = g(S_\cdot). \end{cases} \quad (9.7)$$

This is a coupled Forward-Backward SDE, where the forward component S depends on the backward components (Y, π). Such equations are much more difficult to solve, and we will study them in Chap. 11.

9.2 Linear Backward SDEs

In this section we study the case when f is linear. We first have the following simple result.

Proposition 9.2.1 *Assume $\xi \in L^2(\mathcal{F}_T)$ and $f^0 \in L^{1,2}(\mathbb{F})$. Then, the following linear BSDE has a unique solution in $L^2(\mathbb{F}) \times L^2(\mathbb{F})$:*

$$Y_t = \xi + \int_t^T f_s^0 ds - \int_t^T Z_s dB_s. \quad (9.8)$$

Proof It is obvious that

$$Y_t = E_t\left[\xi + \int_t^T f_s^0 ds\right].$$

Note that

$$\tilde{Y}_t := Y_t + \int_0^t f_s^0 ds = E_t\left[\xi + \int_0^T f_s^0 ds\right]$$

is a P-square integrable martingale. By the Martingale Representation Theorem, there exists a unique $Z \in L^2(\mathbb{F})$ such that

$$d\tilde{Y}_t = Z_t dB_t.$$

One can check straightforwardly that the above pair (Y, Z) satisfies (9.8), and, from the above derivation, it is the unique solution. □

We next consider the general linear BSDE:

$$Y_t = \xi + \int_t^T [\alpha_s Y_s + \beta_s Z_s + f_s^0]ds - \int_t^T Z_s dB_s \quad (9.9)$$

where α and β are bounded processes. Introduce the adjoint process

$$\Gamma_t := \exp\left(\int_0^t \beta_s dB_s + \int_0^t \left[\alpha_s - \frac{1}{2}\beta_s^2\right]ds\right) \quad (9.10)$$

which is the solution to the following SDE:

$$d\Gamma_t = \Gamma_t[\alpha_t dt + \beta_t dB_t], \qquad \Gamma_0 = 1. \tag{9.11}$$

Applying Itô's rule we have

$$d(\Gamma_t Y_t) = -\Gamma_t f_t dt + \Gamma_t[\beta_t Y_t + Z_t]dB_t.$$

Denote

$$\hat{Y}_t := \Gamma_t Y_t; \qquad \hat{Z}_t := \Gamma_t[\beta_t Y_t + Z_t]; \qquad \tilde{\xi} := \Gamma_T \xi; \qquad \hat{f}_t^0 := \Gamma_t f_t^0. \tag{9.12}$$

Then, one may rewrite (9.9) as

$$\hat{Y}_t = \hat{\xi} + \int_t^T \hat{f}_s^0 ds - \int_t^T \hat{Z}_s dB_s. \tag{9.13}$$

This is a linear BSDE in the form (9.8). However, $\hat{\xi}$ and $\hat{f}^0$ are in general not square integrable, and on the other hand, the integrability of $(\hat{Y}, \hat{Z})$ does not lead to the same integrability of (Y, Z). We establish the integrability below, as part of the general theory, see Proposition 9.3.6. We emphasize that the results in the next section rely only on Proposition 9.2.1, and that we will not use the well-posedness of BSDE (9.9).

9.3 Well-Posedness of BSDEs

We now investigate nonlinear BSDE (9.1). For notational simplicity we omit ω in f and simply write it as $f(t, y, z)$. We impose the following assumptions.

Assumption 9.3.1 (i) $\xi \in L^2(\mathcal{F}_T)$ and $f(\cdot, 0, 0) \in L^{1,2}(\mathbb{F})$;
(ii) f is uniformly Lipschitz continuous on (y, z). That is,

$$|f(t, y_1, z_1) - f(t, y_2, z_2)| \le C[|y_1 - y_2| + |z_1 - z_2|], \quad \forall (\omega, t, y_i, z_i). \tag{9.14}$$

Theorem 9.3.2 (A priori estimate) *Assume Assumption 9.3.1. If $(Y, Z) \in L^2(\mathbb{F}) \times L^2(\mathbb{F})$ is a solution to BSDE (9.1), then*

$$\|(Y, Z)\|^2 := E\left\{\sup_{0 \le t \le T} |Y_t|^2 + \int_0^T |Z_t|^2 dt\right\}$$
$$\le CE\left\{|\xi|^2 + \left(\int_0^T |f(t, 0, 0)|dt\right)^2\right\}, \tag{9.15}$$

where the constant C depends only on T, the dimension, and the Lipschitz constant of f.

Remark 9.3.3 In the standard literature, it is required that $E\{\int_0^T |f(t, 0, 0)|^2 dt\} < \infty$. Our condition here is slightly weaker.

9.3 Well-Posedness of BSDEs

To prove the theorem, we need a simple lemma.

Lemma 9.3.4 *Let $p, q > 1$ be conjugates. Assume Y, Z are $\mathbb{F}$-adapted and*

$$E\left\{\sup_{0\le t\le T} |Y_t|^p + \left(\int_0^T |Z_t|^2 dt\right)^{\frac{q}{2}}\right\} < \infty.$$

Then, $\int_0^t Y_s Z_s dB_s$ is a true martingale.

Proof Applying the Burkholder–Davis–Gundy inequality, we have

$$E\left\{\sup_{0\le t\le T} \left|\int_0^t Y_s Z_s dB_s\right|\right\}$$

$$\le CE\left\{\left[\int_0^T |Y_t Z_t|^2 dt\right]^{\frac{1}{2}}\right\} \le CE\left\{\sup_{0\le t\le T} |Y_t| \left(\int_0^T |Z_t|^2 dt\right)^{\frac{1}{2}}\right\}$$

$$\le C\left(E\left\{\sup_{0\le t\le T} |Y_t|^p\right\}\right)^{\frac{1}{p}} \left(E\left\{\left(\int_0^T |Z_t|^2 dt\right)^{\frac{q}{2}}\right\}\right)^{\frac{1}{q}} < \infty.$$

This proves the result. □

Proof of Theorem 9.3.2 Denote $f_t^0 := f(t, 0, 0)$. First, note that

$$|Y_t| \le |\xi| + \int_t^T [|f_s^0| + C|Y_s| + C|Z_s|] ds + \left|\int_t^T Z_s dB_s\right|.$$

Then,

$$\sup_{0\le t\le T} |Y_t| \le C\left[|\xi| + \int_0^T [|f_t^0| + |Y_t| + |Z_t|] dt + \sup_{0\le t\le T} \left|\int_0^t Z_s dB_s\right|\right].$$

Applying the Burkholder–Davis–Gundy inequality we have

$$E\left\{\sup_{0\le t\le T} |Y_t|^2\right\} \le CE\left\{|\xi|^2 + \left(\int_0^T |f_t^0| dt\right)^2 + \int_0^T [|Y_t|^2 + |Z_t|^2] dt\right\} < \infty. \tag{9.16}$$

Next, by Itô's rule,

$$d|Y_t|^2 = 2Y_t dY_t + |Z_t|^2 dt = -2Y_t f(t, Y_t, Z_t) dt + 2Y_t Z_t dW_t + |Z_t|^2 dt. \tag{9.17}$$

Thus,

$$|Y_t|^2 + \int_t^T |Z_s|^2 ds = |\xi|^2 + 2\int_t^T Y_s f(s, Y_s, Z_s) ds + 2\int_t^T Y_s Z_s dB_s. \tag{9.18}$$

By (9.16) and Lemma 9.3.4 we know $\int_0^t Y_s Z_s dB_s$ is a true martingale. Now, taking expectation on both sides of (9.18) and noting that $ab \le \frac{1}{2}a^2 + \frac{1}{2}b^2$, we have

$$E\left\{|Y_t|^2 + \int_t^T |Z_s|^2 ds\right\}$$
$$= E\left\{|\xi|^2 + 2\int_t^T Y_s f(s, Y_s, Z_s) ds\right\}$$
$$\leq E\left\{|\xi|^2 + C\int_t^T |Y_s|[|f_s^0| + |Y_s| + |Z_s|] ds\right\}$$
$$\leq E\left\{|\xi|^2 + C \sup_{0\leq s\leq T} |Y_s| \int_0^T |f_s^0| ds + C\int_0^T [|Y_s|^2 + |Y_s Z_s|] ds\right\}$$
$$\leq E\left\{|\xi|^2 + \varepsilon \sup_{0\leq s\leq T} |Y_s|^2 + C\varepsilon^{-1}\left(\int_0^T |f_s^0| ds\right)^2 + C\int_t^T |Y_s|^2 ds\right.$$
$$\left. + \frac{1}{2}\int_t^T |Z_s|^2 ds\right\}, \tag{9.19}$$

for any $\varepsilon > 0$. This leads to

$$E\left\{|Y_t|^2 + \frac{1}{2}\int_t^T |Z_s|^2 ds\right\}$$
$$\leq E\left\{C\int_t^T |Y_s|^2 ds + |\xi|^2 + \varepsilon \sup_{0\leq s\leq T} |Y_s|^2 + C\varepsilon^{-1}\left(\int_0^T |f_s^0| ds\right)^2\right\}, \tag{9.20}$$

which, together with Fubini's theorem, implies that

$$E\{|Y_t|^2\} \leq E\left\{|\xi|^2 + \varepsilon \sup_{0\leq s\leq T} |Y_s|^2 + C\varepsilon^{-1}\left(\int_0^T |f_s^0| ds\right)^2\right\} + C\int_t^T E\{|Y_s|^2\} ds.$$

Applying the Gronwall inequality, we get

$$E\{|Y_t|^2\} \leq CE\left\{|\xi|^2 + \varepsilon \sup_{0\leq s\leq T} |Y_s|^2 + C\varepsilon^{-1}\left(\int_0^T |f_s^0| ds\right)^2\right\}, \quad \forall t \in [0, T]. \tag{9.21}$$

Then, by letting $t = 0$ and plugging (9.21) into (9.20) we have

$$E\left\{\int_0^T |Z_s|^2 ds\right\} \leq CE\left\{|\xi|^2 + \varepsilon \sup_{0\leq s\leq T} |Y_s|^2 + C\varepsilon^{-1}\left(\int_0^T |f_s^0| ds\right)^2\right\}. \tag{9.22}$$

Plugging (9.21) and (9.22) into (9.16), we get

$$E\left\{\sup_{0\leq t\leq T} |Y_t|^2\right\} \leq CE\left\{|\xi|^2 + \varepsilon \sup_{0\leq s\leq T} |Y_s|^2 + C\varepsilon^{-1}\left(\int_0^T |f_s^0| ds\right)^2\right\}.$$

Finally, choosing $\varepsilon = \frac{1}{2C}$ for the constant C above, we obtain

$$E\left\{\sup_{0\leq t\leq T} |Y_t|^2\right\} \leq CE\left\{|\xi|^2 + \left(\int_0^T |f_s^0| ds\right)^2\right\}.$$

This, together with (9.22), proves (9.15). □

9.3 Well-Posedness of BSDEs

We now establish the well-posedness of BSDE (9.1).

Theorem 9.3.5 (Well-posedness) *Assume Assumption 9.3.1. Then, BSDE (9.1) has a unique solution* $(Y, Z) \in L^2(\mathbb{F}) \times L^2(\mathbb{F})$.

Proof Uniqueness. Assume $(Y^i, Z^i) \in L^2(\mathbb{F}) \times L^2(\mathbb{F})$, $i = 1, 2$ are two solutions. Denote $\Delta Y_t := Y_t^1 - Y_t^2$, $\Delta Z_t := Z_t^1 - Z_t^2$. Then,

$$\Delta Y_t = \int_t^T [f(s, Y_s^1, Z_s^1) - f(s, Y_s^2, Z_s^2)]ds - \int_t^T \Delta Z_s dB_s$$

$$= \int_t^T [\alpha_s \Delta Y_s + \beta_s \Delta Z_s]ds - \int_t^T \Delta Z_s dB_s,$$

where, thanks to the Lipschitz condition (9.14),

$$\alpha_t := \begin{cases} \frac{f(t, Y_t^1, Z_t^1) - f(t, Y_t^2, Z_t^1)}{\Delta Y_t}, & \Delta Y_t \neq 0; \\ 0, & \Delta Y_t = 0; \end{cases}$$

$$\beta_t := \begin{cases} \frac{f(t, Y_t^2, Z_t^1) - f(t, Y_t^2, Z_t^2)}{\Delta Z_t}, & \Delta Z_t \neq 0; \\ 0, & \Delta Z_t = 0 \end{cases} \quad (9.23)$$

are bounded. Then, by Theorem 9.3.2 we get $\|(\Delta Y, \Delta Z)\|^2 \leq 0$. That is,

$$Y_t^1 = Y_t^2 \quad \text{for all } t \in [0, T], \, P\text{-a.s.} \quad \text{and} \quad Z_t^1 = Z_t^2, \quad dt \times dP\text{-a.s.} \quad (9.24)$$

Existence. We use Picard iterations. Denote $Y_t^0 := 0$, $Z_t^0 := 0$. For $n = 1, 2, \ldots$, let

$$Y_t^n = \xi + \int_t^T f(s, Y_s^{n-1}, Z_s^{n-1})ds - \int_t^T Z_s^n dB_s. \quad (9.25)$$

Assume $(Y^{n-1}, Z^{n-1}) \in L^2(\mathbb{F}) \times L^2(\mathbb{F})$. Note that

$$|f(t, Y_t^{n-1}, Z_t^{n-1})| \leq C[|f(t, 0, 0)| + |Y_t^{n-1}| + |Z_t^{n-1}|].$$

Then, $f(t, Y_t^{n-1}, Z_t^{n-1}) \in L^{1,2}(\mathbb{F})$. By Proposition 9.2.1, (9.25) uniquely determines $(Y^n, Z^n) \in L^2(\mathbb{F}) \times L^2(\mathbb{F})$. By induction we have $(Y^n, Z^n) \in L^2(\mathbb{F}) \times L^2(\mathbb{F})$ for all $n \geq 0$.

Denote $\Delta Y_t^n := Y_t^n - Y_t^{n-1}$, $\Delta Z_t^n := Z_t^n - Z_t^{n-1}$. Then,

$$\Delta Y_t^n = \int_t^T [\alpha_s^{n-1} \Delta Y_s^{n-1} + \beta_s^{n-1} \Delta Z_s^{n-1}]ds - \int_t^T \Delta Z_s^n dB_s,$$

where α^n, β^n are defined in a similar way as in (9.23) and are bounded. Let $\gamma > 0$ be a constant which will be specified later. Applying Itô's rule on $e^{\gamma t} |\Delta Y_t^n|^2$ we have

$$d(e^{\gamma t} |\Delta Y_t^n|^2) = \gamma e^{\gamma t} |\Delta Y_t^n|^2 dt - 2e^{\gamma t} \Delta Y_t^n [\alpha_t^{n-1} \Delta Y_t^{n-1} + \beta_t^{n-1} \Delta Z_t^{n-1}]dt$$
$$+ 2e^{\gamma t} \Delta Y_t^n \Delta Z_t^n dB_t + e^{\gamma t} |\Delta Z_t^n|^2 dt.$$

By Lemma 9.3.4, $\int_0^t e^{\gamma s} \Delta Y_s^n \Delta Z_s^n dB_s$ is a true martingale. Noting that $\Delta Y_T^n = 0$, we get

$$E\left\{e^{\gamma t}|\Delta Y_t^n|^2 + \gamma \int_t^T e^{\gamma s}|\Delta Y_s^n|^2 ds + \int_t^T e^{\gamma s}|\Delta Z_s^n|^2 ds\right\}$$

$$= E\left\{2\int_t^T e^{\gamma s} \Delta Y_s^n [\alpha_s^{n-1} \Delta Y_s^{n-1} + \beta_s^{n-1} \Delta Z_s^{n-1}] ds\right\}$$

$$\leq E\left\{C \int_t^T e^{\gamma s}|\Delta Y_s^n|[|\Delta Y_s^{n-1}| + |\Delta Z_s^{n-1}|] ds\right\}$$

$$\leq E\left\{C_0 \int_t^T e^{\gamma s}|\Delta Y_s^n|^2 ds + \frac{1}{4T}\int_t^T e^{\gamma s}|\Delta Y_s^{n-1}|^2 ds \right.$$

$$\left. + \frac{1}{4}\int_t^T e^{\gamma s}|\Delta Z_s^{n-1}|^2 ds\right\}.$$

Choosing $\gamma = C_0$, we get

$$E\left\{e^{\gamma t}|\Delta Y_t^n|^2 + \int_t^T e^{\gamma s}|\Delta Z_s^n|^2 ds\right\}$$

$$\leq E\left\{\frac{1}{4T}\int_t^T e^{\gamma s}|\Delta Y_s^{n-1}|^2 ds + \frac{1}{4}\int_t^T e^{\gamma s}|\Delta Z_s^{n-1}|^2 ds\right\}$$

$$\leq \frac{1}{4}\left[\sup_{0\leq s\leq T} e^{\gamma s} E\{|\Delta Y_s^{n-1}|^2\} + \int_0^T e^{\gamma s} E\{|\Delta Z_s^{n-1}|^2\} ds\right].$$

Thus,

$$\sup_{0\leq t\leq T} e^{\gamma t} E\{|\Delta Y_t^n|^2\} \leq \frac{1}{4}\left[\sup_{0\leq t\leq T} e^{\gamma t} E\{|\Delta Y_t^{n-1}|^2\} + \int_0^T e^{\gamma t} E\{|\Delta Z_t^{n-1}|^2\} dt\right];$$

$$\int_0^T e^{\gamma t}|\Delta Z_t^n|^2 dt \leq \frac{1}{4}\left[\sup_{0\leq t\leq T} e^{\gamma t} E\{|\Delta Y_t^{n-1}|^2\} + \int_0^T e^{\gamma t} E\{|\Delta Z_t^{n-1}|^2\} dt\right].$$

Define the following norm:

$$\|(Y,Z)\|_\gamma^2 := \sup_{0\leq t\leq T} e^{\gamma t} E\{|Y_t|^2\} + E\left\{\int_0^T e^{\gamma t}|Z_t|^2 dt\right\}. \quad (9.26)$$

Then,

$$\|(\Delta Y^n, \Delta Z^n)\|_\gamma^2 \leq \frac{1}{2}\|(\Delta Y^{n-1}, \Delta Z^{n-1})\|_\gamma^2. \quad (9.27)$$

By induction, we get

$$\|(\Delta Y^n, \Delta Z^n)\|_\gamma^2 \leq \frac{1}{2^{n-1}}\|(\Delta Y^1, \Delta Z^1)\|_\gamma^2 = \frac{C}{2^n}.$$

Note that

$$\|(Y^1+Y^2, Z^1+Z^2)\|_\gamma \leq \|(Y^1,Z^1)\|_\gamma + \|(Y^2,Z^2)\|_\gamma.$$

Now, for any $n < m$,

$$\left\|(Y_t^n - Y_t^m, Z_t^n - Z_t^m)\right\|_\gamma \le \sum_{j=n+1}^m \left\|(\Delta Y^j, \Delta Z^j)\right\|_\gamma \le \sum_{j=n+1}^m \frac{C}{2^{\frac{j}{2}}} \le \frac{C}{2^{\frac{n}{2}}}.$$

Then,

$$\left\|(Y_t^n - Y_t^m, Z_t^n - Z_t^m)\right\|_\gamma \to 0, \quad \text{as } n, m \to \infty.$$

Thus, there exists $(Y, Z) \in L^2(\mathbb{F}) \times L^2(\mathbb{F})$ such that

$$\sup_{0 \le t \le T} e^{\gamma t} E\{|Y_t^n - Y_t|^2\} + E\left\{\int_0^T e^{\gamma t}|Z_t^n - Z_t|^2 dt\right\} \to 0, \quad \text{as } n \to \infty.$$

Therefore, by letting $n \to \infty$ in BSDE (9.25) we know that (Y, Z) satisfies BSDE (9.1). □

Following the arguments in Sect. 9.2 and applying Theorem 9.3.5 to linear BSDE (9.9), we have

Proposition 9.3.6 *Assume α, β are bounded, $\xi \in L^2(\mathcal{F}_T)$ and $f^0 \in L^{1,2}(\mathbb{F})$. Then, linear BSDE (9.9) has a unique solution $(Y, Z) \in L^2(\mathbb{F}) \times L^2(\mathbb{F})$. Moreover, for Γ defined in (9.10), we have*

$$Y_t = \Gamma_t^{-1} E_t\left\{\Gamma_T \xi + \int_t^T \Gamma_s f_s^0 ds\right\}. \tag{9.28}$$

Proof Applying Theorem 9.3.5, we know that BSDE (9.9) is well-posed. Using the notation of Sect. 9.2, clearly $E\{\sup_{0 \le t \le T} |\Gamma_t|^2\} < \infty$. Then

$$E\left\{\left(\int_0^T |\hat{Z}_t|^2 dt\right)^{\frac{1}{2}}\right\} < \infty \quad \text{and} \quad E\{|\Gamma_t f_t^0| dt\} < \infty.$$

Thus $\int_0^t \hat{Z}_s dB_s$ is a true martingale, and therefore (9.28) follows from (9.13). □

9.4 Comparison Theorem and Stability Properties of BSDEs

Theorem 9.4.1 (Comparison) *Assume $(\xi_i, f_i), i = 1, 2$ satisfy Assumption 9.3.1 and $\dim(\xi_i) = 1$. Let (Y^i, Z^i) be the solution to the following BSDE:*

$$Y_t^i = \xi_i + \int_t^T f_i(s, Y_s^i, Z_s^i) ds - \int_t^T Z_s^i dW_s, \quad i = 1, 2. \tag{9.29}$$

Assume further that $\xi_1 \ge \xi_2$ a.s., and that for any (y, z), $f_1(\cdot, y, z) \ge f_2(\cdot, y, z)$, $dt \times dP$-a.s. Then, $Y_t^1 \ge Y_t^2, 0 \le t \le T$, a.s. In particular, $Y_0^1 \ge Y_0^2$.

Proof First, let

$$A(y,z) := \left\{(t,\omega) : f_1(t,\omega,y,z) \geq f_2(t,\omega,y,z)\right\} \quad \text{and} \quad A := \bigcup_{y,z \in Q} A(y,z).$$

Then, $E\{\int_0^T 1_A(t,\omega)dt\} = 0$. By the Lipschitz continuity of f_1, f_2, we get that

$$f_1(t,\omega,y,z) \geq f_2(t,\omega,y,z) \quad \text{for all } (y,z) \text{ and } (t,\omega) \in A^c.$$

In particular, this implies that

$$f_t^0(\omega) := f_1(t,\omega,Y_t^2(\omega),Z_t^2(\omega)) - f_2(t,\omega,Y_t^2(\omega),Z_t^2(\omega)) \geq 0, \quad dt \times dP\text{-a.s.} \tag{9.30}$$

Now, denote

$$\Delta Y_t := Y_t^1 - Y_t^2; \qquad \Delta Z_t := Z_t^1 - Z_t^2; \qquad \Delta \xi := \xi_1 - \xi_2.$$

Then,

$$\Delta Y_t = \Delta \xi + \int_t^T [f_1(s,Y_s^1,Z_s^1) - f_2(s,Y_s^2,Z_s^2)]ds - \int_t^T \Delta Z_s dB_s$$

$$= \Delta \xi + \int_t^T [f_1(s,Y_s^1,Z_s^1) - f_1(s,Y_s^2,Z_s^2) + f_s^0]ds - \int_t^T \Delta Z_s dB_s$$

$$= \Delta \xi + \int_t^T [\alpha_s \Delta Y_s + \beta_s \Delta Z_s + f_s^0]ds - \int_t^T \Delta Z_s dB_s,$$

where α and β are bounded. Define Γ by (9.10). By (9.28) we have

$$\Delta Y_t = \Gamma_t^{-1} E_t \left\{ \Gamma_T \Delta \xi + \int_t^T \Gamma_s f_s^0 ds \right\}.$$

This, together with (9.30) and the assumption that $\Delta \xi \geq 0$ a.s., implies that $\Delta Y_t \geq 0$. $\square$

Remark 9.4.2 In the Comparison Theorem we require the process Y to be one-dimensional. The comparison property for general high-dimensional BSDEs is an important, but difficult subject.

Theorem 9.4.3 (Stability) (i) *Assume* $(\xi_i, f_i), i = 1, 2$ *satisfy Assumption* 9.3.1, *and let* (Y^i, Z^i) *be the solution to the BSDE* (9.29). *Then,*

$$\|(Y^1 - Y^2, Z^1 - Z^2)\|^2$$
$$\leq CE\left\{|\xi_1 - \xi_2|^2 + \left(\int_0^T |f_1(t,Y_t^2,Z_t^2) - f_2(t,Y_t^2,Z_t^2)|dt\right)^2\right\}. \tag{9.31}$$

(ii) *Assume* (ξ, f) *and* $(\xi_n, f_n), n \geq 1$ *satisfy Assumption* 9.3.1 *with a common Lipschitz constant. Let* (Y, Z) *and* $(Y^n, Z^n), n \geq 1$, *be the solutions to the corresponding BSDEs. Assume further that*

9.4 Comparison Theorem and Stability Properties of BSDEs

$$\lim_{n\to\infty} E\left\{|\xi_n - \xi|^2 + \left(\int_0^T |f_n(t,0,0) - f(t,0,0)|dt\right)^2\right\} = 0, \quad (9.32)$$

and that, for all (y,z), $f_n(\cdot, y, z) \to f(\cdot, y, z)$, in measure $dt \times dP$. Then,

$$\lim_{n\to\infty} \|(Y^n - Y, Z^n - Z)\| = 0. \quad (9.33)$$

Proof (i) Denote $\Delta Y := Y^1 - Y^2$, and similarly for other values. Then,

$$\Delta Y_t = \Delta \xi + \int_t^T [\alpha_s \Delta Y_s + \beta_s \Delta Z_s + \Delta f(s, Y_s^2, Z_s^2)]ds - \int_t^T \Delta Z_s dB_s,$$

where α, β are bounded. Note that the above BSDE has generator $\tilde{f}(t, \omega, y, z) := \alpha_t(\omega)y + \beta_t(\omega)z + \Delta f(t, \omega, Y_t^2(\omega), Z_t^2(\omega))$ which obviously satisfies Assumption 9.3.1. Then, (9.31) follows from Theorem 9.3.2 immediately.

(ii) We first assume $f_n(\cdot, y, z) \to f(\cdot, y, z)$, $dt \times dP$-a.s. for all (y, z). Denote $\Delta Y^n := Y^n - Y$ and similarly for other values. As in (i) we can easily get from (9.15) that

$$\|(\Delta Y^n, \Delta Z^n)\|^2 \leq CE\left\{|\Delta \xi_n|^2 + \left(\int_0^T |\Delta f_n(t, Y_t, Z_t)|dt\right)^2\right\}. \quad (9.34)$$

By the uniform Lipschitz continuity of f_n and f, following the same arguments as in the proof of Theorem 9.4.1, there exists a set $A \subset [0, T] \times \Omega$ such that $E\{\int_0^T 1_A(t, \omega)dt\} = 0$ and

$$\lim_{n\to\infty} \Delta f_n(t, y, z) = 0, \quad \text{for all } (y, z) \text{ and } (t, \omega) \notin A.$$

This implies that

$$\Delta f_n(t, Y_t, Z_t) \to 0, \quad dt \times dP\text{-a.s.}$$

Moreover, by the uniform Lipschitz continuity of f_n and f again, we have

$$|\Delta f_n(t, Y_t, Z_t) - \Delta f_n(t, 0, 0)| \leq C[|Y_t| + |Z_t|].$$

Then, by the Dominated Convergence Theorem we get

$$\lim_{n\to\infty} E\left\{\left(\int_0^T |\Delta f_n(t, Y_t, Z_t) - \Delta f_n(t, 0, 0)|dt\right)^2\right\} = 0.$$

This, together with (9.32) and (9.34), proves (9.33).

We now prove by contradiction the case in which $f_n(\cdot, y, z)$ converges to $f(\cdot, y, z)$ only in measure. If (9.33) does not hold, without loss of generality we assume $\|(\Delta Y^n, \Delta Z^n)\| \geq c > 0$ for all $n \geq 1$. For each (y, z), there exists a subsequence $\{n_k, k \geq 1\}$ such that $f_{n_k}(t, y, z) \to f(t, y, z)$, $dt \times dP$-a.s. Using the standard diagonalizing argument, there exists a common subsequence $\{n_k\}$ such that $f_{n_k}(t, y, z) \to f(t, y, z)$, $dt \times dP$-a.s. for all rational (y, z). As discussed above, by the uniform continuity of f_n and f we get $f_{n_k}(t, y, z) \to f(t, y, z)$, $dt \times dP$-a.s. for all real numbers (y, z). Then, by the above proof we have $\lim_{k\to\infty} \|(\Delta Y^{n_k}, \Delta Z^{n_k})\| = 0$. This contradicts the assumption that $\|(\Delta Y^n, \Delta Z^n)\| \geq c > 0$ for all $n \geq 1$. Therefore, (9.33) holds. □

We conclude this subsection by extending the estimates to L^p for $p \geq 2$.

Proposition 9.4.4 *Assume Assumption* 9.3.1 *holds and*

$$I_p^p := E\left\{|\xi|^p + \left(\int_0^T |f(t,0,0)|dt\right)^p\right\} < \infty$$

for some $p \geq 2$. Then,

$$\|(Y,Z)\|_p^p := E\left\{\sup_{0\leq t\leq T} |Y_t|^p + \left(\int_0^T |Z_t|^2 dt\right)^{\frac{p}{2}}\right\} \leq C_p I_p^p, \qquad (9.35)$$

where the constant C_p may depend on p as well.

Proof We first assume ξ and f are bounded. Then, Y is bounded and thus

$$E\left\{\sup_{0\leq t\leq T} |Y_t|^p + \int_0^T |Y_t|^{p-2}|Z_t|^2 dt\right\} < \infty.$$

Applying Itô's rule on $|Y_t|^p = (|Y_t|^2)^{\frac{p}{2}}$, we obtain

$$d|Y_t|^p = p|Y_t|^{p-2}Y_t\left[-f(t,Y_t,Z_t)dt + Z_t dB_t\right] + \frac{1}{2}p(p-1)|Y_t|^{p-2}|Z_t|^2 dt.$$

Note that $|Y_t|^{p-1}|Z_t| \leq C_p[|Y_t|^p + |Y_t|^{p-2}|Z_t|^2]$. Following the arguments in the proof of Theorem 9.3.2 one can easily show that

$$\sup_{0\leq t\leq T} E\{|Y_t|^p\} + E\left\{\int_0^T |Y_t|^{p-2}|Z_t|^2 dt\right\} \leq C_p I_p^p. \qquad (9.36)$$

Moreover,

$$\sup_{0\leq t\leq T} |Y_t|^p \leq |Y_0|^p + \int_0^T \left[|Y_t|^{p-1}|f(t,Y_t,Z_t)| + |Y_t|^{p-2}|Z_t|^2\right]dt$$

$$+ \sup_{0\leq t\leq T} \left|\int_0^t |Y_s|^{p-2}Y_s Z_s dB_s\right|.$$

By applying the Burkholder–Davis–Gundy inequality we have

$$E\left\{\sup_{0\leq t\leq T} \left|\int_0^t |Y_s|^{p-2}Y_s Z_s dB_s\right|\right\}$$

$$\leq CE\left\{\left(\int_0^T |Y_t|^{2p-2}|Z_t|^2 dt\right)^{\frac{1}{2}}\right\} \leq CE\left\{\sup_{0\leq t\leq T} |Y_t|^{\frac{p}{2}}\left(\int_0^T |Y_t|^{p-2}|Z_t|^2 dt\right)^{\frac{1}{2}}\right\}$$

$$\leq E\left\{\frac{1}{2}\sup_{0\leq t\leq T} |Y_t|^p + C\left(\int_0^T |Y_t|^{p-2}|Z_t|^2 dt\right)^{\frac{1}{2}}\right\} < \infty.$$

9.4 Comparison Theorem and Stability Properties of BSDEs

Then,

$$E\left\{\sup_{0\le t\le T}|Y_t|^p\right\} \le \frac{1}{2}E\left\{\sup_{0\le t\le T}|Y_t|^p\right\}$$
$$+ C_p E\left\{|Y_0|^p + \left(\int_0^T |f(t,0,0)|dt\right)^p\right.$$
$$+ \int_0^T \left[|Y_t|^p + |Y_t|^{p-2}|Z_t|^2\right]dt\right\},$$

which, combined with (9.36), leads immediately to

$$E\left\{\sup_{0\le t\le T}|Y_t|^p\right\} \le C_p I_p^p. \qquad (9.37)$$

Furthermore, let $\delta := \frac{T}{n} > 0$ be a constant which will be specified later. Denote $t_i := i\delta, i = 0, \ldots, n$. Note that

$$\int_{t_i}^{t_{i+1}} Z_t dB_t = Y_{t_i} - Y_{t_{i+1}} - \int_{t_i}^{t_{i+1}} f(t, Y_t, Z_t) dt.$$

Then, by Doob's maximum inequality,

$$E\left\{\left(\int_{t_i}^{t_{i+1}} |Z_t|^2 dt\right)^{\frac{p}{2}}\right\}$$
$$\le C_p E\left\{\left|\int_{t_i}^{t_{i+1}} Z_t dB_t\right|^p\right\}$$
$$\le C_p E\left\{\sup_{0\le t\le T}|Y_t|^p + \left(\int_0^T |f(t,0,0)|dt\right)^p + \delta^{\frac{p}{2}}\left(\int_{t_i}^{t_{i+1}} |Z_t|^2 dt\right)^{\frac{p}{2}}\right\}.$$

Choose δ small enough so that $C_p \delta^{\frac{p}{2}} \le \frac{1}{2}$ for the constant C_p above. We get

$$E\left\{\left(\int_{t_i}^{t_{i+1}} |Z_t|^2 dt\right)^{\frac{p}{2}}\right\} \le C_p E\left\{\sup_{0\le t\le T}|Y_t|^p + \left(\int_0^T |f(t,0,0)|dt\right)^p\right\} \le C_p I_p^p.$$

Since n is determined by p as well, we get

$$E\left\{\left(\int_0^T |Z_t|^2 dt\right)^{\frac{p}{2}}\right\} \le C_p \sum_{i=1}^n E\left\{\left(\int_{t_i}^{t_{i+1}} |Z_t|^2 dt\right)^{\frac{p}{2}}\right\} \le C_p I_p^p.$$

This, together with (9.37), proves (9.35).

In the general case, let $\xi_n := (-n) \vee \xi \wedge n$, $f_n := (-n) \vee f \wedge n$. Note that f_n, ξ_n satisfy Assumption 9.3.1 uniformly. Let (Y^n, Z^n) denote the solution to BSDE (9.1) with coefficients (f_n, ξ_n). By the above arguments, we obtain

$$\left\|(Y^n, Z^n)\right\|_p^p \le C_p E\left\{|\xi_n|^p + \left(\int_0^T |f_n(t,0,0)|dt\right)^p dt\right\} \le C_p I_p^p.$$

We emphasize that constant C_p does not depend on n. Sending $n \to \infty$, and applying Theorem 9.4.3 and Fatou's lemma, we prove (9.35). □

9.5 Markovian BSDEs and PDEs

In this section we study the BSDEs for which the randomness in the coefficients f and g comes from a Markov diffusion X. More precisely, we focus on the following decoupled FBSDE:

$$\begin{cases} X_t = x + \int_0^t b(s, X_s)ds + \int_0^t \sigma(s, X_s)dB_s; \\ Y_t = g(X_T) + \int_t^T f(s, X_s, Y_s, Z_s)ds - \int_t^T Z_s dB_s. \end{cases} \quad (9.38)$$

Here, the coefficients b, σ, f, g are deterministic functions. This FBSDE is decoupled in the sense that the forward SDE for X does not depend on the backward components (Y, Z). For notational simplicity, we denote $\Theta := (X, Y, Z)$, and define

$$\|\Theta\|^2 := E\left\{\sup_{0 \le t \le T}\left[|X_t|^2 + |Y_t|^2\right] + \int_0^T |Z_t|^2 dt\right\}. \quad (9.39)$$

Our standing assumption is:

Assumption 9.5.1 Deterministic functions b, σ, f, g are Lebesgue measurable in all variables and are uniformly Lipschitz continuous in (x, y, z). Moreover,

$$I_0^2 := \left(\int_0^T \left[|b(t,0)| + |f(t,0,0,0)|\right]dt\right)^2 + \int_0^T |\sigma(t,0)|^2 dt + |g(0)|^2 < \infty.$$

By standard results on SDEs and Theorems 9.3.2 and 9.3.5, we get

Proposition 9.5.2 *Under Assumption 9.5.1, FBSDE (9.38) is well-posed and*

$$\|\Theta\|^2 \le C\left[1 + |x|^2\right].$$

In the estimate above and in the sequel, we use a generic constant C which depends only on T, the dimensions, the Lipschitz constant, and the value I_0.

The most important feature of this FBSDE is its connection with the following semi-linear parabolic PDE:

$$\mathcal{L}u(t, x) = 0, \quad t \in [0, T), \quad \text{and} \quad u(T, x) = g(x);$$
$$\text{where } \mathcal{L}u(t, x) := u_t + \frac{1}{2}\sigma^2 u_{xx} + u_x b + f(t, x, u, u_x \sigma). \quad (9.40)$$

The following theorem can be viewed as a nonlinear Feynman–Kac formula.

Theorem 9.5.3 *Assume Assumption 9.5.1 holds. If PDE (9.40) has a classical solution $u \in C^{1,2}$ such that $|u_x| \le C$, then*

$$Y_t = u(t, X_t), \quad Z_t = u_x \sigma(t, X_t). \quad (9.41)$$

9.5 Markovian BSDEs and PDEs

Proof First, the forward SDE in (9.38) has a unique solution X. Since $u \in C^{1,2}$, we may apply Itô's rule and get

$$du(t, X_t) = \left[u_t + u_x b + \frac{1}{2} u_{xx} \sigma^2\right] dt + u_x \sigma dB_t$$
$$= -f(t, X_t, u, u_x \sigma) dt + u_x \sigma dB_t.$$

Let (Y, Z) be defined by (9.41). Since $|u_x| \leq C$, we have

$$|Y_t| \leq C[1 + |X_t|], \qquad |Z_t| \leq C|\sigma(t, X_t)| \leq C[1 + |X_t|].$$

Then, $\|\Theta\| < \infty$. One can easily check that

$$dY_t = -f(t, X_t, Y_t, Z_t) dt + Z_t dB_t.$$

Moreover, $Y_T := u(T, X_T) = g(X_T)$. Thus, (X, Y, Z) is a solution to (9.38). Since the solution is unique, we have proved the theorem. □

Remark 9.5.4 (i) Actually, it can be shown that Assumption 9.5.1 implies $|u_x| \leq C$.

(ii) Clearly, process (X, Y, Z) is Markovian, and thus we call (9.38) a Markovian FBSDE.

(iii) In the higher dimensional case, the PDE is as follows. Let B, X, and Y take values in $\mathbb{R}^{d \times 1}$, $\mathbb{R}^{n \times 1}$ and $\mathbb{R}^{m \times 1}$, respectively. Then, b, σ, f, g, Z take values in $\mathbb{R}^{n \times 1}$, $\mathbb{R}^{n \times d}$, $\mathbb{R}^{m \times 1}$, $\mathbb{R}^{m \times 1}$ and $\mathbb{R}^{m \times d}$, respectively. Let u be a function of (t, x) taking values in $\mathbb{R}^{m \times 1}$, ∇u the gradient of u with respect to x taking values in $\mathbb{R}^{m \times n}$, and for each $i = 1, \ldots, m$, $D^2 u$ the Hessian of u^i with respect to x taking values in $\mathbb{R}^{n \times n}$. We then have

$$Y_t = u(t, X_t), \qquad Z_t = \nabla u \sigma(t, X_t),$$

where u satisfies the following system of PDEs, by denoting σ^* the transpose of σ:

$$u_t^i + \frac{1}{2} \text{tr}(\sigma \sigma^* D^2 u^i) + \nabla u^i b^i + f^i(t, x, u, \nabla u \sigma) = 0, \quad 1 \leq i \leq m;$$
$$u(T, x) = g(x).$$

Remark 9.5.5 For the European call option in the Black–Scholes model (see Sect. 9.1.1), PDE (9.40) is linear and the Black–Scholes formula is obtained via the solution to (9.40), that is, $Y_0 = u(0, x)$ gives the option price. Moreover, recall that $Z_t \sigma^{-1}$ represents the hedging portfolio. By (9.41), $Z_t \sigma^{-1} = u_x(t, X_t)$ is the sensitivity of the option price Y_t with respect to the stock price X_t. This is exactly the so-called delta-hedging in the option pricing theory.

In general, under Assumption 9.5.1, (9.40) does not have a classical solution. We instead provide a probabilistic representation for the so-called *viscosity solution* to PDE (9.40). Such a representation is important for numerically solving the PDE, in particular when the dimension of X is high.

In order to do that, we first define function u in a different way, by using the FBSDE:

$$u(t,x) := Y_0^t, \qquad (9.42)$$

where $\Theta^t := (X^t, Y^t, Z^t)$ denotes the solution to the following FBSDE on $[0, T-t]$ with shifted coefficients:

$$\begin{cases} X_s^t = x + \int_0^s b(r+t, X_r^t) dr + \int_0^s \sigma(r+t, X_r^t) dB_r; \\ Y_s^t = g(X_{T-t}^t) + \int_s^{T-t} f(r+t, \Theta_r^t) dr - \int_s^{T-t} Z_r^t dB_r. \end{cases} \qquad (9.43)$$

By the Blumenthal 0-1 law, $u(t, x)$ is deterministic. The following result is important.

Theorem 9.5.6 *Assume Assumption 9.5.1 holds. Then, $Y_t = u(t, X_t)$, a.s. for all $t \in [0, T]$.*

Proof Denote $\tilde{B}_s := B_{t+s} - B_t$, $\tilde{\Theta}_s := \Theta_{s+t}$, $s \in [0, T-t]$. Note that $\tilde{\Theta}$ satisfies:

$$\begin{cases} \tilde{X}_s = X_t + \int_0^s b(r+t, \tilde{X}_r) dr + \int_0^s \sigma(r+t, \tilde{X}_r) d\tilde{B}_r; \\ \tilde{Y}_s = g(\tilde{X}_{T-t}) + \int_s^{T-t} f(r+t, \tilde{\Theta}_r) dr - \int_s^{T-t} \tilde{Z}_r d\tilde{B}_r. \end{cases} \qquad (9.44)$$

Since $\tilde{B}$ is a Brownian motion and is independent of X_t, comparing (9.44) and (9.43) we see that $\tilde{Y}_0 = u(t, X_t)$, a.s. That is, $Y_t = u(t, X_t)$, a.s. □

There is another convenient way to define u. Given (t, x), let $\Theta^{t,x} := (X^{t,x}, Y^{t,x}, Z^{t,x})$ be the solution to the following FBSDE over $[t, T]$:

$$\begin{cases} X_s^{t,x} = x + \int_t^s b(r, X_r^{t,x}) dr + \int_t^s \sigma(r, X_r^{t,x}) dB_r; \\ Y_s^{t,x} = g(X_T^{t,x}) + \int_s^T f(r, X_r^{t,x}, Y_r^{t,x}, Z_r^{t,x}) dr - \int_s^T Z_r^{t,x} dB_r. \end{cases} \qquad (9.45)$$

Following similar arguments, we have

$$Y_s^{t,x} = u(s, X_s^{t,x}) \quad \text{and in particular} \quad Y_t^{t,x} = u(t, x). \qquad (9.46)$$

We conclude this section by noting that it can be shown that function u defined by (9.42) is a viscosity solution to PDE (9.40).

9.5.1 Numerical Methods

There are typically two approaches to solve numerically FBSDE (9.38) and related PDE (9.40). One is the PDE approach, e.g., the finite difference methods. While working efficiently in low dimensions, the PDE methods in general do not work in

9.6 BSDEs with Quadratic Growth

high dimensions, due to the so-called curse of dimensionality. The other method is a probabilistic approach, which solves the FBSDE directly, thus also the associated PDE, by using Monte Carlo simulation. A lot of recent and ongoing research has been done on this topic.

9.6 BSDEs with Quadratic Growth

In this section we study BSDE (9.1) whose generator f has quadratic growth in Z. In particular, in this case f is not uniformly Lipschitz continuous in Z. This kind of BSDE appears often in applications, for example, when the utility function is exponential, or when the cost function is quadratic. The main technique to establish the well-posedness of such BSDEs is a nonlinear transformation of Y. To focus on the main idea, we will study only the case in which the solution Y is bounded.

Assumption 9.6.1 (i) $\dim(\xi) = 1$ and $\xi \in L^\infty(\mathcal{F}_T)$.
(ii) There exists a constant C such that, for any (ω, t, y, z),
$$|f(t, y, z)| \leq C[1 + |y| + |z|^2]. \tag{9.47}$$
(iii) There exists a constant C such that, for any (ω, t, y_i, z_i), $i = 1, 2$,
$$\begin{aligned}|f(t, y_1, z_1) - f(t, y_2, z_2)| \\ \leq C[|y_1 - y_2| + (|y_1| + |y_2| + |z_1| + |z_2|)|z_1 - z_2|]. \end{aligned} \tag{9.48}$$

We remark that, when f is differentiable in (y, z), (9.48) is equivalent to
$$|f_y| \leq C, \qquad |f_z| \leq C[1 + |y| + |z|]. \tag{9.49}$$

Remark 9.6.2 We will use the comparison theorem to establish the well-posedness of such BSDEs, and thus it is important to assume ξ and Y are one-dimensional. The well-posedness of higher dimensional BSDEs with quadratic generator is an important and difficult subject.

Theorem 9.6.3 *Assume Assumption* 9.6.1 *holds. Then, BSDE* (9.1) *admits a unique solution* $(Y, Z) \in L^\infty(\mathbb{F}) \times L^2(\mathbb{F})$.

We first establish some a priori estimates.

Theorem 9.6.4 *Assume Assumption* 9.6.1 *holds. If* $(Y, Z) \in L^\infty(\mathbb{F}) \times L^2(\mathbb{F})$ *is a solution to BSDE* (9.1), *then*
$$|Y_t| \leq C \quad \text{and} \quad E_t\left\{\int_t^T |Z_s|^2 ds\right\} \leq C,$$
where constant C depends only on T, the dimensions, and the constant in (9.47) *and* $\|\xi\|_\infty$.

Proof Let $\varphi(y)$ be a smooth function such that $\varphi(y) = |y|$ for $|y| \geq 1$. Denote $C_0 := 2(1 + C)$ for the constant C in (9.47), and

$$\tilde{Y}_t := e^{C_0 e^{C_0 t}[\varphi(Y_t)+1]}, \qquad \tilde{Z}_t := C_0 e^{C_0 t} \varphi_y(Y_t) \tilde{Y}_t Z_t. \tag{9.50}$$

Applying Itô's rule, we obtain:

$$d\left(C_0 e^{C_0 t}[\varphi(Y_t)+1]\right)$$
$$= C_0^2 e^{C_0 t}[\varphi(Y_t)+1]dt + C_0 e^{C_0 t}\left[-\varphi_y(Y_t)f(t, Y_t, Z_t) + \frac{1}{2}\varphi_{yy}(Y_t)Z_t^2\right]dt$$
$$+ C_0 e^{C_0 t}\varphi_y(Y_t)Z_t dB_t,$$

and then

$$d\tilde{Y}_t = C_0 e^{C_0 t}\tilde{Y}_t\left[C_0[\varphi(Y_t)+1] - \varphi_y(Y_t)f(t, Y_t, Z_t) + \frac{1}{2}\varphi_{yy}(Y_t)Z_t^2\right]dt$$
$$+ \frac{1}{2}\left|C_0 e^{C_0 t}\varphi_y(Y_t)Z_t\right|^2 \tilde{Y}_t dt + \tilde{Z}_t dB_t.$$

By the property of φ and (9.47) with $C = \frac{1}{2}C_0 - 1$, we get

$$1_{\{|Y_t|\geq 1\}} d\tilde{Y}_t$$
$$= 1_{\{|Y_t|\geq 1\}} C_0 e^{C_0 t}\tilde{Y}_t\left[C_0[|Y_t|+1] - \text{sign}(Y_t)f(t, Y_t, Z_t) + \frac{1}{2}C_0 e^{C_0 t}|Z_t|^2\right]dt$$
$$+ 1_{\{|Y_t|\geq 1\}}\tilde{Z}_t dB_t$$
$$\geq 1_{\{|Y_t|\geq 1\}} C_0 e^{C_0 t}\tilde{Y}_t\left[C_0[|Y_t|+1] - \left[\frac{C_0}{2}-1\right][1+|Y_t|+|Z_t|^2]\right.$$
$$\left. + \frac{1}{2}C_0 e^{C_0 t}|Z_t|^2\right]dt + 1_{\{|Y_t|\geq 1\}}\tilde{Z}_t dB_t$$
$$\geq 1_{\{|Y_t|\geq 1\}}\tilde{Z}_t dB_t. \tag{9.51}$$

Now, for any t and on the event $\{|Y_t| \geq 1\}$, define a stopping time

$$\tau_t := \inf\{s \geq t : |Y_s| \leq 1\} \wedge T.$$

Then,

$$|Y_s| \geq 1, \quad s \in [t, \tau_t] \quad \text{and} \quad |Y_{\tau_t}| = 1_{\{\tau_t < T\}} + |\xi|1_{\{\tau_t = T\}} \leq |\xi| \vee 1.$$

As Y is bounded, $\tilde{Z}$ is square integrable. Thus, (9.51) leads to

$$0 \leq e^{C_0 e^{C_0 t}[|Y_t|+1]} = \tilde{Y}_t \leq E_t\{\tilde{Y}_{\tau_t}\} = E_t\{e^{C_0 e^{C_0 \tau_t}[|Y_{\tau_t}|+1]}\} \leq e^{C_0 e^{C_0 T}[\|\xi\|_\infty \vee 1+1]}.$$

This implies that

$$|Y_t| \leq e^{C_0(T-t)}[\|\xi\|_\infty \vee 1 + 1] - 1 \quad \text{on} \quad \{|Y_t| \geq 1\}.$$

Therefore,

9.6 BSDEs with Quadratic Growth

$$|Y_t| \leq \left[e^{C_0 T}\left[\|\xi\|_\infty \vee 1 + 1\right] - 1\right] \vee 1.$$

This proves the estimate for Y.

Next, applying Itô's rule on $e^{C_0 Y_t}$ we obtain

$$de^{C_0 Y_t} = -C_0 e^{C_0 Y_t} f(t, Y_t, Z_t) dt + C_0 e^{C_0 Y_t} Z_t dB_t + \frac{1}{2} C_0^2 e^{C_0 Y_t} Z_t^2 dt$$

$$\geq -C_0 e^{C_0 Y_t}\left[\frac{C_0}{2} - 1\right]\left[1 + |Y_t| + |Z_t|^2\right] dt + C_0 e^{C_0 Y_t} Z_t dB_t$$

$$+ \frac{1}{2} C_0^2 e^{C_0 Y_t} Z_t^2 dt$$

$$= C_0 e^{C_0 Y_t}\left[|Z_t|^2 - C_0(1 + |Y_t|)\right] dt + C_0 e^{C_0 Y_t} Z_t dB_t.$$

Then,

$$e^{C_0 Y_t} + E_t\left\{\int_t^T C_0 e^{C_0 Y_s} |Z_s|^2 ds\right\} \leq E_t\left\{e^{C_0 \xi} + \int_t^T C_0 e^{C_0 Y_s} C_0[1 + |Y_s|] ds\right\}.$$

Thus,

$$C_0 e^{-C_0 \|Y\|_\infty} E_t\left\{\int_t^T |Z_s|^2 dt\right\} \leq e^{C_0 \|\xi\|_\infty} + T C_0 e^{C_0 \|Y\|_\infty} C_0[1 + \|Y\|_\infty].$$

This implies the estimate for Z. $\square$

The above estimate for Z implies that the martingale $\int_0^t Z_s dB_s$ is a so-called BMO martingale, for which we have the following important lemma.

Lemma 9.6.5 *Assume an $\mathbb{F}$-adapted process θ satisfies*

$$E_t\left\{\int_t^T |\theta_s|^2 ds\right\} \leq C_0, \quad t \in [0, T]. \tag{9.52}$$

Then

$$E\left\{\exp\left(\varepsilon \int_0^T |\theta_s|^2 ds\right)\right\} \quad \text{for } 0 < \varepsilon < \frac{1}{C_0} \tag{9.53}$$

and the process

$$M_t^\theta := \exp\left(\int_0^t \theta_s dB_s - \frac{1}{2}\int_0^t |\theta_s|^2 ds\right) \quad \text{is an uniformly integrable martingale.} \tag{9.54}$$

Proof We first show that

$$E_t\left\{\left(\int_t^T |\theta_s|^2 ds\right)^n\right\} \leq n! C_0^n. \tag{9.55}$$

Without loss of generality, we prove it only for $t = 0$. In fact, denote $\Gamma_t := \int_0^t |\theta_s|^2 ds$. Then, by applying Itô's rule repeatedly,

$$|\Gamma_T|^n = n! \int_0^T \int_0^{t_1} \cdots \int_0^{t_{n-1}} \theta_{t_1}^2 \cdots \theta_{t_n}^2 dt_n \cdots dt_1$$
$$= n! \int_0^T \int_{t_n}^T \cdots \int_{t_2}^T \theta_{t_1}^2 \cdots \theta_{t_n}^2 dt_1 \cdots dt_n.$$

Thus,

$$E\{|\Gamma_T|^n\} = n! E\left\{\int_0^T \int_{t_n}^T \cdots \int_{t_2}^T \theta_{t_1}^2 \cdots \theta_{t_n}^2 dt_1 \cdots dt_n\right\}$$
$$= n! E\left\{\int_0^T \int_{t_n}^T \cdots \int_{t_3}^T E_{t_2}\left\{\int_{t_2}^T \theta_{t_1}^2 dt_1\right\} \theta_{t_2}^2 \cdots \theta_{t_n}^2 dt_2 \cdots dt_n\right\}$$
$$\leq C_0 n! E\left\{\int_0^T \int_{t_n}^T \cdots \int_{t_3}^T \theta_{t_2}^2 \cdots \theta_{t_n}^2 dt_2 \cdots dt_n\right\}$$
$$\leq \cdots \leq n! C_0^n.$$

Next, for $0 < \varepsilon < \frac{1}{C_0}$,

$$E\left\{\exp\left(\varepsilon \int_0^T |\theta_s|^2 ds\right)\right\} = \sum_{n=0}^\infty \frac{1}{n!} \varepsilon^n E\{|\Gamma_T|^n\} \leq \sum_{n=0}^\infty (\varepsilon C_0)^n = \frac{1}{1 - C_0 \varepsilon} < \infty.$$

This proves (9.53). Moreover, by Novikov's theorem, $M^{\varepsilon\theta}$ is a true martingale for $\varepsilon < \sqrt{\frac{2}{C_0}}$. Now, fixing $0 < \varepsilon < \frac{2}{C_0} \wedge \sqrt{\frac{2}{C_0}}$, we have

$$1 = E_t\left\{\exp\left(\varepsilon \int_t^T \theta_s dB_s - \frac{\varepsilon^2}{2} \int_t^T |\theta_s|^2 ds\right)\right\}$$
$$= E_t\left\{\exp\left(\varepsilon \int_t^T \theta_s dB_s - \frac{\varepsilon}{2} \int_t^T |\theta_s|^2 ds\right) \exp\left(\frac{\varepsilon - \varepsilon^2}{2} \int_t^T |\theta_s|^2 ds\right)\right\}$$
$$\leq \left(E_t\left\{\exp\left(\int_t^T \theta_s dB_s - \frac{1}{2} \int_t^T |\theta_s|^2 ds\right)\right\}\right)^\varepsilon \left(E_t\left\{\exp\left(\frac{\varepsilon}{2} \int_t^T |\theta_s|^2 ds\right)\right\}\right)^{1-\varepsilon}$$
$$\leq ((M_t^\theta)^{-1} E_t\{M_T^\theta\})^\varepsilon \left(\frac{2}{2 - C_0 \varepsilon}\right)^{1-\varepsilon}.$$

This implies that

$$M_t^\theta \leq E_t\{M_T^\theta\} \left(\frac{2}{2 - C_0 \varepsilon}\right)^{\frac{1-\varepsilon}{\varepsilon}}.$$

Note that M^θ is a positive local martingale and thus a supermartingale, so M_T^θ is integrable. Therefore, M^θ is uniformly integrable and thus is a true martingale. □

Proof of Theorem 9.6.3 (Uniqueness) Let (Y^i, Z^i), $i = 1, 2$, be two solutions such that Y^i are bounded. By Theorem 9.6.4,

$$|Y_t^i|^2 + E_t\left\{\int_t^T |Z_s^i|^2 ds\right\} \leq C, \quad i = 1, 2, \tag{9.56}$$

9.6 BSDEs with Quadratic Growth

for some constant C. Denote $\Delta Y := Y^1 - Y^2$, $\Delta Z := Z^1 - Z^2$. Then,

$$d\Delta Y_t = \left[f(t, Y_t^1, Z_t^1) - f(t, Y_t^2, Z_t^2)\right]dt + \Delta Z_t dB_t$$
$$= -[\alpha_t \Delta Y_t + \theta_t \Delta Z_t]dt + \Delta Z_t dB_t,$$

where α is bounded, and $|\theta_t| \leq C[1 + |Y_t^1| + |Y_t^2| + |Z_t^1| + |Z_t^2|]$. By (9.56), θ satisfies (9.52). Then, by Lemma 9.6.5, M^θ is a true martingale. Denote $\Gamma_t := \exp(\int_0^t \alpha_s ds)$. Then,

$$d(\Gamma_t M_t^\theta \Delta Y_t) = \Gamma_t M_t^\theta \Delta Z_t dB_t.$$

That is, $\Gamma M^\theta \Delta Y$ is a local martingale. Note that Γ and ΔY are bounded. Then, $\Gamma M^\theta \Delta Y$ is uniformly integrable and thus is a true martingale. Since $\Delta Y_T = 0$, we obtain $\Gamma_t M_t^\theta \Delta Y_t = 0$ and therefore, $\Delta Y_t = 0$. This clearly implies also $\Delta Z_t = 0$. □

To prove the existence, we need another lemma.

Lemma 9.6.6 *Assume ξ is bounded, $\{f_n, n \geq 1\}$ satisfy Assumption 9.6.1 uniformly, and $f_n(\cdot, y, z) \to f(\cdot, y, z)$ in measure $dt \times dP$ as $n \to \infty$, for any (y, z). If (Y^n, Z^n) is a solution to BSDE (9.1) with coefficient (ξ, f_n) such that Y^n is bounded, and $Y^n \uparrow Y$ or $Y^n \downarrow Y$, then there exists a process Z such that (Y, Z) is a solution to BSDE (9.1) with coefficient (ξ, f).*

Proof Without loss of generality, we assume $Y^n \uparrow Y$. By Theorem 9.6.4,

$$|Y_t^n| \leq C, \qquad E\left\{\int_0^T |Z_t^n|^2 dt\right\} \leq C, \quad n \geq 1. \tag{9.57}$$

Then, Y is also bounded, and $\{Z^n, n \geq 1\}$ has a weak limit point Z in $L^2(\mathbb{F})$. By otherwise choosing a subsequence, we may assume that the whole sequence Z^n converges to Z weakly in $L^2(\mathbb{F})$. Moreover, following similar arguments as in Theorem 9.4.3(ii), we may assume without loss of generality that

$$f_n(\cdot, y, z) \to f(\cdot, y, z), \quad \text{for all } (y, z), \ dt \times dP\text{-a.s.} \tag{9.58}$$

Let $\varphi(y)$ be a convex function such that $\varphi(0) = 0$, $\varphi'(0) = 0$, which will be specified later. For $n < m$, denoting $\Delta Y_t^{m,n} := Y_t^m - Y_t^n$, $\Delta Z_t^{m,n} := Z_t^m - Z_t^n$ and applying Itô's rule, we have

$$d\varphi(\Delta Y_t^{m,n}) = -\varphi'(\Delta Y_t^{m,n})\left[f_m(t, Y_t^m, Z_t^m) - f_n(t, Y_t^n, Z_t^n)\right]dt$$
$$+ \frac{1}{2}\varphi''(\Delta Y_t^{m,n})|\Delta Z_t^{m,n}|^2 dt + \varphi'(\Delta Y_t^{m,n})\Delta Z_t^{m,n} dB_t$$
$$\geq -C|\varphi'(\Delta Y_t^{m,n})|\left[1 + |Y_t^m| + |Y_t^n| + |Z_t^m|^2 + |Z_t^n|^2\right]dt$$
$$+ \frac{1}{2}\varphi''(\Delta Y_t^{m,n})|\Delta Z_t^{m,n}|^2 dt + \varphi'(\Delta Y_t^{m,n})\Delta Z_t^{m,n} dB_t$$
$$\geq -C_0|\varphi'(\Delta Y_t^{m,n})|\left[1 + |\Delta Z_t^{m,n}|^2 + |Z_t^n - Z_t|^2 + |Z_t|^2\right]dt$$
$$+ \frac{1}{2}\varphi''(\Delta Y_t^{m,n})|\Delta Z_t^{m,n}|^2 dt + \varphi'(\Delta Y_t^{m,n})\Delta Z_t^{m,n} dB_t.$$

Define

$$\varphi(y) := \frac{1}{8C_0^2}\left[e^{4C_0 y} - 4C_0 y - 1\right], \quad \text{and thus}$$

$$\varphi'(y) = \frac{1}{2C_0}\left[e^{4C_0 y} - 1\right], \quad \varphi''(y) = 2e^{4C_0 y}.$$

Then,

$$\varphi(0) = 0, \quad \varphi'(0) = 0, \quad \varphi(y) > 0, \quad \varphi'(y) > 0 \text{ for } y > 0, \text{ and}$$
$$\varphi''(y) = 4C_0 \varphi'(y) + 2.$$

Note that $Y^m \geq Y^n$ for $m > n$. Then,

$$d\varphi(\Delta Y_t^{m,n}) \geq -C_0 \varphi'(\Delta Y_t^{m,n})\left[1 + |\Delta Z_t^{m,n}|^2 + |Z_t^n - Z_t|^2 + |Z_t|^2\right]dt$$
$$+ \left[2C_0 \varphi'(\Delta Y_t^{m,n}) + 1\right]|\Delta Z_t^{m,n}|^2 dt + \varphi'(\Delta Y_t^{m,n})\Delta Z_t^{m,n} dB_t.$$

Thus, noting that $\varphi(\Delta Y_T^{m,n}) = \varphi(0) = 0$,

$$E\left\{\int_0^T \left[C_0 \varphi'(\Delta Y_t^{m,n}) + 1\right]|\Delta Z_t^{m,n}|^2 dt\right\}$$
$$\leq C_0 E\left\{\int_0^T \varphi'(\Delta Y_t^{m,n})\left[1 + |Z_t^n - Z_t|^2 + |Z_t|^2\right]dt\right\}. \qquad (9.59)$$

We now fix n. Note that $\varphi'(\Delta Y_t^{m,n})$ is bounded and $\varphi'(\Delta Y_t^{m,n}) \to \varphi'(Y_t - Y_t^n)$, a.s. Then, one can easily see that

$$\sqrt{C_0 \varphi'(\Delta Y_t^{m,n}) + 1}\,\Delta Z_t^m \text{ converges to } \sqrt{C_0 \varphi'(Y_t - Y_t^n) + 1}\left[Z_t - Z_t^n\right] \text{ weakly in } L^2(\mathbb{F}),$$

as $m \to \infty$, and therefore,

$$E\left\{\int_0^T \left[C_0 \varphi'(Y_t - Y_t^n) + 1\right]|Z_t^n - Z_t|^2 dt\right\}$$
$$\leq \limsup_{m\to\infty} E\left\{\int_0^T \left[C_0 \varphi'(\Delta Y_t^{m,n}) + 1\right]|\Delta Z^{m,n}|^2 dt\right\}.$$

Thus, by (9.59)

$$E\left\{\int_0^T \left[C_0 \varphi'(Y_t - Y_t^n) + 1\right]|Z_t^n - Z_t|^2 dt\right\}$$
$$\leq \limsup_{m\to\infty} C_0 E\left\{\int_0^T \varphi'(\Delta Y_t^{m,n})\left[1 + |Z_t^n - Z_t|^2 + |Z_t|^2\right]dt\right\}$$
$$= C_0 E\left\{\int_0^T \varphi'(Y_t - Y_t^n)\left[1 + |Z_t^n - Z_t|^2 + |Z_t|^2\right]dt\right\}.$$

This implies that

9.6 BSDEs with Quadratic Growth

$$E\left\{\int_0^T |Z_t^n - Z_t|^2 dt\right\} \leq C_0 E\left\{\int_0^T \varphi'(Y_t - Y_t^n)[1 + |Z_t|^2] dt\right\}.$$

Sending $n \to \infty$ and applying the Dominated Convergence Theorem, we have

$$\lim_{n \to \infty} E\left\{\int_0^T |Z_t^n - Z_t|^2 dt\right\} = 0. \tag{9.60}$$

Finally, note that

$$E\left\{\int_0^T |f_n(t, Y_t^n, Z_t^n) - f(t, Y_t, Z_t)| dt\right\}$$

$$\leq E\left\{\int_0^T |f_n(t, Y_t, Z_t) - f(t, Y_t, Z_t)| dt\right\}$$

$$+ C_0 E\left\{\int_0^T [|Y_t^n - Y_t| + [1 + |Y_t^n| + |Z_t^n| + |Y_t| + |Z_t|]|Z_t^n - Z_t|] dt\right\}$$

$$\leq E\left\{\int_0^T |f_n(t, Y_t, Z_t) - f(t, Y_t, Z_t)| dt\right\} + C_0 E\left\{\int_0^T |Y_t^n - Y_t| dt\right\}$$

$$+ C_0 \left(E\left\{\int_0^T [1 + |Y_t^n| + |Z_t^n| + |Y_t| + |Z_t|]^2 dt\right\}\right)^{\frac{1}{2}}$$

$$\times \left(E\left\{\int_0^T |Z_t^n - Z_t|^2 dt\right\}\right)^{\frac{1}{2}}$$

$$\leq E\left\{\int_0^T |f_n(t, Y_t, Z_t) - f(t, Y_t, Z_t)| dt\right\} + C_0 E\left\{\int_0^T |Y_t^n - Y_t| dt\right\}$$

$$+ C\left(E\left\{\int_0^T |Z_t^n - Z_t|^2 dt\right\}\right)^{\frac{1}{2}}.$$

Sending $n \to \infty$, we get from (9.58) and (9.60) that

$$\lim_{n \to \infty} E\left\{\int_0^T |f_n(t, Y_t^n, Z_t^n) - f(t, Y_t, Z_t)| dt\right\} = 0.$$

Then, it is straightforward to check that (Y, Z) satisfies BSDE (9.1) with coefficient (ξ, f). □

Proof of Theorem 9.6.3 (Existence) For any $n, m, k \geq 1$, define

$$f_n := f \wedge n, \qquad f_{n,m} := f_n \vee (-m),$$
$$f_{n,m,k}(t, y, z) := \inf_{z'}[f_{n,m}(t, y, z') + k|z - z'|].$$

Then, as n, m, k increase,

$$f_n \uparrow f, \qquad f_{n,m} \downarrow f_n, \qquad f_{n,m,k} \uparrow f_{n,m};$$
$$-m \leq f_{n,m}, \qquad f_{n,m,k} \leq n \quad \text{and} \quad |f_n|, |f_{n,m}| \leq C[1 + |y| + |z|^2]$$
for all (n, m, k);

$f_{n,m,k}$ is uniformly Lipschitz continuous in (y, z) for each (n, m, k).
$$\tag{9.61}$$

Let $(Y^{n,m,k}, Z^{n,m,k})$ denote the unique solution of the BSDE

$$Y_t^{n,m,k} = \xi + \int_t^T f_{n,m,k}(s, Y_s^{n,m,k}, Z_s^{n,m,k})ds - \int_t^T Z_s^{n,m,k} dB_s.$$

By Theorem 9.4.1,

$Y^{n,m,k}$ is increasing in k, decreasing in m, and increasing in n.

Note that

$$Y_t^{n,m,k} = E_t\left\{\xi + \int_t^T f_{n,m,k}(s, Y_s^{n,m,k}, Z_s^{n,m,k})ds\right\}.$$

Clearly,

$$|Y_t^{n,m,k}| \leq \|\xi\|_\infty + (n \vee m)T, \quad \text{for all } k.$$

Denote $Y_t^{n,m} := \lim_{k\to\infty} Y_t^{n,m,k}$.

$Y^{n,m}$ is decreasing in m, increasing in n, and $|Y_t^{n,m}| \leq \|\xi\|_\infty + (n \vee m)T$.

For any k_1, k_2, applying Itô's rule on $|Y_t^{n,m,k_1} - Y_t^{n,m,k_2}|^2$ we obtain

$$E\left\{\int_0^T |Z_t^{n,m,k_1} - Z_t^{n,m,k_2}|^2 dt\right\}$$

$$\leq E\left\{2\int_0^T |Y_t^{n,m,k_1} - Y_t^{n,m,k_2}| |f_{n,m,k_1}(t, Y_t^{n,m,k_1}, Z_t^{n,m,k_1})\right.$$

$$\left. - f_{n,m,k_2}(t, Y_t^{n,m,k_2}, Z_t^{n,m,k_2})|dt\right\}$$

$$\leq 2(n+m)E\left\{\int_0^T |Y_t^{n,m,k_1} - Y_t^{n,m,k_2}|dt\right\} \to 0, \quad \text{as } k_1, k_2 \to \infty,$$

thanks to the Dominated Convergence Theorem. Then, there exists $Z^{n,m}$ such that

$$\lim_{k\to\infty} E\left\{\int_0^T |Z_t^{n,m,k} - Z_t^{n,m}|^2 dt\right\} = 0.$$

It is straightforward to check that $(Y^{n,m}, Z^{n,m})$ satisfies

$$Y_t^{n,m} = \xi + \int_t^T f_{n,m}(s, Y_s^{n,m}, Z_s^{n,m})ds - \int_t^T Z_s^{n,m} dB_s.$$

Applying Theorem 9.6.4,

$$|Y_t^{n,m}|^2 + E_t\left\{\int_t^T |Z_s^{n,m}|^2 ds\right\} \leq C \quad \text{for all } n, m.$$

Now, denote $Y^n := \lim_{m\to\infty} Y^{n,m}$. Then,

Y^n is increasing in n, and $|Y_t^n| \leq C$.

Applying Lemma 9.6.6, there exists Z^n such that

$$Y_t^n = \xi + \int_t^T f_n(s, Y_s^n, Z_s^n) ds - \int_t^T Z_s^n dB_s.$$

Now, denote $Y := \lim_{n \to \infty} Y^n$. Then, Y is bounded. Applying Lemma 9.6.6 again, there exists Z such that (Y, Z) is a solution to BSDE (9.1). □

By the construction of the solution in the above existence proof of Theorem 9.6.3, one gets immediately

Theorem 9.6.7 *Assume (ξ_i, f^i), $i = 1, 2$, satisfy Assumption 9.6.1, and let (Y^i, Z^i) be the corresponding solutions such that Y^i is bounded. If $\xi_1 \leq \xi_2$ a.s. and, for any (y, z), $f_1(\cdot, y, z) \leq f_2(\cdot, y, z)$, $dt \times dP$-a.s. then $Y_t^1 \leq Y_t^2$, $0 \leq t \leq 1$, a.s. In particular, $Y_0^1 \leq Y_0^2$.*

9.7 Further Reading

The theory of BSDEs started from the seminal paper Pardoux and Peng (1990), and its connection with PDEs was established in Peng (1991) and Pardoux and Peng (1992). The basic theory and financial applications of BSDE's can be found in the survey paper El Karoui et al. (1997). For probabilistic numerical methods for Markovian decoupled FBSDEs, we refer to Zhang (2004), Bouchard and Touzi (2004), Gobet et al. (2005), Bender and Denk (2007), and Crisan and Manolarakis (2012). Quadratic BSDEs with bounded terminal conditions was studied by Kobylanski (2000), and Briand and Hu (2006, 2008) extended the results to the case with unbounded terminal conditions. A much more comprehensive presentation of BSDEs and FBSDEs is provided in the book of Zhang (2011).

References

Bender, C., Denk, R.: A forward scheme for backward SDEs. Stoch. Process. Appl. **117**, 1793–1812 (2007)
Bouchard, B., Touzi, N.: Discrete-time approximation and Monte-Carlo simulation of backward stochastic differential equations. Stoch. Process. Appl. **111**, 175–206 (2004)
Briand, P., Hu, Y.: BSDE with quadratic growth and unbounded terminal value. Probab. Theory Relat. Fields **136**, 604–618 (2006)
Briand, P., Hu, Y.: Quadratic BSDEs with convex generators and unbounded terminal conditions. Probab. Theory Relat. Fields **141**, 543–567 (2008)
Crisan, D., Manolarakis, K.: Solving backward stochastic differential equations using the cubature method: application to nonlinear pricing. SIAM J. Financ. Math. **3**, 534–571 (2012)
El Karoui, N., Peng, S., Quenez, M.C.: Backward stochastic differential equations in finance. Math. Finance **7**, 1–71 (1997)
Gobet, E., Lemor, J.-P., Warin, X.: A regression-based Monte-Carlo method to solve backward stochastic differential equations. Ann. Appl. Probab. **15**, 2172–2202 (2005)

Kobylanski, M.: Backward stochastic differential equations and partial differential equations with quadratic growth. Ann. Probab. **28**, 558–602 (2000)

Pardoux, E., Peng, S.: Adapted solution of a backward stochastic differential equation. Syst. Control Lett. **14**, 55–61 (1990)

Pardoux, E., Peng, S.: Backward Stochastic Differential Equations and Quasilinear Parabolic Partial Differential Equations. Lecture Notes in Control and Inform. Sci., vol. 176, pp. 200–217. Springer, New York (1992)

Peng, S.: Probabilistic interpretation for systems of quasilinear parabolic partial differential equations. Stochastics **37**, 61–74 (1991)

Zhang, J.: A numerical scheme for BSDEs. Ann. Appl. Probab. **14**, 459–488 (2004)

Zhang, J.: Backward Stochastic Differential Equations. Book manuscript, University of Southern California (2011, in preparation)

Chapter 10
Stochastic Maximum Principle

Abstract As an important application of BSDEs and FBSDEs, in this chapter we present a classical method of the Stochastic Control Theory, the stochastic maximum principle, the main technical tool in this book. We first present stochastic control of BSDEs and then much more complex stochastic control of FBSDEs. Necessary conditions are obtained in terms of appropriate adjoint processes, which, under some conditions, can be characterized in terms of an FBSDE system. Sufficient conditions are stated in terms of the corresponding Hamiltonian function. Similar results are presented also for the case of the so-called weak formulation, in which the agent controls the distribution of the output process.

10.1 Stochastic Control of BSDEs

Let U be a subset of $\mathbb{R}^k$ and $\mathcal{U}$ be a set of $\mathbb{F}$-adapted processes u taking values in U. For notational simplicity, we assume $k = 1$. For each $u \in \mathcal{U}$, consider the following BSDE:

$$Y_t^u = \xi + \int_t^T f(s, Y_s^u, Z_s^u, u_s) ds - \int_t^T Z_s^u dB_s. \tag{10.1}$$

Our optimization problem is

$$V := \sup_{u \in \mathcal{U}} Y_0^u. \tag{10.2}$$

The goal is to characterize the value V and find the optimal control $u^* \in \mathcal{U}$, if exists.

Remark 10.1.1 (i) When f does not depend on (Y, Z), the optimization problem becomes

$$V := \sup_{u \in \mathcal{U}} Y_0^u = \sup_{u \in \mathcal{U}} E\left\{\xi + \int_0^T f(t, u_t) dt\right\}.$$

Here, f can be interpreted as the running cost or utility.

(ii) In general, Y_0^u can be interpreted as a recursive utility, as introduced by Duffie and Epstein (1992).

We first study sufficient conditions for the optimal control. We adopt the following assumptions.

Assumption 10.1.2 (i) $\xi \in L^2(\mathcal{F}_T)$.
(ii) f is progressively measurable, $\mathbb{F}$-adapted, and uniformly Lipschitz continuous in (y, z).

Assumption 10.1.3 The set $\mathcal{U}$ of admissible controls u satisfies: for each $u \in \mathcal{U}$,

$$E\left\{\left(\int_0^T |f(t,0,0,u_t)|dt\right)^2\right\} < \infty.$$

Theorem 10.1.4 *Assume Assumptions* 10.1.2 *and* 10.1.3 *hold. Suppose that*

$$E\left\{\left(\int_0^T |f^*(t,0,0)|dt\right)^2\right\} < \infty \quad \text{where } f^*(t,y,z) := \sup_{u \in U} f(t,y,z,u),$$
(10.3)

and there exists a progressively measurable and $\mathbb{F}$-*adapted function* $I(t, y, z)$ *such that*

$$f^*(t,y,z) = f\big(t, y, z, I(t, y, z)\big) \quad \text{and} \quad u^* := I\big(\cdot, Y^*, Z^*\big) \in \mathcal{U}, \quad (10.4)$$

where (Y^*, Z^*) *is the solution to the BSDE*

$$Y_t^* = \xi + \int_t^T f^*\big(s, Y_s^*, Z_s^*\big)ds - \int_t^T Z_s^* dB_s. \quad (10.5)$$

Then, $V = Y_0^*$ *and* u^* *is an optimal control for the optimization problem* (10.2).

Proof We first note that (10.3) and Assumption 10.1.2(ii) imply that f^* is finite and uniformly Lipschitz continuous in (y, z). Then, applying Theorem 9.3.5 we know BSDEs (10.1) and (10.5) are well-posed. It is clear that $Y^* = Y^{u^*}$, $Z^* = Z^{u^*}$. Moreover, for each $u \in \mathcal{U}$, since $f(t, y, z, u) \leq f^*(t, y, z)$, applying Comparison Theorem 9.4.1, we have $Y_0^u \leq Y_0^* = Y_0^{u^*}$. That is, u^* is optimal. $\square$

Remark 10.1.5 (i) Two typical cases when (10.4) is satisfied are: U is compact (including the case of discrete U) and f is concave in u.
(ii) In (10.4), we do not require uniqueness or differentiability of function I.

We next look for necessary conditions. We modify the assumptions as follows.

Assumption 10.1.6 (i) $\xi \in L^2(\mathcal{F}_T)$.
(ii) f is progressively measurable, $\mathbb{F}$-adapted, and uniformly Lipschitz continuous in (y, z).
(iii) f is continuously differentiable in u, and $|\partial_u f(t,y,z,u)| \leq C[1 + |\partial_u f(t,0,0,u)| + |y| + |z|]$.

10.1 Stochastic Control of BSDEs

Assumption 10.1.7 Set $\mathcal{U}$ of admissible controls u satisfies:
(i) For each $u \in \mathcal{U}$,

$$E\left\{\left(\int_0^T [|f(t,0,0,u_t)| + |\partial_u f(t,0,0,u_t)|]dt\right)^2\right\} < \infty. \quad (10.6)$$

(ii) $\mathcal{U}$ is locally convex. That is, for any $u, \bar{u} \in \mathcal{U}$, there exist $\tilde{u} \in \mathcal{U}$ and a positive process θ such that $\Delta u := \tilde{u} - u$ is bounded, $\Delta u = \theta(\bar{u} - u)$, and $u^\varepsilon := (1-\varepsilon)u + \varepsilon\tilde{u} \in \mathcal{U}$ for all $\varepsilon \in [0,1]$.

(iii) For any u and u^ε as in (ii) above, $f(t,0,0,u^\varepsilon)$ and $\partial_u f(t,0,0,u^\varepsilon)$ are square integrable uniformly in $\varepsilon \in [0,1]$ in the sense of (10.6). That is,

$$\lim_{R\to\infty} \sup_{\varepsilon\in[0,1]} E\left\{\left(\int_0^T [|f(t,0,0,u_t^\varepsilon)|\mathbf{1}_{\{|f(t,0,0,u_t^\varepsilon)|>R\}}\right.\right.$$
$$\left.\left. + |\partial_u f(t,0,0,u_t^\varepsilon)|\mathbf{1}_{\{|\partial_u f(t,0,0,u_t^\varepsilon)|>R\}}\right]dt\right)^2\right\} = 0.$$

Let $u, \bar{u}, \tilde{u}, \Delta u, u^\varepsilon$ be as in Assumption 10.1.7(ii). Define

$$\Delta Y^\varepsilon := Y^{u^\varepsilon} - Y^u, \quad \Delta Z^\varepsilon := Z^{u^\varepsilon} - Z^u; \quad \nabla Y^\varepsilon := \frac{1}{\varepsilon}\Delta X^\varepsilon, \quad (10.7)$$
$$\nabla Z^\varepsilon := \frac{1}{\varepsilon}\Delta X^\varepsilon,$$

and let $(\nabla Y, \nabla Z) := (\nabla Y^{u,\Delta u}, \nabla Z^{u,\Delta u})$ be the solution to the following linear BSDE:

$$\nabla Y_t = \int_t^T [\partial_y f(s)\nabla Y_s + \partial_z f(s)\nabla Z_s + \partial_u f(s)\Delta u_s]ds - \int_t^T \nabla Z_s dB_s, \quad (10.8)$$

where $\varphi(s) := \varphi(s, Y_s^u, Z_s^u, u_s)$ for an appropriate random field φ.

Lemma 10.1.8 *Assume Assumptions* 10.1.6 *and* 10.1.7 *hold. Then,*

$$\lim_{\varepsilon\to 0} E\left\{\sup_{0\le t\le T} |\Delta Y_t^\varepsilon|^2 + \int_0^T |\Delta Z_t^\varepsilon|^2 dt\right\} = 0; \quad (10.9)$$

$$\lim_{\varepsilon\to 0} E\left\{\sup_{0\le t\le T} |\nabla Y_t^\varepsilon - \nabla Y_t|^2 + \int_0^T |\nabla Z_t^\varepsilon - \nabla Z_t|^2 dt\right\} = 0. \quad (10.10)$$

Proof First, since f is continuous in u, we have $f(t,0,0,u_t^\varepsilon) \to f(t,0,0,u_t)$. Then, Assumption 10.1.7(iii) implies that

$$\lim_{\varepsilon\to 0} E\left\{\left(\int_0^T |f(t,0,0,u_t^\varepsilon) - f(t,0,0,u_t)|dt\right)^2\right\} = 0.$$

Applying Theorem 9.4.3 we obtain (10.9).
Next, by Assumptions 10.1.6 and 10.1.7,

$$E\left\{\left(\int_0^T |\partial_u f(t, Y_t^u, Z_t^u, u_t)\Delta u_t| dt\right)^2\right\} < \infty.$$

Then, it is clear that (10.8) is well-posed. One can easily check that

$$\nabla Y_t^\varepsilon = \int_t^T [\partial_y f^\varepsilon(s) \nabla Y_s^\varepsilon + \partial_z f^\varepsilon(s) \nabla Z_s^\varepsilon + \partial_u f^\varepsilon(s) \Delta u_s] ds - \int_t^T \nabla Z_s^\varepsilon dB_s,$$
(10.11)

where

$$\partial_y f^\varepsilon(s) := \int_0^1 \partial_y f(s, Y_s^u + \theta \Delta Y_s^\varepsilon, Z_s^u + \theta \Delta Z_s^\varepsilon, u_s + \theta \varepsilon \Delta u_s) d\theta,$$

and similarly for $\partial_z f^\varepsilon(s), \partial_u f^\varepsilon(s)$. By (10.9), we have

$$\lim_{\varepsilon \to 0} \partial_y f(s, Y_s^u + \theta \Delta Y_s^\varepsilon, Z_s^u + \theta \Delta Z_s^\varepsilon, u_s + \theta \varepsilon \Delta u_s) = \partial_y f(s, Y_s^u, Z_s^u, u_s), \quad \text{a.s.}$$

Since $\partial_y f, \partial_z f$ are bounded, applying the Dominated Convergence Theorem we get

$$\lim_{\varepsilon \to 0} \partial_y f^\varepsilon(s) = \partial_y f(s), \quad \lim_{\varepsilon \to 0} \partial_z f^\varepsilon(s) = \partial_z f(s), \quad \text{a.s.}$$

Moreover,

$$\lim_{\varepsilon \to 0} \partial_u f(s, Y_s^u + \theta \Delta Y_s^\varepsilon, Z_s^u + \theta \Delta Z_s^\varepsilon, u_s + \theta \varepsilon \Delta u_s) = \partial_u f(s, Y_s^u, Z_s^u, u_s), \quad \text{a.s.}$$

and

$$|\partial_u f(s, Y_s^u + \theta \Delta Y_s^\varepsilon, Z_s^u + \theta \Delta Z_s^\varepsilon, u_s + \theta \varepsilon \Delta u_s)|$$
$$\leq C[1 + |\partial_u f(s, 0, 0, u_s^{\theta \varepsilon})| + |Y_s^u| + |\Delta Y_s^\varepsilon| + |Z_s^u| + |\Delta Z_s^\varepsilon|].$$

Then, by Assumption 10.1.7(iii) we see that

$$\lim_{\varepsilon \to 0} E\left\{\left(\int_0^T |\partial_u f^\varepsilon(t) - \partial_u f(t, Y_t^u, Z_t^u, u_t)||\Delta u_t| dt\right)^2\right\} = 0.$$

Finally, applying Theorem 9.4.3 we prove (10.10). □

We next solve linear BSDE (10.8) by introducing an adjoint process Γ^u:

$$\Gamma_t^u = 1 + \int_0^t \partial_y f(s) \Gamma_s^u ds + \int_0^t \partial_z f(s) \Gamma_s^u dB_s. \qquad (10.12)$$

We emphasize that Γ^u depends only on u, not on Δu. Applying Proposition 9.3.6, we have

Lemma 10.1.9 *Assume Assumptions 10.1.6 and 10.1.7 hold. Then,*

$$\nabla Y_0 = E\left\{\int_0^T \Gamma_t^u \partial_u f(t, Y_t^u, Z_t^u, u_t) \Delta u_t dt\right\}. \qquad (10.13)$$

We now state the main result of this section, which characterizes the necessary condition for the optimal control.

10.1 Stochastic Control of BSDEs

Theorem 10.1.10 *Assume Assumptions 10.1.6 and 10.1.7 hold. If $u^* \in \mathcal{U}$ is an optimal control for the optimization problem (10.2), and u^* is an interior point of $\mathcal{U}$, then*

$$\partial_u f\left(t, Y_t^{u^*}, Z_t^{u^*}, u_t^*\right) = 0. \tag{10.14}$$

Proof Since u^* is an interior point of $\mathcal{U}$, there exists $\tilde{u} \in \mathcal{U}$ such that $\Delta u := \tilde{u} - u$ is bounded and $\Delta u_t = \theta_t \text{sign}(\partial_u f(t, Y_t^{u^*}, Z_t^{u^*}, u_t^*))$ for some positive process θ. Note that $\Gamma^u > 0$. It follows from Lemma 10.1.9 that

$$\nabla Y_0 = E\left\{\int_0^T \Gamma_t^{u^*} \partial_u f\left(t, Y_t^{u^*}, Z_t^{u^*}, u_t^*\right) \Delta u_t dt\right\}$$

$$= E\left\{\int_0^T \Gamma_t^{u^*} \theta_t \left|\partial_u f\left(t, Y_t^{u^*}, Z_t^{u^*}, u_t^*\right)\right| dt\right\} \geq 0.$$

Since u^* is optimal, then $\nabla Y_0^\varepsilon \leq 0$ for all $\varepsilon \in [0, 1]$, and thus by (10.10) we have $\nabla Y_0 \leq 0$. This implies that $\nabla Y_0 = 0$ and therefore, (10.14) holds. $\square$

Remark 10.1.11 (i) When u^* is an interior point of U, clearly the sufficient condition (10.4) implies the necessary condition (10.14).

(ii) The necessary condition (10.14) is called the first order condition. One can do an extension to the second order condition, see e.g. Yong and Zhou (1999).

We next characterize the necessary condition (10.14) via a BSDE.

Proposition 10.1.12 *Assume all the conditions in Theorem 10.1.10 hold. Assume further that there exists a unique function $I(t, y, z)$ such that I is differentiable in (y, z) and*

$$\partial_u f\left(t, y, z, I(t, y, z)\right) = 0. \tag{10.15}$$

Then, the following BSDE is well-posed:

$$Y_t^* = \xi + \int_t^T f^*\left(s, Y_s^*, Z_s^*\right) ds - \int_t^T Z_s^* dB_s$$

$$\text{where } f^*(t, y, z) := f\left(t, y, z, I(t, y, z)\right), \tag{10.16}$$

and

$$V = Y_0^* \quad \text{and} \quad u_t^* = I\left(t, Y_t^*, Z_t^*\right). \tag{10.17}$$

Proof First, by (10.14) we have $u_t^* = I(t, Y_t^{u^*}, Z_t^{u^*})$. It is then straightforward to check that $(Y^*, Z^*) := (Y^{u^*}, Z^{u^*})$ is a solution to BSDE (10.16), and thus $V = Y_0^{u^*} = Y_0^*$.

On the other hand, by (10.15) and the (new) definition of f^* in (10.16), we have

$$\partial_y f^*(t, y, z) = \partial_y f(t, y, z, I(t, y, z)) + \partial_u f(t, y, z, I(t, y, z))\partial_y I(t, y, z)$$
$$= \partial_y f(t, y, z, I(t, y, z));$$
$$\partial_z f^*(t, y, z) = \partial_z f(t, y, z, I(t, y, z)) + \partial_u f(t, y, z, I(t, y, z))\partial_z I(t, y, z)$$
$$= \partial_z f(t, y, z, I(t, y, z)).$$

Thus, f^* is uniformly Lipschitz continuous in (y, z) and BSDE (10.16) has at most one solution. □

10.2 Stochastic Control of FBSDEs

In this section we extend the results of Sect. 10.1 to the following case:

$$V := \sup_{u \in \mathcal{U}} Y_0^u, \tag{10.18}$$

where (X^u, Y^u, Z^u) solve the following decoupled FBSDE:

$$\begin{cases} X_t^u = x + \int_0^t b(s, X_s^u, u_s)ds + \int_0^t \sigma(s, X_s^u, u_s)dB_s; \\ Y_t^u = g(X_T^u) + \int_t^T f(s, X_s^u, Y_s^u, Z_s^u, u_s)ds - \int_t^T Z_s^u dB_s. \end{cases} \tag{10.19}$$

We note that, by considering high-dimensional u, we allow the state process X to have different controls in the drift coefficient b and in the diffusion coefficient σ. However, for notational simplicity we again assume all processes are one-dimensional.

There is no general comparison principle for FBSDEs, so we are not able to obtain a simple sufficient condition as in Theorem 10.1.4 for the optimization problem (10.18). We start with necessary conditions.

Assumption 10.2.1 (i) b, σ, f, g are progressively measurable in all the variables, g is $\mathcal{F}_T$-measurable, and b, σ, f are $\mathbb{F}$-adapted.

(ii) b, σ, f, g are continuously differentiable in (x, y, z) with uniformly bounded derivatives.

(iii) b, σ, f are continuously differentiable in u, and for $\varphi = b, \sigma, f$,

$$|\varphi_u(t, x, y, z, u)| \leq C[1 + |\varphi_u(t, 0, 0, 0, u)| + |x| + |y| + |z|].$$

The next assumption is on the admissible set of controls.

Assumption 10.2.2 (i) $g(0) \in L^2(\mathcal{F}_T)$ and, for each $u \in \mathcal{U}$ and $\varphi = b, f$,

$$E\left\{\left(\int_0^T [|\varphi| + |\partial_u \varphi|](t, 0, 0, 0, u_t)dt\right)^2 + \int_0^T [|\sigma|^2 + |\partial_u \sigma|^2](t, 0, u_t)dt\right\} < \infty. \tag{10.20}$$

10.2 Stochastic Control of FBSDEs

(ii) $\mathcal{U}$ is locally convex, in the sense of Assumption 10.1.7(ii).

(iii) For any u and u^ε as in Assumption 10.1.7(ii) and $\varphi = b, \sigma, f$, we have that $\varphi(t, 0, 0, 0, u^\varepsilon)$ and $\partial_u \varphi(t, 0, 0, 0, u^\varepsilon)$ are uniformly integrable in $\varepsilon \in [0, 1]$ in the sense of (10.20).

Given $u \in \mathcal{U}$ and Δu bounded as in Assumption 10.2.2(ii), let $(\nabla X, \nabla Y, \nabla Z)$ solve the following decoupled linear FBSDE:

$$\begin{cases} \nabla X_t = \int_0^t \big[b_x(s)\nabla X_s + b_u(s)\Delta u_s\big] ds + \int_0^t \big[\sigma_x(s)\nabla X_s + \sigma_u(s)\Delta u_s\big] dB_s; \\ \nabla Y_t = g_x(T)\nabla X_T - \int_t^T \nabla Z_s dB_s \\ + \int_t^T \big[f_x(s)\nabla X_s + f_y(s)\nabla Y_s + f_z(s)\nabla Z_s + f_u(s)\Delta u_s\big] ds, \end{cases} \quad (10.21)$$

where $\varphi(s) := \varphi(s, X_s^u, Y_s^u, Z_s^u, u_s)$.

Under our assumptions, it is straightforward to check that (10.19) and (10.21) are well-posed. Moreover, following the arguments in Lemma 10.1.8, one can easily show that

Lemma 10.2.3 *Assume Assumptions 10.2.1 and 10.2.2 hold. Then,*

$$\lim_{\varepsilon \to 0} E\left\{\sup_{0 \le t \le T}\big[|\nabla X_t^\varepsilon - \nabla X_t^u|^2 + |\nabla Y_t^\varepsilon - \nabla Y_t^u|^2\big] + \int_0^T |\nabla Z_t^\varepsilon - \nabla Z_t^u|^2 dt\right\} = 0, \quad (10.22)$$

where

$$\nabla X^\varepsilon := \frac{1}{\varepsilon}\big[X^{u^\varepsilon} - X^u\big], \qquad \nabla Y^\varepsilon := \frac{1}{\varepsilon}\big[Y^{u^\varepsilon} - Y^u\big], \qquad \nabla Z^\varepsilon := \frac{1}{\varepsilon}\big[Z^{u^\varepsilon} - Z^u\big].$$

In order to solve linear FBSDE (10.21), we introduce the following adjoint processes:

$$\Gamma_t^u = 1 + \int_0^t \alpha_s ds + \int_0^t \beta_s dB_s;$$

$$\bar{Y}_t^u = g_x(X_T^u)\Gamma_T^u + \int_t^T \gamma_s ds - \int_t^T \bar{Z}_s^u dB_s,$$

where α, β, γ will be determined later. Applying Itô's rule, we have

$$d\big(\Gamma_t^u \nabla Y_t^u - \bar{Y}_t^u \nabla X_t^u\big)$$
$$= [\cdots] dB_t + \Big[-\Gamma_t^u\big[f_x(t)\nabla X_t + f_y(t)\nabla Y_t + f_z(t)\nabla Z_t + f_u(t)\Delta u_t\big] + \alpha_t \nabla Y_t$$
$$\quad + \beta_t \nabla Z_t - \bar{Y}_t^u\big[b_x(t)\nabla X_t + b_u(t)\Delta u_t\big] + \gamma_t \nabla X_t$$
$$\quad - \bar{Z}_t^u\big[\sigma_x(t)\nabla X_t + \sigma_u(t)\Delta u_t\big]\Big] dt$$
$$= [\cdots] dB_t + \Big[\nabla X_t\big[-\Gamma_t^u f_x(t) - \bar{Y}_t^u b_x(t) + \gamma_t - \bar{Z}_t^u \sigma_x(t)\big]$$

$$+ \nabla Y_t \left[-\Gamma_t^u f_y(t) + \alpha_t \right] + \nabla Z_t \left[-\Gamma_t^u f_z(t) + \beta_t \right]$$
$$- \Delta u_t \left[\Gamma_t^u f_u(t) + \bar{Y}_t^u b_u(t) + \bar{Z}_t^u \sigma_u(t) \right] \Big] dt. \tag{10.23}$$

Set
$$-\Gamma_t^u f_x(t) - \bar{Y}_t^u b_x(t) + \gamma_t - \bar{Z}_t^u \sigma_x(t) = -\Gamma_t^u f_y(t) + \alpha_t = -\Gamma_t^u f_z(t) + \beta_t = 0.$$

That is,
$$\alpha_t := f_y(t) \Gamma_t^u, \qquad \beta_t := f_z(t) \Gamma_t^u, \qquad \gamma_t := f_x(t) \Gamma_t^u + b_x(t) \bar{Y}_t^u + \sigma_x(t) \bar{Z}_t^u,$$
and thus
$$\Gamma_t^u = 1 + \int_0^t f_y(s, X_s^u, Y_s^u, Z_s^u, u_s) \Gamma_s^u ds + \int_0^t f_z(s, X_s^u, Y_s^u, Z_s^u, u_s) \Gamma_s^u d B_s;$$
$$\bar{Y}_t^u = g_x(X_T^u) \Gamma_T^u - \int_t^T \bar{Z}_s^u d B_s \tag{10.24}$$
$$+ \int_t^T \left[f_x(s, X_s^u, Y_s^u, Z_s^u, u_s) \Gamma_s^u + b_x(s, X_s^u, u_s) \bar{Y}_s^u + \sigma_x(s, X_s^u, u_s) \bar{Z}_s^u \right] ds.$$

We remark again that the adjoint processes $\Gamma^u, \bar{Y}^u, \bar{Z}^u$ depend only on u, but not on Δu. Then, we have

Lemma 10.2.4 *Assume Assumptions* 10.2.1 *and* 10.2.2 *hold. Then,*
$$\nabla Y_0^u = E \left\{ \int_0^T \left[\Gamma_t^u f_u(t, X_t^u, Y_t^u, Z_t^u, u_t) + \bar{Y}_t^u b_u(t, X_t^u, u_t) \right. \right.$$
$$\left. \left. + \bar{Z}_t^u \sigma_u(t, X_t^u, u_t) \right] \Delta u_t dt \right\}. \tag{10.25}$$

Proof By (10.23) we have
$$d \left[\Gamma_t^u \nabla Y_t^u - \bar{Y}_t^u \nabla X_t^u \right] = [\cdots] d B_t - \left[\Gamma_t^u f_u(t) + \bar{Y}_t^u b_u(t) + \bar{Z}_t^u \sigma_u(t) \right] \Delta u_t dt.$$
By standard estimates we know that the term $[\cdots] d B_t$ corresponds to a true martingale. Moreover,
$$\Gamma_0^u \nabla Y_0 - \bar{Y}_0^u \nabla X_0 = \nabla Y_0, \qquad \Gamma_T^u \nabla Y_T - \bar{Y}_T^u \nabla X_T = 0.$$
Thus, we obtain (10.25) immediately. □

Combining the arguments of Theorems 10.1.10 and 10.1.12, Lemma 10.2.4 implies our main result as follows:

Theorem 10.2.5 *Assume Assumptions* 10.2.1 *and* 10.2.2 *hold.*
(i) *If $u^* \in \mathcal{U}$ is an optimal control for the optimization problem* (10.18) *and u^* is an interior point of $\mathcal{U}$, then*
$$\Gamma_t^{u^*} f_u(t, X_t^{u^*}, Y_t^{u^*}, Z_t^{u^*}, u_t^*) + \bar{Y}_t^{u^*} b_u(t, X_t^{u^*}, u_t^*) + \bar{Z}_t^{u^*} \sigma_u(t, X_t^{u^*}, u_t^*) = 0. \tag{10.26}$$

10.2 Stochastic Control of FBSDEs

(ii) *Assume further that there exists a unique function $I(t, x, y, z, \gamma, \bar{y}, \bar{z})$, differentiable in (y, z, γ) and such that*

$$\gamma f_u(t, x, y, z, I) + \bar{y} b_u(t, x, I) + \bar{z} \sigma_u(t, x, I) = 0. \quad (10.27)$$

Denote

$$\varphi^*(t, x, y, z, \gamma, \bar{y}, \bar{z})$$
$$:= \varphi(t, x, y, z, \gamma, \bar{y}, \bar{z}, I(t, x, y, z, \gamma, \bar{y}, \bar{z})) \quad \text{for any function } \varphi. \quad (10.28)$$

Then, $(X^*, \Gamma^*, Y^*, \bar{Y}^*, Z^*, \bar{Z}^*) := (X^{u^*}, \Gamma^{u^*}, Y^{u^*}, \bar{Y}^{u^*}, Z^{u^*}, \bar{Z}^{u^*})$ satisfies the following coupled FBSDE:

$$\begin{cases} X_t^* = x + \int_0^t b^*(s, X_s^*, \Gamma_s^*, Y_s^*, \bar{Y}_s^*, Z_s^*, \bar{Z}_s^*) ds \\ \qquad + \int_0^t \sigma^*(s, X_s^*, \Gamma_s^*, Y_s^*, \bar{Y}_s^*, Z_s^*, \bar{Z}_s^*) dB_s; \\ \Gamma_t^* = 1 + \int_0^t (\partial_y f)^*(s, X_s^*, \Gamma_s^*, Y_s^*, \bar{Y}_s^*, Z_s^*, \bar{Z}_s^*) \Gamma_s^* ds \\ \qquad + \int_0^t (\partial_z f)^*(s, X_s^*, \Gamma_s^*, Y_s^*, \bar{Y}_s^*, Z_s^*, \bar{Z}_s^*) \Gamma_s^* dB_s; \\ Y_t^* = g(X_T^*) + \int_t^T f^*(s, X_s^*, \Gamma_s^*, Y_s^*, \bar{Y}_s^*, Z_s^*, \bar{Z}_s^*) ds - \int_t^T Z_s^* dB_s; \\ \bar{Y}_s^* = g_x(X_T^*) \Gamma_T^* - \int_t^T \bar{Z}_s^* dB_s + \int_t^T \big[(\partial_x f)^*(s, X_s^*, \Gamma_s^*, Y_s^*, \bar{Y}_s^*, Z_s^*, \bar{Z}_s^*) \Gamma_s^* \\ \qquad + (\partial_x b)^*(s, X_s^*, \Gamma_s^*, Y_s^*, \bar{Y}_s^*, Z_s^*, \bar{Z}_s^*) \bar{Y}_s^* \\ \qquad + (\partial_x \sigma)^*(s, X_s^*, \Gamma_s^*, Y_s^*, \bar{Y}_s^*, Z_s^*, \bar{Z}_s^*) \bar{Z}_s^* \big] ds, \end{cases} \quad (10.29)$$

and the optimal control satisfies

$$u_t^* = I(t, X_t^*, \Gamma_t^*, Y_t^*, \bar{Y}_t^*, Z_t^*, \bar{Z}_t^*). \quad (10.30)$$

Remark 10.2.6 (i) Note that

$$\varphi^*(s, X_s^*, \Gamma_s^*, Y_s^*, \bar{Y}_s^*, Z_s^*, \bar{Z}_s^*) = \varphi(s, X_s^*, \Gamma_s^*, Y_s^*, \bar{Y}_s^*, Z_s^*, \bar{Z}_s^*, u_s^*)$$

thanks to (10.28) and (10.30).

(ii) We can, in fact, remove adjoint process Γ in Theorem 10.2.5. Indeed, denote

$$\hat{Y}_t := \bar{Y}_t \Gamma_t^{-1}, \qquad \hat{Z}_t := [\bar{Z}_t - \bar{Y}_t f_z] \Gamma_t^{-1}.$$

Note that $\Gamma > 0$. Then, (10.26) becomes

$$\big[f_u(\cdot) + \hat{Y}_t^{u^*} [b_u(\cdot) + f_z(\cdot) \sigma_u(\cdot)] + \hat{Z}_t^{u^*} \sigma_u(\cdot) \big] (t, X_t^{u^*}, Y_t^{u^*}, Z_t^{u^*}, u_t^*) = 0.$$

Let $\hat{I}(t, x, y, z, \hat{y}, \hat{z})$ be a function differentiable in (y, z) and such that

$$\big[f_u(\cdot) + \hat{y} [b_u(\cdot) + f_z(\cdot) \sigma_u(\cdot)] + \hat{z} \sigma_u(\cdot) \big] (t, x, y, z, \hat{y}, \hat{z}, \hat{I}) = 0.$$

Denote

$$\varphi^*(t,x,y,z,\hat{y},\hat{z}) := \varphi\big(t,x,y,z,\hat{y},\hat{z},\hat{I}(t,x,y,z,\hat{y},\hat{z})\big) \quad \text{for any function } \varphi.$$

Then, $(X^*, Y^*, \hat{Y}^*, Z^*, \hat{Z}^*) := (X^{u^*}, Y^{u^*}, \hat{Y}^{u^*}, Z^{u^*}, \hat{Z}^{u^*})$ satisfies the following coupled FBSDE:

$$\begin{cases} X_t^* = x + \int_0^t b^*\big(s, X_s^*, Y_s^*, \hat{Y}_s^*, Z_s^*, \hat{Z}_s^*\big)ds + \int_0^t \sigma^*\big(s, X_s^*, Y_s^*, \hat{Y}_s^*, Z_s^*, \hat{Z}_s^*\big)dB_s; \\ Y_t^* = g(X_T^*) + \int_t^T f^*\big(s, X_s^*, Y_s^*, \hat{Y}_s^*, Z_s^*, \hat{Z}_s^*\big)ds - \int_t^T Z_s^* dB_s; \\ \hat{Y}_s^* = g_x(X_T^*) - \int_t^T \hat{Z}_s^* dB_s \\ \qquad + \int_t^T \big[\hat{Y}_s^*[(b_x)^* + (f_y)^* + (\sigma_x f_z)^*](\cdot) + \hat{Z}_s^*[(\sigma_x)^* + (f_z)^*](\cdot) \\ \qquad + (f_x)^*(\cdot)\big]\big(s, X_s^*, Y_s^*, \hat{Y}_s^*, Z_s^*, \hat{Z}_s^*\big)ds, \end{cases} \quad (10.31)$$

and the optimal control satisfies

$$u_t^* = I\big(t, X_t^*, Y_t^*, \hat{Y}_t^*, Z_t^*, \hat{Z}_t^*\big).$$

Notice that FBSDE (10.31) does not involve Γ, and is simpler than FBSDE (10.29). However, for the sufficient conditions below, it is more convenient to state the results in terms of adjoint process Γ. We therefore still use the system (10.29).

We now turn to the sufficient conditions. In light of the necessary condition (10.26), we first introduce the following Hamiltonian function:

$$H(t,x,y,z,\gamma,\bar{y},\bar{z},u) := \gamma f(t,x,y,z,u) + \bar{y}b(t,x,u) + \bar{z}\sigma(t,x,u). \quad (10.32)$$

Assumption 10.2.7 (i) The terminal condition g is $\mathcal{F}_T$-measurable, uniformly Lipschitz continuous in x, concave in x, and $E\{|g(0)|^2\} < \infty$;

(ii) For $\varphi = b, \sigma, f$, function φ is progressively measurable and $\mathbb{F}$-adapted, continuously differentiable in (x,y,z) with uniformly bounded derivatives, and both φ and the derivative of φ are continuous in u.

(iii) Set U is convex and admissible set $\mathcal{U}$ is a set of $\mathbb{F}$-adapted processes u taking values in U satisfying:

$$E\left\{\left(\int_0^T [|b|+|f|](t,0,0,0,u_t)dt\right)^2 + \int_0^T |\sigma(t,0,u_t)|^2 dt\right\} < \infty.$$

(iv) The Hamiltonian H is concave in (x,y,z,u) for all $(\gamma,\bar{y},\bar{z})$ in the set of all possible values the adjoint processes $(\Gamma, \bar{Y}, \bar{Z})$ could take, and there exists a function $I(t,x,y,z,\gamma,\bar{y},\bar{z})$ taking values in U such that

$$H\big(t,x,y,z,\gamma,\bar{y},\bar{z},I(t,x,y,z,\gamma,\bar{y},\bar{z})\big) = \sup_{u \in U} H(t,x,y,z,\gamma,\bar{y},\bar{z},u). \quad (10.33)$$

10.2 Stochastic Control of FBSDEs

Remark 10.2.8 The sign of the adjoint processes $(\Gamma, \bar{Y}, \bar{Z})$ is important to ensure the concavity of H, which is the key condition for the sufficiency theorem below. Process Γ is always positive, so we only consider positive γ. Under certain conditions, one can show that $\bar{Y}$ keeps the same sign. In most cases, it is not easy to guarantee that $\bar{Z}$ will preserve the same sign.

Here is one case for which we do have that H is concave for all the possible values of $(\Gamma, \bar{Y}, \bar{Z})$:

(i) f is concave in (x, y, z, u);
(ii) f and g are increasing (resp. decreasing) in x and b is concave (resp. convex) in (x, u);
(iii) σ is linear in (x, u).

We note that condition (ii) implies that $\bar{Y}$ is positive (resp. negative). An alternative to (ii) is to have b also linear in (x, u), in which case we do not need to discuss the values of $\bar{Y}$.

Theorem 10.2.9 *Assume*

(i) *Assumption 10.2.7 holds*;
(ii) *FBSDE* (10.29) *has a solution* $(X^*, \Gamma^*, Y^*, \bar{Y}^*, Z^*, \bar{Z}^*)$, *where* φ^* *is defined by* (10.28) *for the function I in Assumption* 10.2.7(iv);
(iii) *process u^* defined by* (10.30) *is in* $\mathcal{U}$.

Then, $V = Y_0^*$, *and u^* is an optimal control.*

Remark 10.2.10 (i) When I takes values in the interior of U, obviously (10.33) implies (10.27).

(ii) In this theorem, we require neither uniqueness nor differentiability of functions I, neither do we require uniqueness of FBSDE (10.29).

Proof Under Assumption 10.2.7, FBSDE (10.19) is well-posed for each $u \in \mathcal{U}$. Now, for arbitrary $u \in \mathcal{U}$, applying Itô's rule we have

$$\begin{aligned}
&d\bigl(\Gamma_t^* Y_t^u - \bar{Y}_t^* X_t^u\bigr) \\
&= \bigl[-\Gamma_t^* f(t, X_t^u, Y_t^u, Z_t^u, u_t) + \partial_y f(t, X_t^*, Y_t^*, Z_t^*, u_t^*)\Gamma_t^* Y_t^u \\
&\quad + \partial_z f(t, X_t^*, Y_t^*, Z_t^*, u_t^*)\Gamma_t^* Z_t^u - \bar{Y}_t^* b(t, X_t^u, u_t) \\
&\quad + \bigl[\partial_x f(t, X_t^*, Y_t^*, Z_t^*, u_t^*)\Gamma_t^* + \partial_x b(t, X_t^*, u_t^*)\bar{Y}_t^* + \partial_x \sigma(t, X_t^*, u_t^*)\bar{Z}_t^*\bigr]X_t^u \\
&\quad - \bar{Z}_t^* \sigma(t, X_t^u, u_t)\bigr]dt + [\cdots]dB_t \\
&= -H\bigl(t, X_t^u, \Gamma_t^*, Y_t^u, \bar{Y}_t^*, Z_t^u, \bar{Z}_t^*, u_t\bigr)dt + [\cdots]dB_t \\
&\quad + \bigl[H_x\bigl(t, X_t^*, \Gamma_t^*, Y_t^*, \bar{Y}_t^*, Z_t^*, \bar{Z}_t^*, u_t^*\bigr)X_t^u \\
&\quad + H_y\bigl(t, X_t^*, \Gamma_t^*, Y_t^*, \bar{Y}_t^*, Z_t^*, \bar{Z}_t^*, u_t^*\bigr)Y_t^u \\
&\quad + H_z\bigl(t, X_t^*, \Gamma_t^*, Y_t^*, \bar{Y}_t^*, Z_t^*, \bar{Z}_t^*, u_t^*\bigr)Z_t^u\bigr]dt.
\end{aligned}$$

Denote

$$\Delta X_t^u := X_t^u - X_t^*, \qquad \Delta Y_t^u := Y_t^u - Y_t^*, \qquad \Delta Z_t^u := Z_t^u - Z_t^*.$$

Then,

$$\begin{aligned}
d\big(\Gamma_t^* \Delta Y_t^u &- \bar{Y}_t^* \Delta X_t^u\big) \\
&= -\big[H\big(t, X_t^u, \Gamma_t^*, Y_t^u, \bar{Y}_t^*, Z_t^u, \bar{Z}_t^*, u_t\big) - H\big(t, X_t^*, \Gamma_t^*, Y_t^*, \bar{Y}_t^*, Z_t^*, \bar{Z}_t^*, u_t^*\big)\big]dt \\
&\quad + [\cdots]dB_t + \big[H_x\big(t, X_t^*, \Gamma_t^*, Y_t^*, \bar{Y}_t^*, Z_t^*, \bar{Z}_t^*, u_t^*\big)\Delta X_t^u \\
&\quad + H_y\big(t, X_t^*, \Gamma_t^*, Y_t^*, \bar{Y}_t^*, Z_t^*, \bar{Z}_t^*, u_t^*\big)\Delta Y_t^u \\
&\quad + H_z\big(t, X_t^*, \Gamma_t^*, Y_t^*, \bar{Y}_t^*, Z_t^*, \bar{Z}_t^*, u_t^*\big)\Delta Z_t^u\big]dt.
\end{aligned}$$

Noting that $\Gamma_0^* = 1$ and $\Delta X_0^u = 0$ we have

$$\begin{aligned}
\Delta Y_0^u = \Gamma_0^* \Delta Y_0^u - \bar{Y}_0^* \Delta X_0^u &= E\bigg\{\Gamma_T^* \Delta Y_T^u - \bar{Y}_T^* \Delta X_T^u \\
&\quad + \int_0^T \big[H\big(t, X_t^u, \Gamma_t^*, Y_t^u, \bar{Y}_t^*, Z_t^u, \bar{Z}_t^*, u_t\big) \\
&\qquad - H\big(t, X_t^*, \Gamma_t^*, Y_t^*, \bar{Y}_t^*, Z_t^*, \bar{Z}_t^*, u_t^*\big)\big]dt \\
&\quad - \int_0^T \big[H_x\big(t, X_t^*, \Gamma_t^*, Y_t^*, \bar{Y}_t^*, Z_t^*, \bar{Z}_t^*, u_t^*\big)\Delta X_t^u \\
&\qquad + H_y\big(t, X_t^*, \Gamma_t^*, Y_t^*, \bar{Y}_t^*, Z_t^*, \bar{Z}_t^*, u_t^*\big)\Delta Y_t^u \\
&\qquad + H_z\big(t, X_t^*, \Gamma_t^*, Y_t^*, \bar{Y}_t^*, Z_t^*, \bar{Z}_t^*, u_t^*\big)\Delta Z_t^u\big]dt\bigg\}.
\end{aligned}$$

Since g is concave and $\Gamma_T^* > 0$, we have

$$\begin{aligned}
\Gamma_T^* \Delta Y_T^u - \bar{Y}_T^* \Delta X_T^u &= \Gamma_T^*\big[g(X_T^u) - g(X_T^*)\big] - g_x(X_T^*)\Gamma_T^* \Delta X_T^u \\
&= \Gamma_T^*\big[g(X_T^u) - g(X_T^*) - g_x(X_T^*)\Delta X_T^u\big] \le 0.
\end{aligned}$$

Moreover, note that

$$H\big(t, X_t^*, \Gamma_t^*, Y_t^*, \bar{Y}_t^*, Z_t^*, \bar{Z}_t^*, u_t^*\big) = \sup_{u \in U} H\big(t, X_t^*, \Gamma_t^*, Y_t^*, \bar{Y}_t^*, Z_t^*, \bar{Z}_t^*, u\big).$$

Since H is concave in (x, y, z, u), we have

$$\begin{aligned}
H\big(t, X_t^u, \Gamma_t^*, Y_t^u, \bar{Y}_t^*, Z_t^u, \bar{Z}_t^*, u_t\big) &- H\big(t, X_t^*, \Gamma_t^*, Y_t^*, \bar{Y}_t^*, Z_t^*, \bar{Z}_t^*, u_t^*\big) \\
&\le H_x\big(t, X_t^*, \Gamma_t^*, Y_t^*, \bar{Y}_t^*, Z_t^*, \bar{Z}_t^*, u_t^*\big)\Delta X_t^u \\
&\quad + H_y\big(t, X_t^*, \Gamma_t^*, Y_t^*, \bar{Y}_t^*, Z_t^*, \bar{Z}_t^*, u_t^*\big)\Delta Y_t^u \\
&\quad + H_z\big(t, X_t^*, \Gamma_t^*, Y_t^*, \bar{Y}_t^*, Z_t^*, \bar{Z}_t^*, u_t^*\big)\Delta Z_t^u.
\end{aligned}$$

Therefore, $\Delta Y_0^u \le 0$ for all $u \in \mathcal{U}$. That is, u^* is optimal. $\square$

Remark 10.2.11 The well-posedness of FBSDEs will be studied in Chap. 11 below. However, we should point out that the well-posedness of general FBSDEs still remains a very challenging problem. For a system like (10.29), in general none of the methods introduced in Chap. 11 works.

10.3 Stochastic Control of High-Dimensional BSDEs

We conclude this subsection with one example which we can solve completely.

Example 10.2.12 Let

$$b = 0, \quad \sigma = u, \quad f = -\frac{1}{2}u^2, \quad g = -\frac{1}{4}x^2 1_{\{|x| \le 1\}} - \frac{1}{2}\left[|x| - \frac{1}{2}\right] 1_{\{|x| > 1\}}.$$

Assume $U = \mathbb{R}$ and let $\mathcal{U}$ be the set of all $\mathbb{F}$-adapted process u such that

$$E\left\{\left(\int_0^T |u_t|^2 dt\right)^2\right\} < \infty. \tag{10.34}$$

Then, $\Gamma_t^u = 1$, and thus

$$H = -\frac{1}{2}u^2 + \bar{z}u.$$

One can easily check that Assumption 10.2.7 holds. In this case, (10.33) leads to $I = \bar{z}$. Thus, FBSDE (10.29) becomes:

$$\begin{cases} X_t^* = x + \int_0^t \bar{Z}_s^* dB_s; \\ Y_t^* = g(X_T^*) - \int_t^T \frac{1}{2}\bar{Z}_s^{*2} ds - \int_t^T Z_s^* dB_s; \\ \bar{Y}_t^* = g_x(X_T^*) - \int_t^T \bar{Z}_s^* dB_s. \end{cases} \tag{10.35}$$

Note that g_x has a Lipschitz constant less than or equal to $\frac{1}{2}$. Then, the FBSDE for $(X^*, \bar{Y}^*, \bar{Z}^*)$ satisfies condition (11.4) below. Assume further that T is small enough. Then, applying Theorem 11.2.3 below, we see that there exists a unique triple $(X^*, \bar{Y}^*, \bar{Z}^*)$ satisfying (10.35). Moreover, since $|g_x| \le \frac{1}{2}$, by Proposition 9.4.4 we know

$$E\left\{\left(\int_0^T |\bar{Z}_t^*|^2 dt\right)^2\right\} < \infty.$$

This implies that the second equation in (10.35) admits a unique solution (Y^*, Z^*). Finally, $u_t^* = I = \bar{Z}_t^*$ satisfies (10.34), so $u^* \in \mathcal{U}$, and therefore, u^* is indeed the optimal control, which is unique by Theorem 10.2.5 and the uniqueness of solutions for FBSDE (10.35).

10.3 Stochastic Control of High-Dimensional BSDEs

Let A be a subset of $\mathbb{R}$ and a be an $\mathcal{F}_T$-measurable random variable taking values in A. In this subsection we extend the results of Sect. 10.1 to the case when the BSDE under control is associated with another BSDE:

$$V := \sup_{(a,u) \in \mathcal{A}} Y_0^{2,a,u}, \tag{10.36}$$

where $(Y^{a,u}, Z^{a,u}) := (Y^{1,a,u}, Y^{2,a,u}, Z^{1,a,u}, Z^{2,a,u})$ solves the following two-dimensional BSDE:

$$Y_t^{i,a,u} = g_i(a) + \int_t^T f_i(s, Y_s^{a,u}, Z_s^{a,u}, u_s)ds - \int_t^T Z_s^{i,a,u} dB_s, \quad i = 1, 2. \quad (10.37)$$

Such a framework is needed for the principal's problem in Chap. 5. We note that in Sect. 10.1 we did not introduce the additional control a. In fact, in that case if the terminal condition is $g(a)$ with control a, Comparison Theorem implies that the optimization problem is equivalent to one with the terminal condition $\xi := \sup_{a \in A} g(a)$ and thus the control a is redundant in that case.

We do not have a general comparison theorem for high-dimensional BSDEs, so we are not able to obtain a simple sufficient condition as in Theorem 10.1.4 for optimization problem (10.36). We start with the necessary conditions.

Assumption 10.3.1 (i) For $i = 1, 2$, g_i is continuously differentiable in a, and for each $a \in A$, $g_i(a)$ is $\mathcal{F}_T$-measurable.

(ii) For $i = 1, 2$, f_i is progressively measurable and $\mathbb{F}$-adapted; continuously differentiable in (y, z) with uniformly bounded derivatives; and continuously differentiable in u with

$$|\partial_u f_i(t, y, z, u)| \le C\big[1 + |\partial_u f_i(t, 0, 0, u)| + |y| + |z|\big].$$

Assumption 10.3.2 The admissible set $\mathcal{A}$ is a set of pairs (a, u), where a is an $\mathcal{F}_T$-measurable random variable taking values in A, and u is an $\mathbb{F}$-adapted process taking values in U, satisfying:

(i) For $i = 1, 2$, and each $(a, u) \in \mathcal{A}$,

$$E\left\{|g_i(a)|^2 + |g_i'(a)|^2 + \left(\int_0^T [|f_i| + |\partial_u f_i|](t, 0, 0, u_t)dt\right)^2\right\} < \infty. \quad (10.38)$$

(ii) $\mathcal{A}$ is locally convex, in the sense of Assumption 10.1.7(ii).
(iii) For any (a, u) and $(a^\varepsilon, u^\varepsilon)$ as in Assumption 10.1.7(ii), the random variables $g_i(a^\varepsilon)$, $g_i'(a^\varepsilon)$ and the processes $f_i(t, 0, 0, u_t^\varepsilon)$, $\partial_u f_i(t, 0, 0, u_t^\varepsilon)$ are integrable uniformly in $\varepsilon \in [0, 1]$ in the sense of (10.38).

Given $(a, u) \in \mathcal{A}$ and $(\Delta a, \Delta u)$ bounded as in Assumption 10.3.2(ii), for $i = 1, 2$, denote

$$\nabla Y_t^{i,a,u} = g_i'(a)\Delta a - \int_t^T \nabla Z_s^{i,a,u} dB_s + \int_t^T \partial_u f_i(s, Y_s^u, Z_s^u, u_s)\Delta u_s ds$$

$$+ \int_t^T \sum_{j=1}^2 [\partial_{y_j} f_i(s, Y_s^{a,u}, Z_s^{a,u}, u_s)\nabla Y_s^{j,a,u}$$

$$+ \partial_{z_j} f_i(s, Y_s^{a,u}, Z_s^{a,u}, u_s)\nabla Z_s^{j,a,u}]ds. \quad (10.39)$$

Under our assumptions, it is straightforward to check that (10.37) and (10.39) are well-posed.

10.3 Stochastic Control of High-Dimensional BSDEs

Now, denote

$$d\Gamma_t^{i,a,u} = \sum_{j=1}^{2}[\Gamma_t^{j,a,u}\partial_{y_i}f_j(t,Y_t^{a,u},Z_t^{a,u},u_t)dt$$
$$+ \Gamma_t^{j,a,u}\partial_{z_i}f_j(t,Y_t^{a,u},Z_t^{a,u},u_t)dB_t], \quad i=1,2,$$
with $\Gamma_0^{1,a,u} := 0$, $\Gamma_0^{2,a,u} := 1$. \hfill (10.40)

Applying Itô's rule, we have

$$d\left(\sum_{i=1}^{2}\Gamma_t^{i,a,u}\nabla Y_t^{i,a,u}\right)$$
$$= -\sum_{i=1}^{2}\Gamma_t^{i,a,u}\partial_u f_i(s,Y_s^{a,u},Z_s^{a,u},u_s)\Delta u_s ds$$
$$+ \left[\sum_{i=1}^{2}\Gamma_t^{i,a,u}\nabla Z_t^{i,a,u} + \sum_{i,j=1}^{2}\Gamma_t^{j,a,u}\partial_{z_i}f_j(s,Y_s^{a,u},Z_s^{a,u},u_s)\nabla Y_t^{i,a,u}\right]dB_t.$$

Extending the arguments of Lemmas 10.1.8 and 10.1.9 to the high-dimensional case, we obtain immediately

Lemma 10.3.3 *Assume Assumptions* 10.3.1 *and* 10.3.2 *hold. Then,*

$$\lim_{\varepsilon\to 0} E\left\{\sup_{0\le t\le T}\left|\frac{1}{\varepsilon}[Y_t^{a^\varepsilon,u^\varepsilon} - Y_t^{a,u}] - \nabla Y_t^{a,u}\right|^2\right.$$
$$\left. + \int_0^T\left|\frac{1}{\varepsilon}[Z_t^{a^\varepsilon,u^\varepsilon} - Z_t^{a,u}] - \nabla Z_t^{a,u}\right|^2 dt\right\} = 0, \hfill (10.41)$$

and

$$\nabla Y_0^{2,u} = E\left\{\sum_{i=1}^{2}\Gamma_T^{i,a,u}g_i'(u)\Delta u + \int_0^T\sum_{i=1}^{2}\Gamma_t^{i,a,u}\partial_u f_i(t,Y_t^{a,u},Z_t^{a,u},u_t)\Delta u_t dt\right\}. \hfill (10.42)$$

Combining the arguments of Theorems 10.1.10 and 10.1.12, Lemma 10.3.3 implies

Theorem 10.3.4 *Assume Assumptions* 10.3.1 *and* 10.3.2 *hold.*
(i) *If* $(a^*,u^*) \in \mathcal{A}$ *is an optimal control for optimization problem* (10.36) *and* (a^*,u^*) *is an interior point of* $\mathcal{A}$, *then*

$$\sum_{i=1}^{2}\Gamma_T^{i,a^*,u^*}g_i'(a^*) = 0, \quad \sum_{i=1}^{2}\Gamma_t^{i,a^*,u^*}\partial_u f_i(t,Y_t^{a^*,u^*},Z_t^{a^*,u^*},u_t^*) = 0. \hfill (10.43)$$

(ii) *Assume further that there exist unique functions $I_1(\gamma)$ and $I_2(t, y, z, \gamma)$ differentiable in (y, z, γ), and such that*

$$\sum_{i=1}^{2} \gamma_i g_i'(I_1(\gamma)) = 0 \quad \text{and} \quad \sum_{i=1}^{2} \gamma_i \partial_u f_i(t, y, z, I_2(t, y, z, \gamma)) = 0. \quad (10.44)$$

Denote, for $i = 1, 2$,

$$\begin{aligned} g_i^*(\gamma) &:= g_i(I_1(\gamma)), \\ \varphi^*(t, y, z, \gamma) &:= \varphi(t, y, z, I_2(t, y, z, \gamma)) \quad \text{for any function } \varphi. \end{aligned} \quad (10.45)$$

Then, $(\Gamma^*, Y^*, Z^*) := (\Gamma^{a^*, u^*}, Y^{a^*, u^*}, Z^{a^*, u^*})$ satisfies the following coupled FBSDE:

$$\begin{cases} d\Gamma_t^{i,*} = \sum_{j=1}^{2} \Gamma_t^{j,*} \left[(\partial_{y_i} f_j)^*(t, Y_t^*, Z_t^*, \Gamma_t^*) dt + (\partial_{z_i} f_j)^*(t, Y_t^*, Z_t^*, \Gamma_t^*) dB_t \right]; \\ dY_t^{i,*} = -f_i^*(t, Y_t^*, Z_t^*, \Gamma_t^*) dt + Z_t^{i,*} dB_t; \\ \Gamma_0^{1,*} = 0, \quad \Gamma_0^{2,*} = 1, \quad Y_T^{i,*} = g_i^*(\Gamma_T^*). \end{cases} \quad (10.46)$$

Moreover, the optimal control satisfies

$$a^* = I_1(\Gamma_T^*), \quad u_t^* = I_2(t, Y_t^*, Z_t^*, \Gamma_t^*). \quad (10.47)$$

The proof is straightforward. We remark that we have

$$\sum_{j=1}^{2} \gamma_j \partial_{y_i}(f_j^*)(t, y, z, \gamma)$$

$$= \sum_{j=1}^{2} \gamma_j \big[\partial_{y_i} f_j(t, y, z, I_2(t, y, z, \gamma))$$

$$+ \partial_u f_j(t, y, z, I_2(t, y, z, \gamma)) \partial_{y_i} I_2(t, y, z, \gamma) \big]$$

$$= \sum_{j=1}^{2} \gamma_j \partial_{y_i} f_j(t, y, z, I_2(t, y, z, \gamma)) = \sum_{j=1}^{2} \gamma_j (\partial_{y_i} f_j)^*(t, y, z, \gamma),$$

thanks to (10.44). Similarly,

$$\sum_{j=1}^{2} \gamma_j \partial_{z_i}(f_j^*)(t, y, z, \gamma) = \sum_{j=1}^{2} \gamma_j (\partial_{z_i} f_j)^*(t, y, z, \gamma).$$

When f_1 is independent of (y_2, z_2), $\Gamma^2 > 0$. In this case, as in Remark 10.2.6(ii), one can actually remove the adjoint process Γ^2. Indeed, denote $D := \Gamma^1(\Gamma^2)^{-1}$. Apply Itô's formula, we obtain

10.3 Stochastic Control of High-Dimensional BSDEs

$$D_t = \int_0^t \left[(-\partial_{y_2} f_2 + |\partial_{z_2} f_2|^2 + \partial_{y_1} f_1 - \partial_{z_1} f_1 \partial_{z_2} f_2)D_s + (\partial_{y_1} f_2 - \partial_{z_1} f_2 \partial_{z_2} f_2)\right] ds$$
$$+ \int_0^t \left[(-\partial_{z_2} f_2 + \partial_{z_1} f_1)D_s + \partial_{z_1} f_2\right] dB_s. \tag{10.48}$$

As a consequence of Theorem 10.3.4, we obtain

Theorem 10.3.5 *Assume Assumptions* 10.3.1 *and* 10.3.2 *hold, and* f_1 *is independent of* (y_2, z_2).

(i) *If* $(a^*, u^*) \in \mathcal{A}$ *is an optimal control for optimization problem* (10.36) *and* (a^*, u^*) *is an interior point of* $\mathcal{A}$, *then*

$$g_1'(a^*)D_T^{a^*,u^*} + g_2'(a^*) = 0,$$
$$\partial_u f_1(t, Y_t^{1,a^*,u^*}, Z_t^{1,a^*,u^*}, u_t^*)D_t^{a^*,u^*} + \partial_u f_2(t, Y_t^{a^*,u^*}, Z_t^{a^*,u^*}, u_t^*) = 0. \tag{10.49}$$

(ii) *Assume further that there exist unique functions* $I_1(D)$ *and* $I_2(t, y, z, D)$ *differentiable in* (y, z, D), *and such that*

$$g_1'(I_1(D))D + g_2'(I_1(D)) = 0; \tag{10.50}$$
$$\partial_u f_1(t, y_1, z_1, I_2(t, y, z, D))D + \partial_u f_2(t, y, z, I_2(t, y, z, D)) = 0.$$

Denote, for $i = 1, 2$,

$$g_i^*(D) := g_i(I_1(D)),$$
$$\varphi^*(t, y, z, D) := \varphi(t, y, z, I_2(t, y, z, D)) \quad \text{for any function } \varphi. \tag{10.51}$$

Then, $(D^*, Y^*, Z^*) := (D^{a^*,u^*}, Y^{a^*,u^*}, Z^{a^*,u^*})$ *satisfies the following coupled FBSDE:*

$$\begin{cases} dD_t^* = \left[(-\partial_{y_2} f_2 + |\partial_{z_2} f_2|^2 + \partial_{y_1} f_1 - \partial_{z_1} f_1 \partial_{z_2} f_2)^*(t, Y_t^*, Z_t^*, D_t^*)D_t^* \\ \quad + (\partial_{y_1} f_2 - \partial_{z_1} f_2 \partial_{z_2} f_2)^*(t, Y_t^*, Z_t^*, D_t^*)\right] dt \\ \quad + \left[(-\partial_{z_2} f_2 + \partial_{z_1} f_1)^*(t, Y_t^*, Z_t^*, D_t^*)D_t^* \\ \quad + (\partial_{z_1} f_2)^*(t, Y_t^*, Z_t^*, D_t^*)\right] dB_t; \\ dY_t^{i,*} = -f_i^*(t, Y_t^*, Z_t^*, D_t^*) dt + Z_t^{i,*} dB_t; \\ D_0^* = 0, \quad Y_T^{i,*} = g_i^*(D_T^*). \end{cases} \tag{10.52}$$

Moreover, the optimal control satisfies

$$a^* = I_1(D_T^*), \quad u_t^* = I_2(t, Y_t^*, Z_t^*, D_t^*). \tag{10.53}$$

We now turn to the sufficient conditions. As in Sect. 10.2, we introduce the following Hamiltonian functions:

$$H_1(\gamma, a) := \sum_{i=1}^2 \gamma_i g_i(a) \quad \text{and} \quad H_2(t, y, z, \gamma, u) := \sum_{i=1}^2 \gamma_i f_i(t, y, z, u). \tag{10.54}$$

Assumption 10.3.6 (i) For $i = 1, 2$, g_i is progressively measurable; f_i is progressively measurable and $\mathbb{F}$-adapted, continuously differentiable in (y, z) with uniformly bounded derivatives, and continuously differentiable in u.

(ii) The admissible set $\mathcal{A}$ is a set of pairs (a, u), where a is an $\mathcal{F}_T$-measurable random variable taking values in A and u is an $\mathbb{F}$-adapted processes taking values in U, satisfying: for $i = 1, 2$,

$$E\left\{|g_i(a)|^2 + \left(\int_0^T |f_i(t, 0, 0, u_t)| dt\right)^2\right\} < \infty.$$

(iii) The Hamiltonian H_2 is concave in (y, z, u), for all γ in the set of possible values the adjoint process Γ could take.

Theorem 10.3.7 *Assume*

(i) *Assumption 10.3.6 holds;*
(ii) *there exist functions $I_1(g)$, $I_2(t, y, z, \gamma)$ taking values in A and U, respectively, such that*

$$\begin{aligned} H_1(\gamma, I_1(\gamma)) &= \sup_{a \in A} H_1(\gamma, a) \quad \text{and} \\ H_2(t, y, z, \gamma, I_2(t, y, z, \gamma)) &= \sup_{u \in U} H_2(t, y, z, \gamma, u); \end{aligned} \quad (10.55)$$

(iii) *FBSDE (10.46) has a solution (Γ^*, Y^*, Z^*), where f^*, g^* are defined by (10.45) for functions I_1, I_2 in (ii) above, and the pair (a^*, u^*) defined by (10.47) is in $\mathcal{A}$.*

Then, $V = Y_0^*$ and (a^*, u^*) is an optimal control.

Proof The proof is similar to that of Theorem 10.2.9, so we only sketch it. For arbitrary $(a, u) \in \mathcal{A}$, applying Itô's rule we have

$$d\left(\sum_{i=1}^2 \Gamma_t^{i,*} Y_t^{i,a,u}\right)$$
$$= [\cdots] dB_t - H_2(t, Y_t^{a,u}, Z_t^{a,u}, \Gamma_t^*, u_t) dt$$
$$+ \sum_{i,j=1}^2 \Gamma_t^{j,*} [\partial_{y_i} f_j(t, Y_t^*, Z_t^*, u_t^*) Y_t^{i,a,u} + \partial_{z_i} f_j(t, Y_t^*, Z_t^*, u_t^*) Z_t^{i,a,u}] dt.$$

Denote

$$\Delta Y^{a,u} := Y^{a,u} - Y^{a^*, u^*}, \qquad \Delta Z^{a,u} := Z^{a,u} - Z^{a^*, u^*}.$$

Then,

$$\Delta Y_0^{2,a,u} = \sum_{i=1}^2 \Gamma_0^{i,*} \Delta Y_0^{i,a,u}$$

10.3 Stochastic Control of High-Dimensional BSDEs

$$\begin{aligned}= E\bigg\{ &H_1(\Gamma_T^*, a) - H_1(\Gamma_T^*, a^*) + \int_0^T \big[H_2(t, Y_t^{a,u}, Z_t^{a,u}, \Gamma_t^*, u_t) \\&- H_2(t, Y_t^*, Z_t^*, \Gamma_t^*, u_t^*)\big] dt \\&- \int_0^T \sum_{i,j=1}^2 \Gamma_t^{j,*}\big[\partial_{y_i} f_j(t, Y_t^*, Z_t^*, u_t^*)[Y_t^{i,a,u} - Y_t^{i,*}] \\&+ \partial_{z_i} f_j(t, Y_t^*, Z_t^*, u_t^*)[Z_t^{i,a,u} - Z_t^{i,*}]\big] dt \bigg\}.\end{aligned}$$

By the concavity of H_2, together with the maximum condition (10.55), we obtain $\Delta Y_0^{2,a,u} \leq 0$ for any $(a, u) \in \mathcal{A}$. That is, (a^*, u^*) is optimal. □

Remark 10.3.8 Here, the values of γ_i, and in particular γ_1, can be positive or negative. Thus, it is in general difficult to ensure concavity of the function H_2. In addition to the trivial case when both f_1 and f_2 are linear, we note the following conditions under which H_2 satisfies the required concavity:

$$\begin{aligned}&f_1 \text{ is independent of } (y_2, z_2) \text{ and linear in } (y_1, z_1, u), \text{ and} \\&f_2 \text{ is concave in } (y, z, u).\end{aligned} \quad (10.56)$$

In this case Γ^2 is positive and thus H_2 is concave.

The constraint that f_1 is linear in (y_1, z_1, u) is undesirable in applications. We next provide another approach for the sufficient conditions, without assuming the concavity of H_2 explicitly. However, this approach also requires strong technical conditions.

Assumption 10.3.9

(i) Assumption 10.3.1 holds, g_i' and $\partial_u f_i$ are bounded, and

$$E\bigg\{|g_i(0)|^2 + \int_0^T |f_i(t, 0, 0, 0, 0, 0)|^2 dt\bigg\} < \infty; \quad (10.57)$$

(ii) $\mathcal{A} = L^2(\mathcal{F}_T) \times L^2(\mathbb{F})$ (in particular, $A = U = \mathbb{R}$);
(iii) There exist unique functions $\hat{I}_1(\gamma, \xi)$ and $\hat{I}_2(t, y, z, \gamma, \eta)$ such that

$$\sum_{i=1}^2 \gamma_i g_i'(\hat{I}_1(\gamma, \xi)) = \xi, \quad \sum_{i=1}^2 \gamma_i \partial_u f_i(t, y, z, \hat{I}_2(t, y, z, \gamma, \eta)) = \eta. \quad (10.58)$$

Theorem 10.3.10 *Assume*

(i) *Assumption 10.3.9 holds;*
(ii) *For any $(\xi, \eta) \in \mathcal{A}$, the following FBSDE has a unique solution:*

$$\begin{cases} d\hat{\Gamma}_t^{i,\xi,\eta} = \sum_{j=1}^{2} \hat{\Gamma}_t^{j,\xi,\eta} \big[\partial_{y_i} \hat{f}_j(t, \hat{Y}_t^{\xi,\eta}, \hat{Z}_t^{\xi,\eta}, \hat{\Gamma}_t^{\xi,\eta}, \eta_t) dt \\ \qquad\qquad + \partial_{z_i} \hat{f}_j(t, \hat{Y}_t^{\xi,\eta}, \hat{Z}_t^{\xi,\eta}, \hat{\Gamma}_t^{\xi,\eta}, \eta_t) dB_t \big]; \\ d\hat{Y}_t^{i,\xi,\eta} = -\hat{f}_i(t, \hat{Y}_t^{\xi,\eta}, \hat{Z}_t^{\xi,\eta}, \hat{\Gamma}_t^{\xi,\eta}, \eta_t) dt + \hat{Z}_t^{i,\xi,\eta} dB_t; \\ \hat{\Gamma}_0^{1,\xi,\eta} = 0, \qquad \hat{\Gamma}_0^{2,\xi,\eta} = 1, \qquad \hat{Y}_T^{i,\xi,\eta} = \hat{g}_i(\hat{\Gamma}_T^{\xi,\eta}, \xi), \end{cases} \quad (10.59)$$

where

$$\hat{g}(\gamma, \xi) := g(\hat{I}_1(\gamma, \xi)), \qquad \hat{f}(t, y, z, \gamma, \eta) := f(t, y, z, \hat{I}_2(t, y, z, \gamma, \eta)). \tag{10.60}$$

(iii) *The mapping*

$$(\xi, \eta) \mapsto \big(\hat{I}_1(\hat{\Gamma}_T^{\xi,\eta}, \xi), \hat{I}_2(\cdot, \hat{Y}^{\xi,\eta}, \hat{Z}^{\xi,\eta}, \eta) \big) \tag{10.61}$$

is continuous under the inner product norm $\|\cdot\|^2 := \langle \cdot, \cdot \rangle$.

Then, $V = \hat{Y}_0^{2,0,0}$ and the optimal control is

$$a^* := \hat{I}_1(\hat{\Gamma}_T^{0,0}, 0), \qquad u_t^* := \hat{I}_2(t, \hat{Y}_t^{0,0}, \hat{Z}_t^{0,0}, \hat{\Gamma}_t^{0,0}, 0). \tag{10.62}$$

Proof Clearly, $\mathcal{A}$ is a Banach space under the inner product:

$$\langle (a, u), (\tilde{a}, \tilde{u}) \rangle := E\Big\{ a\tilde{a} + \int_0^T u_t \tilde{u}_t dt \Big\}.$$

Define the mapping $F : \mathcal{A} \to \mathbb{R}$ by $F(a, u) := Y_0^{2,a,u}$. Then, F is differentiable and following the arguments in Lemma 10.3.3,

$$\lim_{\varepsilon \to 0} \frac{1}{\varepsilon} \big[F(a + \varepsilon \Delta a, u + \varepsilon \Delta u) - F(a, u) \big] = \langle f(a, u), (a, u) \rangle$$

where $f(a, u) := \Big(\sum_{i=1}^{2} \Gamma_T^{i,a,u} g_i'(a), \sum_{i=1}^{2} \Gamma^{i,a,u} \partial_u f_i(\cdot, Y^{a,u}, Z^{a,u}, u) \Big).$ (10.63)

That is, f is the Frechet derivative of F. We note that mapping (10.61) is exactly the inverse of f. We also note that here $(\Delta a, \Delta u) \in \mathcal{A}$ may not be bounded, and that is why we assumed g_i' and $\partial_u f_i$ are bounded in Assumption 10.3.6(i).

By Ekeland's variational principle, see Ekeland (1974), there exists a sequence $(a^n, u^n) \in \mathcal{A}$ such that

$$\lim_{n \to \infty} \|f(a^n, u^n)\| = 0 \quad \text{and} \quad \lim_{n \to \infty} F(a^n, u^n) = V. \tag{10.64}$$

For the pair (a^*, u^*) defined in (10.62), we have $f(a^*, u^*) = 0$. By Condition (ii), we get $\lim_{n \to \infty} \|(a^n - a^*, u^n - u^*)\| = 0$. This implies further that $\lim_{n \to \infty} F(a^n, u^n) = F(a^*, u^*)$. Thus, $V = F(a^*, u^*) = \hat{Y}_0^{2,0,0}$ and (a^*, u^*) is the optimal control. □

Remark 10.3.11 Consider (ξ, η) as parameters of FBSDE (10.59). Then, condition (iii) is exactly the stability of the FBSDE. This is in general a difficult problem.

10.4 Stochastic Optimization in Weak Formulation

In the second best and third best problems, it is usually assumed that the agent affects the outcome by affecting the underlying distribution. In SDE models this translates into the so-called weak solutions. We present the theory next.

10.4.1 Weak Formulation Versus Strong Formulation

We recall the setup of Sect. 10.2. In general, FSBDE system (10.29) is difficult to solve. In many applications, it is more tractable to use the weak formulation, that we introduce next. In that case, the stochastic maximum principle leads to BSDEs that are easier to solve.

We assume the control is only on the drift b, and $b(t, x, u) = 0$ whenever $\sigma(t, x) = 0$. Let X be a (strong) solution to the following SDE (without control):

$$X_t = x + \int_0^t \sigma(s, X_s) dB_s, \quad P\text{-a.s.} \tag{10.65}$$

For $u \in \mathcal{U}$, denote $\theta_t^u := b(t, X_t, u_t) \sigma^{-1}(t, X_t)$, where $0/0 := 0$, and

$$B_t^u := B_t - \int_0^t \theta_s^u ds, \quad M_t^u := \exp\left(\int_0^t \theta_s^u dB_s - \frac{1}{2} \int_0^t |\theta_s^u|^2 ds\right), \tag{10.66}$$

$$dP^u := M_T^u dP.$$

Assume the Girsanov theorem holds. Then, B^u is a P^u-Brownian motion. Notice that

$$X_t = x + \int_0^t b(s, X_s, u_s) ds + \int_0^t \sigma(s, X_s) dB_s^u, \quad P^u\text{-a.s.} \tag{10.67}$$

Thus, for different u, X is a solution to an SDE with a different Brownian motion. The solutions in which Brownian motion is not fixed are called weak solutions.

In analogy with problem (10.18)–(10.19), our optimization problem in weak formulation is now

$$V := \sup_{u \in \mathcal{U}} Y_0^u \tag{10.68}$$

where

$$Y_t^u = g(X_T) + \int_t^T h(s, X_s, Y_s^u, Z_s^u, u_s) ds - \int_t^T Z_s^u dB_s^u$$

$$= g(X_T) + \int_t^T f(s, X_s, Y_s^u, Z_s^u, u_s) ds - \int_t^T Z_s^u dB_s, \tag{10.69}$$

and

$$f(t,x,y,z,u) := h(t,x,y,z,u) + b(t,x,u)\sigma^{-1}(t,x)z. \quad (10.70)$$

Remark 10.4.1 In the strong formulation, probability P is fixed and the outcomes of state process X^u are affected directly by the control process. In the weak formulation, state process X is fixed, and probability P^u is affected by the control process. In other words, in the weak formulation, the distribution of X is controlled.

Remark 10.4.2 In this remark let us assume that function f in (10.19) is equal to function h in (10.69).

(i) Although the P-distribution of B is equal to the P^u-distribution of B^u, the P-distribution of u is not the same as the P^u-distribution of u. Thus, (10.19) and (10.69) may define different values of Y_0^u, and the two formulations are not equivalent, in general.

(ii) In both formulations we require that u be $\mathbb{F}^B$-adapted. If a control u in the weak formulation happens to be $\mathbb{F}^{B^u}$-adapted, then one can find a mapping v such that $v(B^u) = u(B)$, and thus Y_0^u in (10.69) is equal to Y_0^v in (10.19). On the other hand, if a control u in the strong formulation happens to be $\mathbb{F}^{X^u}$-adapted, then one can find a mapping v such that $v(X^u) = u(B)$, and thus Y_0^u in (10.19) is equal to Y_0^v in (10.69).

(iii) If the optimal control u_W^* in the weak formulation is $\mathbb{F}^{B^{u_W^*}}$-adapted, and the optimal control u_S^* in the strong formulation is $\mathbb{F}^{X^{u_S^*}}$-adapted, then the two formulations define the same value V. This is typically the case in discrete time models and in Markovian models.

Remark 10.4.3 In light of the second equation in (10.69), weak formulation (10.69) can be viewed as a special case of strong formulation (10.19) with coefficients:

$$\tilde{b} := 0, \qquad \tilde{\sigma} := \sigma(t,x), \qquad \tilde{f} := f.$$

If we assume that

$$b(t,x,u)\sigma^{-1}(t,x) \quad \text{is bounded for all } u \in U, \quad (10.71)$$

then, f is uniformly Lipschitz continuous in (y,z), and one can apply the results of Sect. 10.1 to this case.

Condition (10.71) is not appropriate in many applications. We carry out the analysis rigorously under weaker conditions in subsections below. Throughout this section we always assume

Assumption 10.4.4 Process σ is progressively measurable and $\mathbb{F}$-adapted, and SDE (10.65) has a strong solution X.

From now on we fix such a solution X. For notational simplicity, we omit argument X in coefficients b, h, f, and denote $\xi := g(X_T)$.

10.4.2 Sufficient Conditions in Weak Formulation

We impose the following conditions.

Assumption 10.4.5 (i) The coefficients b, h and hence f are progressively measurable and $\mathbb{F}$-adapted, and ξ is $\mathcal{F}_T$-measurable.
  (ii) h is continuously differentiable in (y, z) with uniformly bounded derivatives.
  (iii) For each $u \in \mathcal{U}$, M^u is a true P-martingale and

$$E^u\left\{|\xi|^2 + \left(\int_0^T |h(t, 0, 0, u_t)| dt\right)^2\right\} < \infty.$$

We notice that, in general, $\mathbb{F}^{B^u}$ is smaller than $\mathbb{F}^B$ and thus X may not be $\mathbb{F}^{B^u}$-adapted. Nevertheless we still have the following martingale representation theorem:

Lemma 10.4.6 *Assume M^u is a true martingale and $\xi \in L^2(\mathcal{F}_T^B, P^u)$. There exists a unique $Z \in L^2(\mathbb{F}^B, P^u)$ such that*

$$\xi = E^u\{\xi\} + \int_0^T Z_t dB_t^u.$$

Proof Clearly, it suffices to show that there exists a couple $(Y, Z) \in L^2(\mathbb{F}, P^u) \times L^2(\mathbb{F}, P^u)$ satisfying

$$Y_t = \xi - \int_t^T Z_s dB_s^u. \tag{10.72}$$

We first assume ξ is bounded. Let

$$\tau_n := \inf\{t : M_t^u \geq n\}.$$

Then, $M_{\tau_n}^u E_{\tau_n}\{\xi\} \in L^2(\mathcal{F}_\tau, P)$. By Proposition 9.3.6, there exists $(\tilde{Y}^n, \tilde{Z}^n) \in L^2(\mathbb{F}, P) \times L^2(\mathbb{F}, P)$ such that

$$\tilde{Y}_t^n = M_{\tau_n}^u E_{\tau_n}\{\xi\} - \int_t^{\tau_n} \tilde{Z}^n dB_t, \quad 0 \leq t \leq \tau_n.$$

Denote

$$Y_t^n := \tilde{Y}_{t \wedge \tau_n}^n [M_{t \wedge \tau_n}^u]^{-1}, \quad Z_t^n := [\tilde{Z}_t^n - u_t \tilde{Y}_t^n][M_t^u]^{-1} \mathbf{1}_{[0, \tau_n]}(t), \quad 0 \leq t \leq T.$$

One can check directly that

$$Y_t^n = E_{\tau_n}\{\xi\} - \int_t^T Z_t^n dB_t^u, \quad 0 \leq t \leq T.$$

Note that $E_{\tau_n}\{\xi\} \to \xi$, P-a.s., and thus P^u-a.s. Since ξ is bounded, we have $E^u\{|E_{\tau_n}\{\xi\} - \xi|^2\} \to 0$ as $n \to \infty$. Then, following exactly the same arguments as in Theorem 9.3.2 we have

$$E^u\left\{\sup_{0\leq t\leq T}|Y_t^n-Y_t^m|^2+\int_0^T|Z^n-Z_t^m|^2 dt\right\}\leq CE^u\{|E_{\tau_n}\{\xi\}-E_{\tau_m}\{\xi\}|^2\}\to 0,$$

as $n,m\to\infty$. This proves (10.72).

In general, let $\xi_n:=(-n)\vee\xi\wedge n$ and let (Y^n,Z^n) be the solution to the BSDE

$$Y_t^n=\xi_n-\int_t^T Z_s^n dB_s^u.$$

Since $|\xi_n|\leq\xi$ and $\xi_n\to\xi$, by the Dominated Convergence Theorem we know $\lim_{n\to\infty}E^u\{|\xi_n-\xi|^2\}=0$. Then, following the same arguments as above we see that Y^n,Z^n are Cauchy sequences under appropriate norms, and thus (10.72) holds. □

Combining Lemma 10.4.6 and the arguments in Chap. 9, under Assumption 10.4.5 one can easily see that BSDE (10.69) admits a unique solution $(Y^u,Z^u)\in L^2(\mathbb{F},P^u)\times L^2(\mathbb{F},P^u)$. Similarly to Theorem 10.2.9, we have

Theorem 10.4.7 *Assume Assumptions* 10.4.4 *and* 10.4.5 *hold. If* $u^*\in\mathcal{U}$ *satisfies*

$$f(t,Y_t^{u^*},Z_t^{u^*},u_t^*)=\sup_{u\in U}f(t,Y_t^{u^*},Z_t^{u^*},u) \tag{10.73}$$

and, for each $u\in\mathcal{U}$, *there exists* $\delta>0$ *such that*

$$E^u\left\{\left(\int_0^T[|Y_t^{u^*}|^2+|Z_t^{u^*}|^2]dt\right)^{\frac{1+\delta}{2}}\right\}<\infty, \tag{10.74}$$

then u^* *is an optimal control for the optimization problem* (10.68).

Proof Let $u\in\mathcal{U}$ and denote $\Delta Y^u:=Y^u-Y^{u^*}$, $\Delta Z^u:=Z^u-Z^{u^*}$. By (10.73) we have

$$\begin{aligned}d(\Delta Y_t^u)&=-[f(t,Y_t^u,Z_t^u,u_t)-f(t,Y_t^*,Z_t^*,u_t^*)]dt+\Delta Z_t^u dB_t\\&\geq -[f(t,Y_t^u,Z_t^u,u_t)-f(t,Y_t^*,Z_t^*,u_t)]dt+\Delta Z_t^u dB_t\\&=-[h(t,Y_t^u,Z_t^u,u_t)-h(t,Y_t^*,Z_t^*,u_t)+b(t,u_t)\sigma^{-1}(t)\Delta Z_t^u]dt\\&\quad+\Delta Z_t^u dB_t\\&=-[\alpha_t\Delta Y_t^u+\beta_t\Delta Z_t^u]dt+\Delta Z_t^u dB_t^u,\end{aligned}$$

where α,β are defined in an obvious way, and are bounded due to Assumption 10.4.5. Denote

$$\Gamma_t=1+\int_0^t\alpha_s\Gamma_s ds+\int_0^t\beta_s\Gamma_s dB_s^u.$$

Since $\Delta Y_T=0$, then

$$\Delta Y_0^u=\Gamma_0\Delta Y_0^u\leq\int_0^T\Gamma_t[\beta_t\Delta Y_t^u-\Delta Z_t^u]dB_t^u. \tag{10.75}$$

10.4 Stochastic Optimization in Weak Formulation

It is clear that
$$E^u\left\{\sup_{0\le t\le T}|\Gamma_t|^p\right\}<\infty \quad \text{for all } p\ge 1.$$

Then, for a possibly smaller $\delta > 0$, by (10.74) we have
$$E^u\left\{\left(\int_0^T |\Gamma_t[\beta_t\Delta Y_t^u - \Delta Z_t^u]|^2 dt\right)^{\frac{1+\delta}{2}}\right\}<\infty.$$

This implies that $\int_0^t \Gamma_s[\beta_s\Delta Y_s^u - \Delta Z_s^u]dB_s^u$ is a P^u-martingale. Taking expectation on both sides of (10.75), we get $\Delta Y_0^u \le 0$. That is, u^* is optimal. □

We next provide a tractable set of sufficient conditions. Denote
$$f^*(t,y,z) := \sup_{u\in U} f(t,y,z,u), \tag{10.76}$$

Theorem 10.4.8 *Assume*

(i) *Assumption 10.4.5 holds, and for each $u\in\mathcal{U}$, there exists $\delta := \delta_u > 0$ such that*
$$E\{|M_T^u|^{1+\delta}\}<\infty. \tag{10.77}$$

(ii) *There exists a progressively measurable and $\mathbb{F}$-adapted function $I(t,y,z)$ taking values in U such that*
$$f^*(t,y,z) = f(t,y,z,I(t,y,z)). \tag{10.78}$$

(iii) *Random variable ξ is bounded, and*
$$|f^*(t,y,z)| \le C[1+|y|+|z|^2],$$
$$|b(t,I(t,y,z))\sigma^{-1}(t)| \le C[1+|y|+|z|]. \tag{10.79}$$

(iv) *Process*
$$u^* := I(\cdot,Y^*,Z^*)\in\mathcal{U}, \tag{10.80}$$

where (Y^*,Z^*) is the unique solution of the following BSDE such that Y^* is bounded:
$$Y_t^* = \xi + \int_t^T f^*(s,Y_s^*,Z_s^*)ds - \int_t^T Z_s^* dB_s. \tag{10.81}$$

Then, u^* is an optimal control for the optimization problem (10.68) and $V = Y_0^*$.

Proof We first apply Theorem 9.6.3 to prove well-posedness of BSDE (10.81). It suffices to check condition (9.48). Clearly, f is uniformly Lipschitz continuous in y, thus
$$|f^*(t,y_1,z) - f^*(t,y_2,z)| \le \sup_{u\in U}|f(t,y_1,z,u) - f(t,y_2,z,u)| \le C|y_1-y_2|.$$

Moreover,

$$f^*(t, y, z_1) - f^*(t, y, z_2)$$
$$= f(t, y, I(t, y, z_1)) - f^*(t, y, z_2)$$
$$\leq f(t, y, z_1, I(t, y, z_1)) - f(t, y, z_2, I(t, y, z_1))$$
$$= h(t, y, z_1) - h(t, x, y, z_2) + b(t, I(t, y, z_1))\sigma^{-1}(t)(z_1 - z_2)$$
$$\leq C|z_1 - z_2| + C[1 + |y| + |z_1|]|z_1 - z_2| \leq C[1 + |y| + |z_1|]|z_1 - z_2|.$$

Similarly,
$$f^*(t, y, z_2) - f^*(t, y, z_1) \leq C[1 + |y| + |z_2|]|z_1 - z_2|.$$

Combining the estimates together we prove (9.48). Then, by Theorem 9.6.3, BSDE (10.81) has a unique solution (Y^*, Z^*) such that Y^* is bounded.

Now, denote $u_t^* := I(t, Y_t^*, Z_t^*)$. It is clear that $Y^{u^*} = Y^*$, $Z^{u^*} = Z^*$. By Theorem 9.6.4 and (10.77) we obtain (10.74). Then, applying Theorem 10.4.7 we prove the result. □

Remark 10.4.9 The following result shows that u^* defined by (10.80) indeed satisfies (10.77). Therefore, if we let $\mathcal{U}$ be the set of all those u satisfying (10.77), then (10.80) holds.

Proposition 10.4.10 *Assume the conditions* (i)–(iii) *of Theorem* 10.4.8 *hold. Let* (Y^*, Z^*) *be the solution to BSDE* (10.81) *such that* Y^* *is bounded, and let* u^* *be defined by* (10.80). *Then,* u^* *satisfies* (10.77).

Proof First, by the proof of Theorem 10.4.8 we know BSDE (10.81) is well-posed. By Theorem 9.6.4, we have $E_t\{\int_t^T |Z_s^*|^2 ds\} < \infty$. Then, (10.79) leads to $E_t\{\int_t^T |\theta_s^{u^*}|^2 ds\} < \infty$. Applying Lemma 9.6.5 we know M^{u^*} is a true martingale, and B^{u^*} is a P^{u^*}-Brownian motion.

Note that (10.81) can be rewritten as
$$Y_t^* = \xi + \int_t^T [f^*(s, Y_s^*, Z_s^*) - b(s, I(s, Y_s^*, Z_s^*))\sigma_s^{-1} Z_s^*] ds - \int_t^T Z_s^* dB_s^{u^*}.$$

Although the above BSDE may not satisfy Assumption 9.6.1, by the fact that Y^* is bounded we have the following growth condition:
$$|f^*(s, Y_s^*, Z_s^*) - b(s, I(s, Y_s^*, Z_s^*))\sigma_s^{-1} Z_s^*| \leq C[1 + |Z_s^*|^2].$$

Following the argument in the last part of the proof of Theorem 9.6.4, we see that $E_t^{P^{u^*}}\{\int_t^T |Z_s^*|^2 ds\} < \infty$, and thus $E_t^{P^{u^*}}\{\int_t^T |\theta_s^{u^*}|^2 ds\} < \infty$. By (9.53), there exists $\varepsilon > 0$ such that
$$E^{P^{u^*}}\left\{e^{\varepsilon \int_0^T |\theta_s^{u^*}|^2 ds}\right\} < \infty.$$

That is,
$$E\left\{\exp\left(\int_0^T \theta_t^{u^*} dB_t - \left(\frac{1}{2} - \varepsilon\right)\int_0^T |\theta_s^{u^*}|^2 ds\right)\right\} < \infty.$$

10.4 Stochastic Optimization in Weak Formulation

Let $\delta > 0$ such that $\delta + 2\delta^2 = \varepsilon$. Then,

$$E\{|M_T^{u^*}|^{1+\delta}\}$$
$$= E\left\{\exp\left((1+\delta)\int_0^T \theta_t^{u^*} dB_t - \frac{1+\delta}{2}\int_0^T |\theta_s^{u^*}|^2 ds\right)\right\}$$
$$= E\left\{\exp\left(\left(\frac{1}{2}+\delta\right)\int_0^T \theta_t^{u^*} dB_t - \left(\frac{1}{2}+\delta\right)^2 \int_0^T |\theta_s^{u^*}|^2 ds\right)\right.$$
$$\left.\times \exp\left(\frac{1}{2}\int_0^T \theta_t^{u^*} dB_t - \left[\frac{1+\delta}{2} - \left(\frac{1}{2}+\delta\right)^2\right]\int_0^T |\theta_s^{u^*}|^2 ds\right)\right\}$$
$$\leq \left(E\left\{\exp\left((1+2\delta)\int_0^T \theta_t^{u^*} dB_t - \frac{(1+2\delta)^2}{2}\int_0^T |\theta_s^{u^*}|^2 ds\right)\right\}\right)^{\frac{1}{2}}$$
$$\times \left(E\left\{\exp\left(\int_0^T \theta_t^{u^*} dB_t - \left[1+\delta - \frac{(1+2\delta)^2}{2}\right]\int_0^T |\theta_s^{u^*}|^2 ds\right)\right\}\right)^{\frac{1}{2}}$$
$$\leq \left(E\left\{\exp\left(\int_0^T \theta_t^{u^*} dB_t - \left[\frac{1}{2}-\varepsilon\right]\int_0^T |\theta_s^{u^*}|^2 ds\right)\right\}\right)^{\frac{1}{2}} < \infty.$$

This ends the proof. □

Remark 10.4.11 In this section we have considered the quadratic case, as can be seen in (10.79). This is mainly due to the fact that well-posedness has been shown for such BSDEs. If the BSDE theory could be extended to more general cases, then one could easily extend the stochastic maximum principle accordingly. For example, the assumption that ξ is bounded can be weakened, see e.g. Briand and Hu (2006, 2008).

Example 10.4.12 Assume $\sigma = 1$, $b(t, u) = |u| \wedge 1$, $f = -\frac{u^2}{2}$, ξ is bounded, and $U = \mathbb{R}$. Then, we may set

$$I(t, y, z) := \begin{cases} 0, & z \leq 0; \\ z, & 0 < z \leq 1; \\ 1, & z > 1, \end{cases}$$

and we have

$$f^*(t, x, y, z) = \frac{1}{2}|z|^2 1_{[0,1]}(z) + \left(z - \frac{1}{2}\right) 1_{(1,\infty)}(z).$$

Since $u_t^* := I(t, Y_t^*, Z_t^*)$ is bounded, it is easy to check that u^* is in $\mathcal{U}$ for a reasonably large set $\mathcal{U}$.

Note that function I may not be unique. In fact, in this example, we can also set

$$I(t, y, z) := \begin{cases} 0, & z \leq 0; \\ -z, & 0 < z \leq 1; \\ 1, & z > 1. \end{cases}$$

The next example is somewhat more general.

Example 10.4.13 Assume $\sigma = 1$, b is bounded, $f = -\frac{u^2}{2}$ for $|u| \geq R$ for some fixed constant $R > 0$, ξ is bounded, and $U = \mathbb{R}$. Then, we can easily see that function I still exists. Notice that here we do not require convexity of h on $\mathbb{R}$.

The following simple example can be solved explicitly, and is the basis for Sect. 5.3.

Example 10.4.14 Assume $U = \mathbb{R}$,

$$b = uv, \qquad \sigma = v, \qquad f = f_0(t) - \frac{u^2}{2}, \tag{10.82}$$

where v is a given positive process satisfying $E\{\int_0^T |v_t|^2 dt\} < \infty$. In this case, $I = z$, and thus BSDE (10.81) becomes

$$Y_t = \xi + \int_t^T \left[f_0(s) + \frac{1}{2}|Z_s|^2 \right] ds - \int_t^T Z_s dB_s. \tag{10.83}$$

Assume ξ and f_0 are bounded. By Theorem 9.6.3, BSDE (10.83) is well-posed and such that Y is bounded. We can in fact solve this BSDE explicitly. Denote

$$\hat{Y}_t := e^{Y_t}, \qquad \hat{Z}_t := e^{Y_t} Z_t, \qquad \hat{\xi} := e^{\xi}, \qquad \Gamma_t := e^{\int_0^t f_0(s) ds}. \tag{10.84}$$

Applying Itô's rule, we have

$$\Gamma_t \hat{Y}_t = \Gamma_T \hat{\xi} - \int_t^T \Gamma_s \hat{Z}_s dB_s. \tag{10.85}$$

Then,

$$\Gamma_t \hat{Y}_t = E_t\{\Gamma_T \hat{\xi}\} \quad \text{or equivalently,} \quad e^{Y_t} = E_t\{e^{\int_t^T f_0(s) ds + \xi}\}. \tag{10.86}$$

One important feature of this example is that we can obtain a closed form formula for M^u. Noting that $u^* = I(t, Y_t, Z_t) = Z_t$, we get

$$M_t^u = \exp\left(\int_0^t u_s dB_s - \frac{1}{2} \int_0^t |u_s|^2 ds \right) = \exp\left(\int_0^t Z_s dB_s - \frac{1}{2} \int_0^t |Z_s|^2 ds \right).$$

Note that

$$Y_t = Y_0 - \int_0^t \left[f_0(s) + \frac{1}{2} Z_s^2 \right] ds + \int_0^t Z_s dB_s.$$

Then,

$$M_t^u = \exp\left(Y_t - Y_0 + \int_0^t f_0(s) ds \right) = e^{-Y_0} \Gamma_t \hat{Y}_t = e^{-Y_0} E_t\{e^{\xi + \int_t^T f_0(s) ds}\}. \tag{10.87}$$

10.4.3 Necessary Conditions in Weak Formulation

We impose the following stronger conditions on the coefficients.

Assumption 10.4.15 (i) Coefficients b, h and hence f are progressively measurable and $\mathbb{F}$-adapted, and ξ is $\mathcal{F}_T$-measurable.

(ii) h is continuously differentiable in (y, z) with uniformly bounded derivatives; b and h are continuously differentiable in u, $b_u \sigma^{-1}$ is bounded, and

$$\left|h_u(t, y, z, u)\right| \leq C\bigl[1 + \left|h_u(t, 0, 0, u)\right| + |y| + |z|\bigr]. \tag{10.88}$$

Admissible set $\mathcal{U}$ satisfies the following conditions.

Assumption 10.4.16

(i) For each $u \in \mathcal{U}$, (10.77) holds.

(ii) $\mathcal{U}$ is locally convex, in the sense of Assumption 10.1.7(ii).

(iii) For u, Δu, u^ε as in Assumption 10.1.7(ii), there exists $\delta := \delta(u, \Delta u) > 0$ such that

$$\sup_{0 \leq \varepsilon \leq 1} E^{u^\varepsilon}\left\{|\xi|^{2+\delta} + \left(\int_0^T \bigl[\bigl|h(t, 0, 0, u_t^\varepsilon)\bigr| + \bigl|h_u(t, 0, 0, u_t^\varepsilon)\bigr|\bigr]dt\right)^{2+\delta}\right\} < \infty. \tag{10.89}$$

Remark 10.4.17 Condition (10.77) implies that M^u is a true martingale. Moreover, for any $p > 1$ and each $u \in \mathcal{U}$ and Δu bounded, denoting $\Delta b(t) := b(t, u_t + \Delta u_t) - b(t, u_t)$ and $\tilde{\theta}_t := [b(t, u_t) + p\Delta b(t)]\sigma^{-1}(t)$, one can check straightforwardly that

$$E^u\bigl\{\bigl(M_T^{u+\Delta u}[M_T^u]^{-1}\bigr)^p\bigr\}$$

$$= E\left\{\exp\left(\int_0^T \tilde{\theta}_t dB_t - \frac{1}{2}\int_0^T |\tilde{\theta}_t|^2 dt + \frac{1}{2}(p^2 - p)\int_0^T |\Delta b(t)\sigma^{-1}(t)|^2\right)\right\}$$

$$\leq E\left\{\exp\left(\int_0^T \tilde{\theta}_t dB_t - \frac{1}{2}\int_0^T |\tilde{\theta}_t|^2 dt\right)\right\} \exp\left(\frac{CT}{2}(p^2 - p)\right) \leq C_p < \infty, \tag{10.90}$$

thanks to the assumption that $b_u \sigma^{-1}$ is bounded.

To understand condition (10.77) further, we state

Lemma 10.4.18 *Assume Assumption (10.4.15) holds, u satisfies (10.77) with $\delta_u = \delta$, and Δu is bounded. Then, for any $0 < \tilde{\delta} < \delta$, we have*

$$E\bigl\{\bigl|M_T^{u+\Delta u}\bigr|^{1+\tilde{\delta}}\bigr\} < \infty.$$

Proof Define conjugates $p := \delta/(\delta - \tilde{\delta})$, $q := \delta/\tilde{\delta}$. Then, by (10.90),

$$E\{|M_T^{u+\Delta u}|^{1+\tilde{\delta}}\}$$
$$= E\{|M_T^{u+\Delta u}(M_T^u)^{-1}|^{1+\tilde{\delta}}(M_T^u)^{1+\tilde{\delta}}\}$$
$$= E\{|M_T^{u+\Delta u}(M_T^u)^{-1}|^{1+\tilde{\delta}}(M_T^u)^{(1-\frac{\tilde{\delta}}{\delta})+\frac{\tilde{\delta}}{\delta}(1+\delta)}\}$$
$$\leq (E\{|M_T^{u+\Delta u}(M_T^u)^{-1}|^{p(1+\tilde{\delta})}(M_T^u)^{p(1-\frac{\tilde{\delta}}{\delta})}\})^{\frac{1}{p}} (E\{(M_T^u)^{q\frac{\tilde{\delta}}{\delta}(1+\delta)}\})^{\frac{1}{q}}$$
$$= (E^u\{|M_T^{u+\Delta u}(M_T^u)^{-1}|^{p(1+\tilde{\delta})}\})^{\frac{1}{p}} (E\{(M_T^u)^{(1+\delta)}\})^{\frac{1}{q}} < \infty.$$

This completes the proof. □

As a direct consequence of Lemma 10.4.18, we have

Remark 10.4.19 (i) The set $\mathcal{U}_0$ of all those u satisfying (10.77) is locally convex.

(ii) By (10.89), for such $u, \Delta u, u^\varepsilon$, for a possibly smaller $\delta > 0$, the random variable
$$M_T^{u^\varepsilon}\left[|\xi|^{2+\delta} + \left(\int_0^T [|h(t,0,0,u_t^\varepsilon)| + |h_u(t,0,0,u_t^\varepsilon)|]dt\right)^{2+\delta}\right]$$
is integrable under P, uniformly in $\varepsilon \in [0, 1]$. In fact, denoting that random variable as $M_T^{u^\varepsilon} \eta_\delta^\varepsilon$, we have
$$\sup_{0\leq\varepsilon\leq 1} E\{(M_T^{u^\varepsilon} \eta_\delta^\varepsilon)^{1+\delta^2}\}$$
$$= \sup_{0\leq\varepsilon\leq 1} E\{(|M_T^{u^\varepsilon}|^{\frac{1}{1+\delta}}|\eta_\delta^\varepsilon|^{1+\delta^2})|M_T^{u^\varepsilon}|^{\frac{\delta+\delta^2+\delta^3}{1+\delta}}\}$$
$$\leq \sup_{0\leq\varepsilon\leq 1} (E\{M_T^{u^\varepsilon}|\eta_\delta^\varepsilon|^{(1+\delta^2)(1+\delta)}\})^{\frac{1}{1+\delta}} (E\{|M_T^{u^\varepsilon}|^{1+\delta+\delta^2}\})^{\frac{\delta}{1+\delta}} < \infty,$$
when δ is small enough.

Now, for $u \in \mathcal{U}$, Δu bounded, and $u^\varepsilon = u + \varepsilon \Delta u \in \mathcal{U}$ as in Assumption 10.4.16(ii), denote
$$\Delta Y^\varepsilon := Y^{u^\varepsilon} - Y^u, \qquad \Delta Z^\varepsilon := Z^{u^\varepsilon} - Z^u, \qquad \nabla Y^\varepsilon := \frac{1}{\varepsilon}\Delta Y^\varepsilon,$$
$$\nabla Z^\varepsilon := \frac{1}{\varepsilon}\Delta Z^\varepsilon.$$

Then,
$$\Delta Y_t^\varepsilon = \int_t^T [h_y^\varepsilon(s)\Delta Y_s^\varepsilon + h_z^\varepsilon(s)\Delta Z_s^\varepsilon + h_u^\varepsilon(s)\varepsilon\Delta u_s + b_u^\varepsilon(s)\sigma^{-1}(s)Z_s^{u^\varepsilon}\varepsilon\Delta u_s]ds$$
$$+ \int_t^T b(s,u_s)\sigma^{-1}(s)\Delta Z_s^\varepsilon ds - \int_t^T \Delta Z_s^\varepsilon dB_s$$
$$= \int_t^T [h_y^\varepsilon \Delta Y_s^\varepsilon + h_z^\varepsilon \Delta Z_s^\varepsilon + [h_u^\varepsilon + b_u^\varepsilon \sigma^{-1} Z_s^{u^\varepsilon}]\varepsilon\Delta u_s]ds - \int_t^T \Delta Z_s^\varepsilon dB_s^u;$$
$$\tag{10.91}$$

10.4 Stochastic Optimization in Weak Formulation

$$\nabla Y_t^\varepsilon = \int_t^T \left[h_y^\varepsilon \nabla Y_s^\varepsilon + h_z^\varepsilon \nabla Z_s^\varepsilon + \left[h_u^\varepsilon + b_u^\varepsilon \sigma^{-1} Z_s^{u^\varepsilon} \right] \Delta u_s \right] ds - \int_t^T \nabla Z_s^\varepsilon dB_s^u; \tag{10.92}$$

where

$$h_y^\varepsilon(s) := \int_0^1 h_y\left(s, Y_s^u + \theta \Delta Y_s^\varepsilon, Z_s^u + \theta \Delta Z_s^\varepsilon, u_s + \theta \varepsilon \Delta u_s \right) d\theta,$$

and similarly for $h_z^\varepsilon, h_u^\varepsilon, b_u^\varepsilon$. Consider the BSDE

$$\nabla Y_t^u = \int_t^T \left[h_y^0(s) \nabla Y_s^u + f_z^0(s) \nabla Z_s^u + \left[h_u^0(s) + b_u^0(s) \sigma^{-1}(s) Z_s^u \right] \Delta u_s \right] ds$$
$$- \int_t^T \nabla Z_s^u dB_s^u, \tag{10.93}$$

where $h_y^0(s) = h_y(s, Y_s^u, Z_s^u, u_s)$, and similarly for other terms.

Lemma 10.4.20 *Assume Assumptions* 10.4.4, 10.4.15, *and* 10.4.16 *hold. Then, BSDE* (10.93) *has a unique solution in* $L^2(\mathbb{F}, P^u) \times L^2(\mathbb{F}, P^u)$, *and*

$$\lim_{\varepsilon \to 0} E \left\{ \sup_{0 \le t \le T} \left[|\Delta Y_t^\varepsilon|^2 + |\nabla Y_t^\varepsilon - \nabla Y_t^u|^2 \right] + \int_0^T \left[|\Delta Z_t^\varepsilon|^2 + |\nabla Z_t^\varepsilon - \nabla Z_t^u|^2 \right] dt \right\}$$
$$= 0.$$

Proof In this proof, let $\delta > 0$ denote a generic constant which may vary from line to line.

(i) First, by (10.89) and Proposition 9.4.4, we have

$$\sup_{0 \le \varepsilon \le 1} E^{u^\varepsilon} \left\{ \sup_{0 \le t \le T} |Y_t^{u^\varepsilon}|^{2+\delta} + \left(\int_0^T |Z_t^{u^\varepsilon}|^2 dt \right)^{1+\frac{\delta}{2}} \right\} < \infty, \quad \text{for some } \delta > 0.$$

Then, (10.90) leads to, for a possibly smaller $\delta > 0$,

$$\sup_{0 \le \varepsilon \le 1} E^u \left\{ \sup_{0 \le t \le T} |Y_t^{u^\varepsilon}|^{2+\delta} + \left(\int_0^T |Z_t^{u^\varepsilon}|^2 dt \right)^{1+\frac{\delta}{2}} \right\} < \infty. \tag{10.94}$$

By Assumption 10.4.15(ii), $h_y^\varepsilon, h_z^\varepsilon$ are bounded, and

$$\sup_{0 \le \varepsilon \le 1} E^{u^\varepsilon} \left\{ \left(\int_0^T |h_u^\varepsilon + b_u^\varepsilon \sigma^{-1} Z_t^{u^\varepsilon}| dt \right)^{2+\delta} \right\} < \infty.$$

Since Δu is bounded, applying Proposition 9.4.4 on (10.91) we obtain

$$\lim_{\varepsilon \to 0} E \left\{ \sup_{0 \le t \le T} |\Delta Y_t^\varepsilon|^{2+\delta} + \left(\int_0^T |\Delta Z_t^\varepsilon|^2 dt \right)^{1+\frac{\delta}{2}} \right\} = 0.$$

(ii) Following similar arguments as above, it is clear that BSDE (10.93) is well-posed. Denote $\Delta \nabla Y^\varepsilon := \nabla Y^\varepsilon - \nabla Y^u$, $\Delta \nabla Z^\varepsilon := \nabla Z^\varepsilon - \nabla Z^u$. Then,

$$\Delta \nabla Y_t^\varepsilon = \int_t^T [h_y^\varepsilon \Delta \nabla Y_s^\varepsilon + h_z^\varepsilon \Delta \nabla Z_s^\varepsilon] ds - \int_t^T \Delta \nabla Z_s^\varepsilon dB_s^u$$

$$+ \int_t^T [(h_y^\varepsilon - h_y^0) \nabla Y_s^u + (h_z^\varepsilon - h_z^0) \nabla Z_s^u$$

$$+ [(h_u^\varepsilon - h_u^0) + b_u^\varepsilon \sigma^{-1} \Delta Z_s^\varepsilon + (b_u^\varepsilon - b_u^0) \sigma^{-1} Z_s^u] \Delta u_s] ds. \quad (10.95)$$

By Dominated Convergence Theorem, we have

$$\lim_{\varepsilon \to 0} E^u \left\{ \left(\int_0^T [|(h_y^\varepsilon - h_y^0) \nabla Y_s^u| + |(h_z^\varepsilon - h_z^0) \nabla Z_s^u| \right. \right.$$

$$\left. \left. + |(b_u^\varepsilon - b_u^0) \sigma^{-1} Z_s^u \Delta u_s|] ds \right)^2 \right\} = 0.$$

By (i) and Assumption 10.4.16(iii), we see that

$$\lim_{\varepsilon \to 0} E^u \left\{ \left(\int_0^T [|h_u^\varepsilon - h_u^0| + |b_u^\varepsilon \sigma^{-1} \Delta Z_s^\varepsilon|] ds \right)^2 \right\} = 0.$$

Then, the result follows from (10.95). □

We now introduce adjoint process Γ^u for BSDE (10.93), that depends only on u, but not on Δu:

$$\Gamma_t^u = 1 - \int_0^t h_y^0 \Gamma_s^u ds - \int_0^T h_z^0 \Gamma_s^u dB_s^u. \quad (10.96)$$

Applying Itô's rule, one obtains immediately that

$$\nabla Y_0^u = E^u \left\{ \int_0^T \Gamma_t^u [h_u^0(t) + b_u^0(t) \sigma^{-1}(t) Z_t^u] \Delta u_t dt \right\}. \quad (10.97)$$

Then, following the same argument as in Theorem 10.1.10 we have

Theorem 10.4.21 *Assume Assumptions 10.4.4, 10.4.15 and 10.4.16 hold. If $u^* \in \mathcal{U}$ is an optimal control for the optimization problem (10.68) and u^* is an interior point of $\mathcal{U}$, then*

$$h_u(t, Y_t^{u^*}, Z_t^{u^*}, u_t^*) + b_u(t, u_t^*) \sigma^{-1}(t) Z_t^{u^*} = 0. \quad (10.98)$$

Moreover, we have

Theorem 10.4.22 *Assume:*

(i) *all the conditions in Theorem 10.4.21 hold;*
(ii) *there exists a unique function $u = I(t, y, z)$ taking values in U such that*

$$h_u(t, y, z, I(t, y, z)) + b_u(t, I(t, y, z)) \sigma^{-1}(t) z = 0, \quad (10.99)$$

and I is differentiable in (y, z);

10.4 Stochastic Optimization in Weak Formulation

(iii) ξ and Y^{u^*} are bounded, and (10.79) holds, with f^* redefined by

$$f^*(t, y, z) := f(t, y, z, I(t, y, z)). \qquad (10.100)$$

Then, BSDE (10.81) is well-posed with a solution $(Y^*, Z^*) := (Y^{u^*}, Z^{u^*})$, and optimal control u^* satisfies (10.80).

Proof First, note that $u_t^* = I(t, Y_t^{u^*}, Z_t^{u^*})$. Then, one can check straightforwardly that (Y^{u^*}, Z^{u^*}) is a solution to (10.81). Next, by (10.99),

$$\begin{aligned}
f_y^*(t, y, z) &= h_y(t, y, z, I(t, y, z)) + h_u(t, y, z, I(t, y, z))I_y(t, y, z) \\
&\quad + b_u I_y(t, y, z)\sigma^{-1}(t)z \\
&= h_y(t, y, z, I(t, y, z)); \\
f_z^*(t, y, z) &= h_z(t, y, z, I(t, y, z)) + h_u(t, y, z, I(t, y, z))I_z(t, y, z) \\
&\quad + b_u I_z(t, y, z)\sigma^{-1}(t)z + b(t, I(t, y, z))\sigma^{-1}(t) \\
&= h_z(t, y, z, I(t, y, z)) + b(t, I(t, y, z))\sigma^{-1}(t).
\end{aligned} \qquad (10.101)$$

Then, (10.79) implies (9.48). Applying Theorem 9.6.3, we know BSDE (10.81) is well-posed among those solutions whose component Y is bounded. $\square$

Remark 10.4.23 By imposing stronger conditions on $\mathcal{U}$ so as to make $\mathcal{U}$ smaller, it will be easier to check the necessary conditions for the optimal control in Theorems 10.4.21 and 10.4.22. However, that will make it more difficult to check that u^* defined by (10.80) in Theorem 10.4.8 is in $\mathcal{U}$. In the case in which $h = -\frac{u^2}{2}$, we will actually consider a larger set $\mathcal{U}$.

10.4.4 Stochastic Optimization for High-Dimensional BSDEs

As in Sect. 10.3, we consider the case where the controlled BSDE is associated with another BSDE:

$$V := \sup_{(a, u) \in \mathcal{A}} Y_0^{2, a, u} \qquad (10.102)$$

where, for $i = 1, 2$,

$$\begin{aligned}
Y_t^{i, a, u} &= g_i(a) + \int_t^T h_i(s, Y_s^{a, u}, Z_s^{a, u}, u_s) ds - \int_t^T Z_s^{i, a, u} dB_s^u \\
&= g_i(a) + \int_t^T f_i(s, Y_s^{a, u}, Z_s^{a, u}, u_s) ds - \int_t^T Z_s^{i, a, u} dB_s, \quad (10.103)
\end{aligned}$$

and

$$f_i(t, y, z, u) := h_i(t, y, z, u) + b(t, u)\sigma^{-1}(t)z_i. \qquad (10.104)$$

We recall that there is no general theory for high-dimensional BSDEs with quadratic growth. However, by assuming h_i are uniformly Lipschitz continuous in (y, z), even though f_i are not uniformly Lipschitz continuous in z, we will be able to get the desired results.

Similarly to Sect. 10.3, we could formally derive some sufficient conditions here. However, since they are not that useful, we study necessary conditions only.

Note that, for $i, j = 1, 2$,

$$\partial_{y_j} f_i = \partial_{y_j} h_i, \qquad \partial_{z_j} f_i = \partial_{z_j} h_i + b(t, u)\sigma^{-1}(t) 1_{\{i=j\}},$$
$$\partial_u f_i = \partial_u h_i + \partial_u b(t, u)\sigma_t^{-1} z_i.$$

Given (a, u) and $(\Delta a, \Delta u)$, (10.39) becomes

$$\nabla Y_t^{i,a,u} = g_i'(a)\Delta a - \int_t^T \nabla Z_s^{i,a,u} dB_s^u$$
$$+ \int_t^T \left[\partial_u h_i(s, Y_s^u, Z_s^u, u_s) + \partial_u b(s, u_s)\sigma^{-1}(s) Z_s^{i,a,u} \right] \Delta u_s ds$$
$$+ \int_t^T \sum_{j=1}^2 \left[\partial_{y_j} h_i(s, Y_s^{a,u}, Z_s^{a,u}, u_s) \nabla Y_s^{j,a,u} \right.$$
$$\left. + \partial_{z_j} h_i(s, Y_s^{a,u}, Z_s^{a,u}, u_s) \nabla Z_s^{j,a,u} \right] ds. \tag{10.105}$$

Similarly to (10.40), we introduce the following adjoint processes

$$d\Gamma_t^{i,a,u} = \sum_{j=1}^2 [\Gamma_t^{j,a,u} \partial_{y_i} h_j(t, Y_t^{a,u}, Z_t^{a,u}, u_t) dt$$
$$+ \Gamma_t^{j,a,u} \partial_{z_i} h_j(t, Y_t^{a,u}, Z_t^{a,u}, u_t) dB_t^u], \quad i = 1, 2,$$
$$\text{with } \Gamma_0^{1,a,u} := 0, \ \Gamma_0^{2,a,u} := 1. \tag{10.106}$$

Applying Itô's rule, we have

$$d\left(\sum_{i=1}^2 \Gamma_t^{i,a,u} \nabla Y_t^{i,a,u} \right)$$
$$= \left[\sum_{i=1}^2 \Gamma_t^{i,a,u} \nabla Z_t^{i,a,u} + \sum_{i,j=1}^2 \Gamma_t^{j,a,u} \partial_{z_i} h_j(t, Y_t^{a,u}, Z_t^{a,u}, u_t) \nabla Y_t^{i,a,u} \right] dB_t^u$$
$$- \sum_{i=1}^2 \Gamma_t^{i,a,u} \left[\partial_u h_i(t, Y_t^{a,u}, Z_t^{a,u}, u_t) + \partial_u b(t, u_t)\sigma_t^{-1} Z_t^{i,a,u} \right] \Delta u_t dt.$$

Then,

10.4 Stochastic Optimization in Weak Formulation

$$\nabla Y_0^{2,a,u} = E^u \left\{ \sum_{i=1}^{i} \Gamma_T^{i,a,u} g_i'(a) \Delta a \right.$$

$$+ \int_0^T \sum_{i=1}^{2} \Gamma_t^{i,a,u} \left[\partial_u h_i(t, Y_t^{a,u}, Z_t^{a,u}, u_t) \right.$$

$$\left. \left. + \partial_u b(t, u_t) \sigma_t^{-1} Z_t^{i,a,u} \right] \Delta u_t dt \right\}. \tag{10.107}$$

We now specify the technical conditions.

Assumption 10.4.24

(i) Coefficient b is progressively measurable and $\mathbb{F}$-adapted, continuously differentiable in u, and $\partial_u b \sigma^{-1}$ is bounded.
(ii) For $i = 1, 2$, g_i is progressively measurable and $\mathcal{F}_T$-measurable, and continuously differentiable in a.
(iii) For $i = 1, 2$, h_i is progressively measurable and $\mathbb{F}$-adapted, continuously differentiable in (y, z) with uniformly bounded derivatives, and continuously differentiable in u with

$$\left| \partial_u h_i(t, y, z, u) \right| \le C \left[1 + \left| \partial_u h_i(t, 0, 0, u) \right| + |y| + |z| \right].$$

Assumption 10.4.25 Admissible set $\mathcal{A}$ is a set of pairs (a, u), where a is an $\mathcal{F}_T$-measurable random variable taking values in A and u is an $\mathbb{F}$-adapted processes taking values in U, satisfying:

(i) For each $(a, u) \in \mathcal{A}$, (10.77) holds.
(ii) $\mathcal{A}$ is locally convex, in the sense of Assumption 10.1.7(ii).
(iii) For (a, u), $(\Delta a, \Delta u)$, and $(a^\varepsilon, u^\varepsilon)$ as in Assumption 10.1.7(ii), there exists $\delta > 0$ such that, for $i = 1, 2$,

$$\sup_{0 \le \varepsilon \le 1} E^{u^\varepsilon} \left\{ \left| g_i(a^\varepsilon) \right|^{2+\delta} + \left| g_i'(a^\varepsilon) \right|^{2+\delta} \right.$$

$$\left. + \left(\int_0^T [|h_i| + |\partial_u h_i|](t, 0, 0, u_t^\varepsilon) dt \right)^{2+\delta} \right\} < \infty.$$

Combining the arguments of Theorem 10.3.4 and those in Sect. 10.4.3, we obtain

Theorem 10.4.26 *Assume Assumptions* 10.4.4, 10.4.24 *and* 10.4.25 *hold.*

(i) *If* $(a^*, u^*) \in \mathcal{A}$ *is an optimal control for the optimization problem* (10.102) *and* (a^*, u^*) *is an interior point of* $\mathcal{A}$, *then*

$$\sum_{i=1}^{2} \Gamma_T^{i,a^*,u^*} g_i'(a^*) = 0,$$

$$\sum_{i=1}^{2} \Gamma_t^{i,a^*,u^*} \left[\partial_u h_i(t, Y_t^{a^*,u^*}, Z_t^{a^*,u^*}, u_t^*) + \partial_u b(t, u_t^*) \sigma_t^{-1} Z_t^{i,a^*,u^*} \right] = 0.$$
(10.108)

(ii) *Assume further that there exist unique functions $I_1(\gamma)$ and $I_2(t, y, z, \gamma)$ such that they are differentiable in (y, z, γ) and*

$$\sum_{i=1}^{2} \gamma_i g_i'(I_1(\gamma)) = 0,$$

$$\sum_{i=1}^{2} \gamma_i \left[\partial_u h_i(t, y, z, I_2(t, y, z, \gamma)) + \partial_u b(t, I_2(t, y, z, \gamma)) \sigma_t^{-1} z_i \right] = 0.$$
(10.109)

Let g_i^* and φ^* be defined as in (10.45). Then, $(\Gamma^*, Y^*, Z^*) := (\Gamma^{a^*,u^*}, Y^{a^*,u^*}, Z^{a^*,u^*})$ satisfies the following coupled FBSDE:

$$\begin{cases} d\Gamma_t^{i,*} = \sum_{j=1}^{2} \Gamma_t^{j,*} \left[(\partial_{y_i} h_j)^*(t, Y_t^*, Z_t^*, \Gamma_t^*) dt + (\partial_{z_i} h_j)^*(t, Y_t^*, Z_t^*, \Gamma_t^*) dB_t^{u^*} \right]; \\ dY_t^{i,*} = -h_i^*(t, Y_t^*, Z_t^*, \Gamma_t^*) dt + Z_t^{i,*} dB_t^{u^*}; \\ \Gamma_0^{1,*} = 0, \qquad \Gamma_0^{2,*} = 1, \qquad Y_T^{i,*} = g_i^*(\Gamma_T^*), \end{cases}$$
(10.110)

and the optimal control satisfies (10.47).

10.4.5 Stochastic Optimization for FBSDEs

As in Sect. 10.2, we consider the case where the controlled BSDE is associated with an SDE:

$$V := \sup_{u \in \mathcal{A}} Y_0^u \qquad (10.111)$$

where

$$\begin{cases} \tilde{X}_t^u = \tilde{x} + \int_0^t \tilde{b}(s, \tilde{X}_s^u, u_s) ds + \int_0^t \tilde{\sigma}(s, \tilde{X}_s^u, u_s) dB_s^u; \\ Y_t^u = g(\tilde{X}_T^u) + \int_t^T h(s, \tilde{X}_s^u, Y_s^u, Z_s^u, u_s) ds - \int_t^T Z_s^u dB_s^u. \end{cases} \qquad (10.112)$$

We emphasize that $\tilde{X}^u$ here is different from the fixed underlying process X in (10.65). By allowing $\tilde{b}$ and $\tilde{\sigma}$ to be random, they actually depend on both $\tilde{X}^u$ and X, but the variable X is omitted for notational simplification.

BSDE (10.112) can be rewritten in the form of BSDE (10.19) with coefficients

10.4 Stochastic Optimization in Weak Formulation

$$\bar{b} := \tilde{b}(t,\tilde{x},u) - \tilde{\sigma}(t,\tilde{x},u)b(t,u)\sigma_t^{-1}, \qquad \bar{\sigma} := \tilde{\sigma}(t,\tilde{x},u),$$
$$\bar{f} := h(t,\tilde{x},y,z,u) + b(t,u)\sigma_t^{-1}z.$$

Given u and Δu, (10.21) becomes

$$\begin{aligned}
\nabla \tilde{X}_t &= \int_0^t \big[\bar{b}_x(s)\nabla \tilde{X}_s + [\bar{b}_u(s) - \tilde{\sigma}(s)b_u(s)\sigma_s^{-1}]\Delta u_s\big]ds \\
&\quad + \int_0^t \big[\tilde{\sigma}_x(s)\nabla \tilde{X}_s + \tilde{\sigma}_u(s)\Delta u_s\big]dB_s^u; \\
\nabla Y_t &= g_x(\tilde{X}_T^u)\nabla \tilde{X}_T - \int_t^T \nabla Z_s dB_s^u \qquad (10.113) \\
&\quad + \int_t^T \big[h_x(s)\nabla \tilde{X}_s + h_y(s)\nabla Y_s + h_z(s)\nabla Z_s \\
&\quad + [h_u(s) + b_u(s)\sigma_s^{-1}]\Delta u_s\big]ds,
\end{aligned}$$

where, for any function φ,

$$\varphi(s) := \varphi\big(s, \tilde{X}_s^u, Y_s^u, Z_s^u, u_s\big). \qquad (10.114)$$

Similarly to (10.24), we introduce adjoint processes

$$\begin{aligned}
\Gamma_t^u &= 1 + \int_0^t h_y(s)\Gamma_s^u ds + \int_0^t h_z(s)\Gamma_s^u dB_s^u; \\
\bar{Y}_t^u &= g_x(X_T^u)\Gamma_T^u + \int_t^T \big[h_x(s)\Gamma_s^u + \bar{b}_x(s)\bar{Y}_s^u + \tilde{\sigma}_x(s)\bar{Z}_s^u\big]ds - \int_t^T \bar{Z}_s^u dB_s^u.
\end{aligned} \qquad (10.115)$$

By applying Itô's rule on $\Gamma_t^u \nabla Y - \bar{Y}^u \nabla \tilde{X}$, we have

$$\begin{aligned}
\nabla Y_0 = E^u \bigg\{ \int_0^T &\big[\Gamma_t^u [h_u(t) + b_u(t)\sigma_t^{-1}] \\
&+ \bar{Y}_t^u [\bar{b}_u(t) - \tilde{\sigma}(t)b_u(t)\sigma_t^{-1}] + \bar{Z}_t^u \tilde{\sigma}_u(t)\big]\Delta u_t dt \bigg\}.
\end{aligned}$$

We now specify the technical conditions we need.

Assumption 10.4.27

(i) b is progressively measurable and $\mathbb{F}$-adapted, continuously differentiable in u, and $\partial_u b \sigma^{-1}$ is bounded.
(ii) $\tilde{b}, \tilde{\sigma}, h, g$ are progressively measurable, g is $\mathcal{F}_T$-measurable, and $\tilde{b}, \tilde{\sigma}, h$ are $\mathbb{F}$-adapted.
(iii) $\tilde{b}, \tilde{\sigma}, h, g$ are continuously differentiable in (x,y,z) with uniformly bounded derivatives.
(iv) $\tilde{b}, \tilde{\sigma}, h$ are continuously differentiable in u, and for $\varphi = \tilde{b}, \tilde{\sigma}, h$,

$$\big|\varphi_u(t,\tilde{x},y,z,u)\big| \leq C\big[1 + \big|\varphi_u(t,0,0,0,u)\big| + |\tilde{x}| + |y| + |z|\big].$$

Assumption 10.4.28 Agent's admissible set $\mathcal{U}$ is a set of $\mathbb{F}$-adapted processes u taking values in U satisfying:

(i) For each $u \in \mathcal{U}$, (10.77) holds.
(ii) $\mathcal{U}$ is locally convex, in the sense of Assumption 10.1.7(ii).
(iii) For $u, \Delta u, u^\varepsilon$ as in Assumption 10.1.7(ii), there exists $\delta > 0$ such that, for $\varphi = \tilde{b}, \tilde{\sigma}, h$,

$$\sup_{0 \leq \varepsilon \leq 1} E^{u^\varepsilon} \left\{ |g(0)|^{2+\delta} + \left(\int_0^T [|\varphi| + |\partial_u f|](t, 0, 0, 0, u_t^\varepsilon) dt \right)^{2+\delta} \right\} < \infty.$$

Combining the arguments of Theorem 10.2.5 and those in Sect. 10.4.3, we get

Theorem 10.4.29 *Assume Assumptions* 10.4.4, 10.4.27 *and* 10.4.28 *hold.*

(i) *If $u^* \in \mathcal{U}$ is an optimal control for the optimization problem* (10.111) *and u^* is an interior point of $\mathcal{U}$, then*

$$\Gamma_t^{u^*} \left[h_u(t, \tilde{X}_t^{u^*}, Y_t^{u^*}, Z_t^{u^*}, u_t^*) + b_u(t, u_t^*) \sigma_t^{-1} \right]$$
$$+ \bar{Y}_t^{u^*} \left[\tilde{b}_u(t, \tilde{X}_t^{u^*}, u_t^*) - \tilde{\sigma}(t, \tilde{X}_t^{u^*}, u_t^*) b_u(t, u_t^*) \sigma_t^{-1} \right] + \bar{Z}_t^{u^*} \tilde{\sigma}_u(t, \tilde{X}_t^{u^*}, u_t^*) = 0.$$

(ii) *Assume further that there exists a unique function $I(t, x, y, z, \gamma, \bar{y}, \bar{z})$ differentiable in (y, z, γ) and such that*

$$\gamma \left[h_u(t, \tilde{x}, y, z, I) + b_u(t, u) \sigma_t^{-1} \right] + \bar{y} \left[\tilde{b}_u(t, \tilde{x}, I) - \tilde{\sigma}(t, \tilde{x}, u) b_u(t, u) \sigma_t^{-1} \right]$$
$$+ \bar{z} \tilde{\sigma}_u(t, \tilde{x}, I) = 0.$$

Then, $(\tilde{X}^*, \Gamma^*, Y^*, \bar{Y}^*, Z^*, \bar{Z}^*) := (\tilde{X}^{u^*}, \Gamma^{u^*}, Y^{u^*}, \bar{Y}^{u^*}, Z^{u^*}, \bar{Z}^{u^*})$ satisfies the following coupled FBSDE, defining φ^* as in (10.28) *for the above function I,*

$$\begin{cases}
\tilde{X}_t^* = x + \int_0^t \tilde{b}^*(s, \tilde{X}_s^*, \Gamma_s^*, Y_s^*, \bar{Y}_s^*, Z_s^*, \bar{Z}_s^*) ds \\
\qquad + \int_0^t \tilde{\sigma}^*(s, \tilde{X}_s^*, \Gamma_s^*, Y_s^*, \bar{Y}_s^*, Z_s^*, \bar{Z}_s^*) dB_s^{u^*}; \\
\Gamma_t^* = 1 + \int_0^t (\partial_y h)^*(s, \tilde{X}_s^*, \Gamma_s^*, Y_s^*, \bar{Y}_s^*, Z_s^*, \bar{Z}_s^*) \Gamma_s^* ds \\
\qquad + \int_0^t (\partial_z h)^*(s, \tilde{X}_s^*, \Gamma_s^*, Y_s^*, \bar{Y}_s^*, Z_s^*, \bar{Z}_s^*) \Gamma_s^* dB_s^{u^*}; \\
Y_t^* = g(\tilde{X}_T^*) + \int_t^T h^*(s, \tilde{X}_s^*, \Gamma_s^*, Y_s^*, \bar{Y}_s^*, Z_s^*, \bar{Z}_s^*) ds - \int_t^T Z_s^* dB_s^{u^*}; \\
\bar{Y}_s^* = g_x(\tilde{X}_T^*) \Gamma_T^* - \int_t^T \bar{Z}_s^* dB_s^{u^*} + \int_t^T \left[(\partial_x h)^*(s, \tilde{X}_s^*, \Gamma_s^*, Y_s^*, \bar{Y}_s^*, Z_s^*, \bar{Z}_s^*) \Gamma_s^* \right. \\
\qquad + (\partial_x \tilde{b})^*(s, \tilde{X}_s^*, \Gamma_s^*, Y_s^*, \bar{Y}_s^*, Z_s^*, \bar{Z}_s^*) \bar{Y}_s^* \\
\qquad + \left. (\partial_x \tilde{\sigma})^*(s, X_s^*, \Gamma_s^*, Y_s^*, \bar{Y}_s^*, Z_s^*, \bar{Z}_s^*) \bar{Z}_s^* \right] ds,
\end{cases} \qquad (10.116)$$

and the optimal control u^* satisfies

$$u_t^* = I(t, \tilde{X}_t^*, \Gamma_t^*, Y_t^*, \bar{Y}_t^*, Z_t^*, \bar{Z}_t^*). \tag{10.117}$$

10.5 Some Technical Proofs

In this section we provide the proofs for some results of Chaps. 4 and 5.

10.5.1 Heuristic Derivation of the Results of Sect. 4.7

We follow the arguments in Sect. 10.2. Given a perturbation $(\Delta C_T, \Delta u, \Delta v, \Delta c, \Delta e)$, we have

$$\nabla X_t = \int_0^t \Delta v_s dB_s + \int_0^t \left[\partial_x b(s)\nabla X_s + \partial_u b(s)\Delta u_s + \partial_v b(s)\Delta v_s + \partial_c b(s)\Delta c_s \right. \\ \left. + \partial_e b(s)\Delta e_s\right]ds;$$

$$\nabla G_t = \int_0^t \left[\partial_x g(s)\nabla X_s + \partial_u g(s)\Delta u_s + \partial_v g(s)\Delta v_s + \partial_c g(s)\Delta c_s \right. \\ \left. + \partial_e g(s)\Delta e_s\right]ds;$$

$$\nabla W_t^A = \partial_X U_A(T)\nabla X_T + \partial_C U_A(T)\Delta C_T + \partial_G U_A(T)\nabla G_T - \int_t^T \nabla Z_s^A dB_s$$
$$+ \int_t^T \left[\partial_x u_A(s)\nabla X_s + \partial_u u_A(s)\Delta u_s + \partial_v u_A(s)\Delta v_s + \partial_c u_A(s)\Delta c_s \right.$$
$$\left. + \partial_e u_A(s)\Delta e_s + \partial_y u_A(s)\nabla W_s^A + \partial_z u_A(s)\nabla Z_s^A\right]ds;$$

$$\nabla W_t^P = \partial_X U_P(T)\nabla X_T + \partial_C U_P(T)\Delta C_T - \int_t^T \nabla Z_s^P dB_s$$
$$+ \int_t^T \left[\partial_x u_P(s)\nabla X_s + \partial_v u_P(s)\Delta v_s + \partial_c u_P(s)\Delta c_s + \partial_y u_P(s)\nabla W_s^P \right.$$
$$\left. + \partial_z u_P(s)\nabla Z_s^P\right]ds.$$

Recall the adjoint processes defined in (4.48). Then, we have

$$d\big(\Gamma_t^A \nabla W_t^A\big) = [\cdots]dB_t - \Gamma_t^A\big[\partial_x u_A(t)\nabla X_t + \partial_u u_A(t)\Delta u_t + \partial_v u_A(t)\Delta v_t$$
$$+ \partial_c u_A(t)\Delta c_t + \partial_e u_A(t)\Delta e_t\big]dt;$$

$$d\big(\Gamma_t^P \nabla W_t^P\big) = [\cdots]dB_t - \Gamma_t^P\big[\partial_x u_P(t)\nabla X_t + \partial_v u_P(s)\Delta v_s + \partial_c u_P(t)\Delta c_t\big]dt;$$

$$d\big(\bar{Y}_t^1 \nabla X_t\big) = [\cdots]dB_t + \big[\bar{Z}_t^1 \Delta v_t - \big[\Gamma_t^P \partial_x u_P(t) + \lambda \Gamma_t^A \partial_x u_A(t)$$
$$+ \bar{Y}_t^2 \partial_x g(t)\big]\nabla X_t\big]dt + \bar{Y}_t^1\big[\partial_u b(t)\Delta u_t + \partial_v b(t)\Delta v_t + \partial_c b(t)\Delta c_t$$
$$+ \partial_e b(t)\Delta e_t\big]dt;$$

$$d\big(\bar{Y}_t^2 \nabla G_t\big) = [\cdots]dB_t$$
$$+ \bar{Y}_t^2\big[\partial_x g(t)\nabla X_t + \partial_u g(t)\Delta u_t + \partial_v g(t)\Delta v_t + \partial_c g(t)\Delta c_t$$
$$+ \partial_e g(t)\Delta e_t\big]dt;$$

and
$$d\left(\left[\Gamma_t^P \nabla W_t^P + \lambda \Gamma_t^A \nabla W_t^A\right] - \left[\bar{Y}_t^1 \nabla X_t + \bar{Y}_t^2 \nabla G_t\right]\right)$$
$$= [\cdots]dB_t - \Big[\big[\lambda\Gamma_t^A \partial_u u_A(t) + \bar{Y}_t^1 \partial_u b(t) + \bar{Y}_t^2 \partial_u g(t)\big]\Delta u_t$$
$$+ \big[\Gamma_t^P \partial_v u_P(t) + \lambda\Gamma_t^A \partial_v u_A(t) + \bar{Y}_t^1 \partial_v b(t) + \bar{Y}_t^2 \partial_v g(t) + \bar{Z}_t^1\big]\Delta v_t$$
$$+ \big[\Gamma_t^P \partial_c u_P(t) + \lambda\Gamma_t^A \partial_c u_A(t) + \bar{Y}_t^1 \partial_c b(t) + \bar{Y}_t^2 \partial_c g(t)\big]\Delta c_t$$
$$+ \big[\lambda\Gamma_t^A \partial_e u_A(t) + \bar{Y}_t^1 \partial_e b(t) + \bar{Y}_t^2 \partial_e g(t)\big]\Delta e_t\Big]dt.$$

Thus,
$$\nabla W_0^P + \lambda \nabla W_0^A$$
$$= \Gamma_0^P \nabla W_0^P + \lambda \Gamma_0^A \nabla W_0^A - \bar{Y}_0 \nabla X_0$$
$$= E\bigg\{\big[\Gamma_T^P \partial_C U_P(T) + \lambda \Gamma_T^A \partial_C U_A(T)\big]\Delta C_T$$
$$+ \int_0^T \Big[\big[\lambda\Gamma_t^A \partial_u u_A(t) + \bar{Y}_t^1 \partial_u b(t) + \bar{Y}_t^2 \partial_u g(t)\big]\Delta u_t$$
$$+ \big[\Gamma_t^P \partial_v u_P(t) + \lambda\Gamma_t^A \partial_v u_A(t) + \bar{Y}_t^1 \partial_v b(t) + \bar{Y}_t^2 \partial_v g(t) + \bar{Z}_t^1\big]\Delta v_t$$
$$+ \big[\Gamma_t^P \partial_c u_P(t) + \lambda\Gamma_t^A \partial_c u_A(t) + \bar{Y}_t^1 \partial_c b(t) + \bar{Y}_t^2 \partial_u g(t)\big]\Delta c_t$$
$$+ \big[\lambda\Gamma_t^A \partial_e u_A(t) + \bar{Y}_t^1 \partial_e b(t) + \bar{Y}_t^2 \partial_u g(t)\big]\Delta e_t\Big]dt\bigg\}. \quad (10.118)$$

The statements from Sect. 4.7 follow from this.

10.5.2 Heuristic Derivation of the Results of Sect. 5.5

We proceed similar to Sect. 10.5.1. We first prove the necessary conditions (5.111) for the agent's problem. Given a perturbation $(\Delta u, \Delta e)$, we have
$$\nabla G_t = \int_0^t \big[\partial_u g(s)\Delta u_s + \partial_e g(s)\Delta e_s\big]ds;$$
$$\nabla W_t^A = \partial_G U_A(T)\nabla G_T - \int_t^T \nabla Z_s^A dB_s^u + \int_t^T \big[\partial_u u_A(s)\Delta u_s$$
$$+ \partial_e u_A(s)\Delta e_s + \partial_y u_A(s)\nabla W_s^A + \partial_z u_A(s)\nabla Z_s^A + Z_s^A \Delta u_s\big]ds.$$

Recall the adjoint processes defined in (5.110). Then, for the W^A in (5.107) we have
$$d\big(\Gamma_t^A \nabla W_t^A\big) = [\cdots]dB_t^u - \Gamma_t^A\big[\big[\partial_u u_A(t) + Z_t^A\big]\Delta u_t + \partial_e u_A(t)\Delta e_t\big]dt;$$
$$d\big(\bar{Y}_t^A \nabla G_t\big) = [\cdots]dB_t^u + \bar{Y}_t^A\big[\partial_u g(t)\Delta u_t + \partial_e g(t)\Delta e_t\big]dt;$$

and
$$d\big(\Gamma_t^A \nabla W_t^A - \bar{Y}_t^A \nabla G_t\big) = [\cdots]dB_t^u - \Big[\big[\Gamma_t^A \partial_u u_A(t) + \Gamma_t^A Z_t^A + \bar{Y}_t^A \partial_u g(t)\big]\Delta u_t$$
$$+ \big[\Gamma_t^A \partial_e u_A(t) + \bar{Y}_t^A \partial_e g(t)\big]\Delta e_t\Big]dt.$$

Thus,

10.5 Some Technical Proofs

$$\nabla W_0^A = \Gamma_0^A \nabla W_0^A - \bar{Y}_0^A \nabla G_0$$
$$= E\left\{\int_0^T \left[[\Gamma_t^A \partial_u u_A(t) + \Gamma_t^A Z_t^A + \bar{Y}_t^A \partial_u g(t)]\Delta u_t \right.\right.$$
$$\left.\left. + [\Gamma_t^A \partial_e u_A(t) + \bar{Y}_t^A \partial_e g(t)]\Delta e_t\right]dt\right\}. \quad (10.119)$$

This implies (5.111).

We now derive formally necessary conditions (5.119) for the principal's problem. Given perturbation $(\Delta u, \Delta v, \Delta c)$, by (5.116) we have

$$\nabla X_t = \int_0^t \Delta v_s dB_s = \int_0^t u_s \Delta v_s ds + \int_0^t \Delta v_s dB_s^u;$$

$$\nabla G_t = \int_0^t \left[\partial_x \hat{g}(s)\nabla X_s + \partial_u \hat{g}(s)\Delta u_s + \partial_v \hat{g}(s)\Delta v_s + \partial_c \hat{g}(s)\Delta c_s + \partial_\Gamma \hat{g}(s)\nabla \Gamma_s^A \right.$$
$$\left. + \partial_{\bar{y}}\hat{g}(s)\nabla \bar{Y}_s^A\right]ds;$$

$$\nabla \Gamma_t^A = \int_0^t \left[\nabla \Gamma_s^A \widehat{\partial_y u_A}(s) + \Gamma_s^A \left[\partial_x \widehat{\partial_y u_A}(s)\nabla X_s + \partial_u \widehat{\partial_y u_A}(s)\Delta u_s\right.\right.$$
$$+ \partial_v \widehat{\partial_y u_A}(s)\Delta v_s + \partial_c \widehat{\partial_y u_A}(s)\Delta c_s$$
$$\left.\left. + \partial_\Gamma \widehat{\partial_y u_A}(s)\nabla \Gamma_s^A + \partial_{\bar{y}}\widehat{\partial_y u_A}(s)\nabla \bar{Y}_s^A + \partial_y \widehat{\partial_y u_A}(s)\nabla W_s^A\right]\right.$$
$$\left. - \Gamma_s^A \widehat{\partial_y u_A}(s)\Delta u_s\right]ds$$
$$+ \int_0^t \left[\nabla \Gamma_s^A \widehat{\partial_z u_A}(s) + \Gamma_s^A \left[\partial_x \widehat{\partial_z u_A}(s)\nabla X_s + \partial_u \widehat{\partial_z u_A}(s)\Delta u_s\right.\right.$$
$$+ \partial_v \widehat{\partial_z u_A}(s)\Delta v_s + \partial_c \widehat{\partial_z u_A}(s)\Delta c_s$$
$$\left.\left. + \partial_\Gamma \widehat{\partial_z u_A}(s)\nabla \Gamma_s^A + \partial_{\bar{y}}\widehat{\partial_z u_A}(s)\nabla \bar{Y}_s^A + \partial_y \widehat{\partial_z u_A}(s)\nabla W_s^A\right]\right]dB_s^u;$$

$$\nabla W_t^A = -\int_0^t \left[\partial_x \hat{u}_A(s)\nabla X_s + \partial_u \hat{u}_A(s)\Delta u_s + \partial_v \hat{u}_A(s)\Delta v_s + \partial_c \hat{u}_A(s)\Delta c_s\right.$$
$$\left. + \partial_\Gamma \hat{u}_A(s)\nabla \Gamma_s^A + \partial_{\bar{y}}\hat{u}_A(s)\nabla \bar{Y}_s^A + \partial_y \hat{u}_A(s)\nabla W_s^A - J_1^A(s)\Delta u_s\right]ds$$
$$+ \int_0^t \left[\partial_x J_1^A(s)\nabla X_s + \partial_u J_1^A(s)\Delta u_s + \partial_v J_1^A(s)\Delta v_s + \partial_c J_1^A(s)\Delta c_s\right.$$
$$\left. + \partial_\Gamma J_1^A(s)\nabla \Gamma_s^A + \partial_{\bar{y}}J_1^A(s)\nabla \bar{Y}_s^A\right]dB_s^u;$$

$$\nabla W_t^P = \partial_x \hat{U}_P(T)\nabla X_T + \partial_y \hat{U}_P(T)\nabla W_T^A + \partial_G \hat{U}_P(T)\nabla G_T - \int_t^T \nabla Z_s^P dB_s^u$$
$$+ \int_t^T \left[\partial_x u_P(s)\nabla X_s + \partial_v u_P(s)\Delta v_s + \partial_c u_P(s)\Delta c_s\right.$$
$$\left. + \partial_y u_P(s)\nabla W_s^P + \partial_z u_P(s)\nabla Z_s^P + Z_s^P \Delta u_s\right]ds;$$

$$\nabla \bar{Y}_t^A = \partial_x \widehat{\partial_G U_A}(T)\nabla X_T + \partial_y \widehat{\partial_G U_A}(T)\nabla W_T^A + \partial_G \widehat{\partial_G U_A}(T)\nabla G_T$$
$$+ \widehat{\partial_G U_A}(T)\nabla \Gamma_T^A + \int_t^T \bar{Z}_s^A \Delta u_s ds - \int_t^T \nabla \bar{Z}_s^A dB_s^u.$$

Recall the adjoint processes defined in (5.118). Then, we have

$$d\big([\Gamma_t^1 \nabla W_t^P + \Gamma_t^2 \nabla \bar{Y}_t^A] - [\bar{Y}_t^1 \nabla X_t + \bar{Y}_t^2 \nabla G_t + \bar{Y}_t^3 \nabla \Gamma_t^A + \bar{Y}_t^4 \nabla W_t^A]\big)$$
$$= [\cdots] dB_t^u - \big[\Lambda_u(t)\Delta u_t + \Lambda_v(t)\Delta v_t + \Lambda_c(t)\Delta c_t\big]dt;$$

where

$$\Lambda_u := \Gamma^1 Z^P + \Gamma^2 \bar{Z}^A + \bar{Y}^2 \partial_u \hat{g} + \bar{Y}^3 \Gamma^A [\partial_u \widehat{\partial_y u_A} - \widehat{\partial_y u_A}] + \bar{Z}^3 \Gamma^A \partial_u \widehat{\partial_z u_A}$$
$$\quad - \bar{Y}^4 [\partial_u \hat{u}_A - J_1^A] + \bar{Z}^4 \partial_u J_1^A;$$

$$\Lambda_v := \Gamma^1 \partial_v u_P + \bar{Y}^1 u + \bar{Z}^1 + \bar{Y}^2 \partial_v \hat{g} + \bar{Y}^3 \Gamma^A \partial_v \widehat{\partial_y u_A} + \bar{Z}^3 \Gamma^A \partial_v \widehat{\partial_z u_A} - \bar{Y}^4 \partial_v \hat{u}_A$$
$$\quad + \bar{Z}^4 \partial_v J_1^A;$$

$$\Lambda_c := \Gamma^1 \partial_c u_P + \bar{Y}^2 \partial_c \hat{g} + \bar{Y}^3 \Gamma^A \partial_c \widehat{\partial_y u_A} + \bar{Z}^3 \Gamma^A \partial_c \widehat{\partial_z u_A} - \bar{Y}^4 \partial_c \hat{u}_A + \bar{Z}^4 \partial_c J_1^A.$$

Thus,

$$\nabla W_0^P = [\Gamma_0^1 \nabla W_0^P + \Gamma_0^2 \nabla \bar{Y}_0^A] - [\bar{Y}_0^1 \nabla X_0 + \bar{Y}_0^2 \nabla G_0 + \bar{Y}_0^3 \nabla \Gamma_0^A + \bar{Y}_0^4 \nabla W_0^A]$$
$$= E\bigg\{\int_0^T \big[\Lambda_u(t)\Delta u_t + \Lambda_v(t)\Delta v_t + \Lambda_c(t)\Delta c_t\big]dt\bigg\}. \qquad (10.120)$$

The statements from Sect. 5.5 follow from this.

10.5.3 Sketch of Proof for Theorem 5.2.12

Since the technical setting of Sect. 5.2.2 is slightly different from that in the previous section, we here sketch the proofs and point out the differences. We follow the arguments of Sect. 10.3.

Let $(C_T, c) \in \mathcal{A}$ be given and $(C_T^\varepsilon, c^\varepsilon) \in \mathcal{A}$ be as in Assumption 5.2.11(iii). Recall that $\Delta C_T = C_T^1 - C_T$ and $\Delta c = c^1 - c$ are bounded. Denote

$$\Delta W_t^{A,\varepsilon} := W^{A,C_T^\varepsilon,c^\varepsilon} - W^{A,C_T,c}, \qquad \Delta Z_t^{A,\varepsilon} := Z^{A,C_T^\varepsilon,c^\varepsilon} - Z^{A,C_T,c},$$
$$\Delta W^{P,\lambda,\varepsilon} := W^{P,\lambda,C_T^\varepsilon,c^\varepsilon} - W^{P,\lambda,C_T,c}, \qquad \Delta Z^{P,\lambda,\varepsilon} := Z^{P,\lambda,C_T^\varepsilon,c^\varepsilon} - Z^{P,\lambda,C_T,c},$$
$$\nabla W^{A,\varepsilon} := \frac{1}{\varepsilon}\Delta W^{A,\varepsilon}, \quad \nabla Z^{A,\varepsilon} := \frac{1}{\varepsilon}\Delta Z^{A,\varepsilon}, \quad \nabla W^{P,\lambda,\varepsilon} := \frac{1}{\varepsilon}\Delta W^{P,\lambda,\varepsilon},$$
$$\nabla Z^{P,\lambda,\varepsilon} := \frac{1}{\varepsilon}\Delta Z^{P,\lambda,\varepsilon},$$

and define, recalling (5.17), (5.19), and (10.101),

$$\nabla W_t^A = U_A'(C_T)\Delta C_T + \int_t^T \nabla Z_s^A I_A(t,c,Z_s^A)ds - \int_t^T \nabla Z_s^A dB_s$$
$$\quad + \int_t^T \partial_c u_A(t,c,I_A(t,c,Z_s^A))\Delta c_s ds$$

10.5 Some Technical Proofs

$$\nabla W_t^{P,\lambda} = U_P'(C_T)\Delta C_T + \int_t^T \nabla Z_s^{P,\lambda} I_A(t, c, Z_s^A) ds - \int_t^T \nabla Z_s^{P,\lambda} dB_s \quad (10.121)$$
$$+ \int_t^T \left[\partial_c u_P(t, c)\Delta c_s + \lambda_s \nabla W_s^A + Z_s^{P,\lambda} \partial_c I_A(s, c_s, Z_s^A)\Delta c_s \right.$$
$$\left. + Z_s^{P,\lambda} \partial_z I_A(s, c_s, Z_s^A) \nabla Z_s^A \right] ds.$$

Lemma 10.5.1 *Assume Assumptions 5.2.10 and 5.2.11 hold. Then, BSDE (10.121) is well-posed and*

$$\lim_{\varepsilon \to 0} E \left\{ \sup_{0 \le t \le T} \left[|\nabla W_t^{A,\varepsilon} - \nabla W_t^A|^2 + |\nabla W_t^{P,\lambda,\varepsilon} - \nabla W_t^{P,\lambda}|^2 \right] \right.$$
$$\left. + \int_0^T \left[|\nabla Z_t^{A,\varepsilon} - \nabla Z_t^A|^2 + |\nabla Z_t^{P,\lambda,\varepsilon} - \nabla Z_t^{P,\lambda}|^2 \right] dt \right\} = 0. \quad (10.122)$$

Proof First, recalling (10.101) and the assumption that $I_A \in U$ is bounded, we know BSDE (5.13) and the first equation in (10.121) are well-posed. Moreover, by (5.22) and applying Proposition 9.4.4, we have

$$E \left\{ \sup_{0 \le t \le T} \left[|W_t^A|^4 + |\nabla W_t^A|^4 \right] + \left(\int_0^T \left[|Z_t^A|^2 + |\nabla Z_t^A|^2 \right] dt \right)^2 \right\} < \infty.$$

This implies further that the last equation in (5.21) is well-posed and

$$E \left\{ \sup_{0 \le t \le T} |W_t^{P,\lambda}|^4 + \left(\int_0^T |Z_t^{P,\lambda}|^2 dt \right)^2 \right\} < \infty.$$

Then,

$$E \left\{ \left(\int_0^T |Z_t^{P,\lambda} \nabla Z_t^A| dt \right)^2 \right\} < \infty,$$

and therefore, the second equation in (10.121) is well-posed with solution $(\nabla W^{P,\lambda}, \nabla Z^{P,\lambda}) \in L^2(\mathbb{F}, P) \times L^2(\mathbb{F}, P)$.

Following the arguments in Lemma 10.1.8 and using Proposition 9.4.4, one can easily show that

$$\lim_{\varepsilon \to 0} E \left\{ \sup_{0 \le t \le T} \left[|\Delta W_t^{A,\varepsilon}|^4 + |\Delta W_t^{P,\lambda,\varepsilon}|^4 \right] + \left(\int_0^T \left[|\Delta Z_t^{A,\varepsilon}|^2 + |\Delta Z_t^{P,\lambda,\varepsilon}|^2 \right] dt \right)^2 \right\}$$
$$= 0; \quad (10.123)$$
$$\lim_{\varepsilon \to 0} E \left\{ \sup_{0 < t < T} |\nabla W_t^{A,\varepsilon} - \nabla W_t^A|^4 + \left(\int_0^T |\nabla Z_t^{A,\varepsilon} - \nabla Z_t^A|^2 dt \right)^2 \right\} = 0.$$

To prove the convergence of $\nabla W^{P,\lambda,\varepsilon}$ and $\nabla Z^{P,\lambda,\varepsilon}$ in (10.122), we again follow the arguments in Lemma 10.1.8. In particular, denoting

$$\partial_z I_A^\varepsilon(t) := \int_0^1 \partial_z I_A(t, c_s, Z_t^A + \theta \Delta Z_t^{A,\varepsilon}) d\theta,$$

we have

$$E\left\{\left(\int_0^T |Z_t^{P,\lambda}\partial_z I_A^\varepsilon(t)\nabla Z_t^{A,\varepsilon} - Z_t^{P,\lambda}\partial_z I_A(t,c_t,Z_t^A)\nabla Z_t^A|dt\right)^2\right\}$$

$$\leq CE\left\{\left(\int_0^T [|Z_t^{P,\lambda}\nabla Z_t^A[\partial_z I_A^\varepsilon(t) - \partial_z I_A(t,c_t,Z_t^A)]|\right.\right.$$
$$\left.\left.+ |Z_t^{P,\lambda}\partial_z I_A(t,c_t,Z_t^A)[\nabla Z_t^{A,\varepsilon} - \nabla Z_t^A]|dt\right)^2\right\}$$

$$\leq CE\left\{\left(\int_0^T |Z_t^{P,\lambda}\nabla Z_t^A[\partial_z I_A^\varepsilon(t) - \partial_z I_A(t,c_t,Z_t^A)]|dt\right)^2\right.$$
$$\left.+ \int_0^T |Z_t^{P,\lambda}|^2 dt \int_0^T |\nabla Z_t^{A,\varepsilon} - \nabla Z_t^A|^2 dt\right\}$$
$$\to 0 \quad \text{as } \varepsilon \to 0,$$

where the first term converges to zero thanks to the Dominated Convergence Theorem, and the second term thanks to (10.123). Then, one can easily prove (10.122). □

Notice that in this case the adjoint processes in (10.40) become

$$\Gamma_t^1 := \int_0^t \lambda_s \Gamma_s^2 ds + \int_0^t [I_A(s,c_s,Z_s^A)\Gamma_s^1 + Z_s^P \partial_z I_A(s,c_s,Z_s^A)\Gamma_s^2]dB_s;$$
$$\Gamma_t^2 = 1 + \int_0^t I_A(s,c_s,Z_s^A)\Gamma_s^2 dB_s. \qquad (10.124)$$

Combining the arguments in Lemma 10.3.3 and the estimates in Lemma 10.5.1, we obtain

Lemma 10.5.2 *Assume Assumptions 5.2.10 and 5.2.11 hold. Then,*

$$\nabla W_0^{P,\lambda} = E\left\{\Gamma_T^2[D_T U_A'(C_T) + U_P'(C_T)]\Delta C_T\right.$$
$$+ \int_0^T \Gamma_t^2[D_t \partial_c u_A(t,c_t,I_A(t,c,Z_t^A)) + \partial_c u_P(t,c_t)$$
$$\left.+ Z_t^{P,\lambda}\partial_c I_A(t,c,Z_t^A)]\Delta c_t dt\right\}.$$

Finally, the proof of Theorem 5.2.12 follows exactly the same arguments as the proofs of Theorems 10.3.4 and 10.3.5.

10.6 Further Reading

The stochastic maximum principle was the original motivation for studying BSDEs, see Bismut (1973), Peng (1990), and El Karoui et al. (2001) for related work. The book Yong and Zhou (1999) is an excellent reference for the subject. We also refer to Wu (1998) and Yong (2010) for stochastic control on coupled FBSDEs.

References

Bismut, J.M.: Conjugate convex functions in optimal stochastic control. J. Math. Anal. Appl. **44**, 384–404 (1973)

El Karoui, N., Peng, S., Quenez, M.C.: A dynamic maximum principle for the optimization of recursive utilities under constraints. Ann. Appl. Probab. **11**, 664–693 (2001)

Peng, S.: A general stochastic maximum principle for optimal control problems. SIAM J. Control **28**, 966–979 (1990)

Wu, Z.: Maximum principle for optimal control problem of fully coupled forward-backward stochastic systems. J. Syst. Sci. Math. Sci. **11**, 249–259 (1998)

Yong, J.: Optimality variational principle for optimal controls of forward-backward stochastic differential equations. SIAM J. Control Optim. **48**, 4119–4156 (2010)

Yong, J., Zhou, X.Y.: Stochastic Controls: Hamiltonian Systems and HJB Equations. Springer, New York (1999)

Chapter 11
Forward-Backward SDEs

Abstract The theory of existence and uniqueness of FBSDEs is not as satisfactorily developed as the one for BSDEs. Even linear FBSDEs do not necessarily have a solution. We present several approaches to establishing existence and uniqueness under specific conditions.

11.1 FBSDE Definition

We study coupled FBSDEs of the form

$$\begin{cases} X_t = x + \int_0^t b(s, \omega, X_s, Y_s, Z_s)ds + \int_0^t \sigma(s, \omega, X_s, Y_s, Z_s)dB_s; \\ Y_t = g(\omega, X_T) + \int_t^T f(s, \omega, X_s, Y_s, Z_s)ds - \int_t^T Z_s dB_s, \end{cases} \quad (11.1)$$

where b, σ, f, g are progressively measurable, and b, σ, f are $\mathbb{F}$-adapted for any (x, y, z). We omit ω in the coefficients, and for simplicity we assume all the processes are one-dimensional. A solution to (11.1) is a triplet of $\mathbb{F}$-adapted processes $\Theta := (X, Y, Z)$ such that

$$\|\Theta\|^2 := E\left\{\sup_{0 \le t \le T}\left[|X_t|^2 + |Y_t|^2\right] + \int_0^T |Z_t|^2 dt\right\} < \infty.$$

We say that (11.1) is coupled because one cannot solve the FSDE separately as one did for (9.38). FBSDEs of type (11.1) appear naturally in many applications, as we see in Chap. 9 and in the main body of the book. In this chapter we focus on well-posedness of FBSDEs. We adopt the following standard assumptions.

Assumption 11.1.1 (i) $g(0) \in L^2(\mathcal{F}_T)$, $\varphi(\cdot, 0, 0, 0) \in L^{1,2}(\mathbb{F})$ for $\varphi = b, f$, and $\sigma(\cdot, 0, 0, 0) \in L^2(\mathbb{F})$.

(ii) b, σ, f, g are uniformly Lipschitz continuous in (x, y, z).

We note that FBSDE (11.1) is much more complicated than BSDE (9.1) and decoupled FBSDE (9.38), and Assumption 11.1.1 alone is not enough to ensure well-posedness of (11.1). In fact, even a linear FBSDE may have no solution.

Example 11.1.2 For $\xi \in L^2(\mathcal{F}_T)$, but not deterministic, the following FBSDE has no $\mathbb{F}$-adapted solution:

$$\begin{cases} X_t = \int_0^t Z_s dB_s; \\ Y_t = X_T + \xi + \int_t^T Z_s dB_s. \end{cases}$$

Proof Assume (X, Y, Z) is an $\mathbb{F}$-adapted solution. Denote $\bar{Y}_t := Y_t - X_t$. Note that $X_t = X_T - \int_t^T Z_s dB_s$, thus $\bar{Y}_t = \xi$, which is not $\mathbb{F}$-adapted. A contradiction. □

There are typically three approaches in the literature, as outlined below. However, each approach has its limits. In many applications, including the Principal–Agent problem studied in this book, none of the three approaches works for general existence results. Thus, a more fundamental understanding of well-posedness of general FBSDEs is still needed.

11.2 Fixed Point Approach

The fixed point approach works very well for BSDEs, however, for FBSDEs one needs additional assumptions. For notational simplicity, for $\varphi = b, \sigma, f, g$, let

$\varphi_x, \varphi_y, \varphi_z$ denote the Lipschitz constant of φ with respect to x, y, z.

We emphasize that they are constants, not partial derivatives. Moreover, let k_b, k_f be two constants (possibly negative) such that

$$[b(t, x_1, y, z) - b(t, x_2, y, z)][x_1 - x_2] \le k_b |x_1 - x_2|^2; \quad (11.2)$$
$$[f(t, x, y_1, z) - f(t, x, y_2, z)][y_1 - y_2] \le k_f |y_1 - y_2|^2. \quad (11.3)$$

Assumption 11.2.1 Assume

$$\sigma_z g_x < 1; \quad (11.4)$$

and one of the following five conditions holds true:

(i) T is small enough;
(ii) The forward SDE is weakly coupled to (y, z), that is, $b_y, b_z, \sigma_y, \sigma_z$ are small enough;
(iii) The backward SDE is weakly coupled to x, that is, f_x, g_x are small enough;
(iv) The forward SDE is sharply mean reverting, that is, k_b is negative enough;
(v) The backward SDE is sharply mean reverting, that is, k_f is negative enough;

We also denote

$$K := \max(b_y, b_z, \sigma_x, \sigma_y, \sigma_z, f_x, f_y, f_z, k_b, k_f). \quad (11.5)$$

We emphasize that when $k_b \le 0$ (resp. $k_f \le 0$), K does not depend on b_x (resp. f_y).

11.2 Fixed Point Approach

Remark 11.2.2 From the arguments below we can see that the next theorem holds even without assuming that b is Lipschitz continuous in x and that f is Lipschitz continuous in y.

Theorem 11.2.3 *Assume Assumptions 11.1.1 and 11.2.1 hold true. Then, FBSDE (11.1) admits a unique solution Θ and we have*

$$\|\Theta\|^2 \leq C I_0^2 \quad \text{where}$$

$$I_0^2 := E\left\{\left(\int_0^T [|b|+|f|](t,0,0,0)dt\right)^2 + \int_0^T |\sigma(t,0,0,0)|^2 dt \right.$$
$$\left. + |g(0)|^2 + |x|^2\right\}. \tag{11.6}$$

Proof. We emphasize that in this proof generic constant C depends only on K and the dimensions, but not on the values k_b, k_f and $\varphi_x, \varphi_y, \varphi_z$. Let $\mathcal{L}$ denote the space of all $\mathbb{F}$-adapted processes (Y, Z) such that

$$\sup_{0\leq t\leq T} E\{|Y_t|^2\} + E\left\{\int_0^T |Z_t|^2 dt\right\} < \infty.$$

For any $(y, z) \in \mathcal{L}$, let $\Theta := \Theta^{y,z}$ be the unique solution to the following decoupled FBSDE:

$$\begin{cases} X_t = x + \int_0^t b(s, X_s, y_s, z_s) ds + \int_0^t \sigma(s, X_s, y_s, y_s) dB_s; \\ Y_t = g(X_T) + \int_t^T f(s, X_s, y_s, z_s) ds - \int_t^T Z_s dB_s. \end{cases} \tag{11.7}$$

By Assumption 11.1.1 and applying Theorem 9.3.5, it is clear that $(Y, Z) \in \mathcal{L}$. Define a mapping $F : \mathcal{L} \to \mathcal{L}$ by $F(y, z) := (Y, Z)$. Now, for $(y^i, z^i) \in \mathcal{L}$, $i = 1, 2$, let Θ^i be the solution to (11.7). Denote $\Delta y := y^1 - y^2, \Delta z := z^1 - z^2, \Delta\Theta := \Theta^1 - \Theta^2$. Then,

$$\begin{cases} \Delta X_t = \int_0^t [\alpha_s^1 \Delta X_s + \beta_s^1 \Delta y_s + \gamma_s^1 \Delta z_s] ds \\ \qquad + \int_0^t [\alpha_s^2 \Delta X_s + \beta_s^2 \Delta y_s + \gamma_s^2 \Delta z_s] dB_s; \\ \Delta Y_t = \lambda \Delta X_T + \int_t^T [\alpha_s^3 \Delta X_s + \beta_s^3 \Delta y_s + \gamma_s^3 \Delta z_s] ds - \int_t^T \Delta Z_s dB_s, \end{cases} \tag{11.8}$$

where, by Assumption 11.1.1, $|\alpha^j|, |\beta^j|, |\gamma^j| \leq K$, $j = 1, 2, 3$, and λ are bounded. Now, in our notation, we have, for any $\varepsilon, \delta > 0$,

$$E\{|\Delta X_t|^2\} = E\left\{\int_0^t [2\Delta X_s[\alpha_s^1 \Delta X_s + \beta_s^1 \Delta y_s + \gamma_s^1 \Delta z_s]\right.$$
$$\left. + [\alpha_s^2 \Delta X_s + \beta_s^2 \Delta y_s + \gamma_s^2 \Delta z_s]^2] ds\right\}$$

$$\leq E\left\{\int_0^t [2b_x|\Delta X_s|^2 + 2|\Delta X_s|[b_y|\Delta y_s| + b_z|\Delta z_s|]\right.$$
$$\left. + [K|\Delta X_s| + \sigma_y|\Delta y_s| + \sigma_z|\Delta z_s|]^2 ds\right\}$$
$$\leq E\left\{\int_0^t \left[(2b_x + C\varepsilon^{-1})|\Delta X_s|^2 + \varepsilon(b_y^2|\Delta y_s|^2 + b_z^2|\Delta z_s|^2)\right.\right.$$
$$\left.\left. + (1+\varepsilon)(\sigma_y|\Delta y_s| + \sigma_z|\Delta z_s|)^2\right]ds\right\}$$
$$\leq E\left\{\int_0^t \left[(2b_x + C\varepsilon^{-1})|\Delta X_s|^2 + (\varepsilon b_y^2 + (1+\varepsilon)(1+\delta^{-1})\sigma_y^2)|\Delta y_s|^2\right.\right.$$
$$\left.\left. + (\varepsilon b_z^2 + (1+\varepsilon)(1+\delta)\sigma_z^2)|\Delta z_s|^2\right]ds\right\};$$

and

$$E\left\{|\Delta Y_t|^2 + \int_t^T |\Delta Z_s|^2 ds\right\}$$
$$= E\left\{\lambda^2|\Delta X_T|^2 + \int_t^T 2\Delta Y_s[\alpha_s^3 \Delta X_s + \beta_s^3 \Delta y_s + \gamma_s^3 \Delta z_s]ds\right\}$$
$$\leq E\left\{g_x^2|\Delta X_T|^2 + \int_t^T \left[(2f_y + C\varepsilon^{-1})|\Delta Y_s|^2 + \varepsilon[f_x^2|\Delta X_s|^2\right.\right.$$
$$\left.\left. + |f_z|^2|\Delta Z_s|^2]\right]ds\right\}.$$

Let

$$\Gamma_x(t) := e^{-(2b_x + C\varepsilon^{-1})t}, \qquad \Gamma_y(t) := e^{(2f_y + C\varepsilon^{-1})t}, \qquad \Gamma_t := (\Gamma_x(t))^{-1}\Gamma_y(t), \tag{11.9}$$

and assume

$$\varepsilon f_z^2 < 1. \tag{11.10}$$

We get

$$E\{\Gamma_x(t)|\Delta X_t|^2\} \leq E\left\{\int_0^t \Gamma_x(s)\left[(\varepsilon b_y^2 + (1+\varepsilon)(1+\delta^{-1})\sigma_y^2)|\Delta y_s|^2\right.\right.$$
$$\left.\left. + (\varepsilon b_z^2 + (1+\varepsilon)(1+\delta)\sigma_z^2)|\Delta z_s|^2\right]ds\right\}$$
$$= E\left\{\int_0^t (\Gamma_s)^{-1}\left[(\varepsilon b_y^2 + (1+\varepsilon)(1+\delta^{-1})\sigma_y^2)\Gamma_y(s)|\Delta y_s|^2\right.\right.$$
$$\left.\left. + (\varepsilon b_z^2 + (1+\varepsilon)(1+\delta)\sigma_z^2)\Gamma_y(s)|\Delta z_s|^2\right]ds\right\};$$

and

11.2 Fixed Point Approach

$$E\left\{\Gamma_y(t)|\Delta Y_t|^2 + (1-\varepsilon f_z^2)\int_t^T \Gamma_y(s)|\Delta Z_s|^2 ds\right\}$$
$$\leq E\left\{g_x^2 \Gamma_T \Gamma_x(T)|\Delta X_T|^2 + \varepsilon \int_t^T f_x^2 \Gamma_s \Gamma_x(s)|\Delta X_s|^2 ds\right\}.$$

Thus,

$$E\left\{\Gamma_y(t)|\Delta Y_t|^2 + (1-\varepsilon f_z^2)\int_t^T \Gamma_y(s)|\Delta Z_s|^2 ds\right\}$$
$$\leq E\Bigg\{\int_0^T \left(\varepsilon b_y^2 + (1+\varepsilon)(1+\delta^{-1})\sigma_y^2\right)(\Gamma_s)^{-1}$$
$$\times \left(g_x^2 \Gamma_T + \varepsilon f_x^2 \int_{s\vee t}^T \Gamma_r dr\right)\Gamma_y(s)|\Delta y_s|^2 ds$$
$$+ \int_0^T \left(\varepsilon b_z^2 + (1+\varepsilon)(1+\delta)\sigma_z^2\right)(\Gamma_s)^{-1}$$
$$\times \left(g_x^2 \Gamma_T + \varepsilon f_x^2 \int_{s\vee t}^T \Gamma_r dr\right)\Gamma_y(s)|\Delta z_s|^2 ds\Bigg\}. \quad (11.11)$$

We claim that, under Assumption 11.2.1, one can choose appropriate ε and δ such that there exists a constant $c < 1$ satisfying:

$$\|(\Delta Y, \Delta Z)\|_w^2 \leq c\|(\Delta y, \Delta z)\|_w^2,$$
where $\|(Y, Z)\|_w^2 := \sup_{0\leq t\leq T} E\left\{\Gamma_y(t)|Y_t|^2 + (1-\varepsilon f_z^2)\int_t^T \Gamma_y(t)|Z_t|^2 dt\right\}.$
$$(11.12)$$

Clearly, $(\mathcal{L}, \|\cdot\|_w)$ is a Banach space. By the contraction mapping theorem, the mapping F has a unique fixed point (Y, Z). Let X be the solution to the SDE with the given fixed point (Y, Z):

$$X_t = x + \int_0^t b(s, X_s, Y_s, Z_s)ds + \int_0^t \sigma(s, X_s, Y_s, Z_s)dB_s.$$

Then, clearly (X, Y, Z) is the unique solution to FBSDE (11.1).

Moreover, let $\Theta^0 := \Theta^{0,0}$ be the solution to (11.7) with $y = 0, z = 0$. Then,

$$\|(Y - Y^0, Z - Z^0)\|_w \leq \sqrt{c}\|(Y - 0, Z - 0)\|_w = \sqrt{c}\|(Y, Z)\|_w^2.$$

Thus,

$$\|(Y, Z)\|_w \leq \|(Y - Y^0, Z - Z^0)\|_w + \|(Y^0, Z^0)\|_w$$
$$\leq \sqrt{c}\|(Y, Z)\|_w^2 + \|(Y^0, Z^0)\|_w.$$

This implies

$$\|(Y, Z)\|_w \leq \frac{1}{1-\sqrt{c}}\|(Y^0, Z^0)\|_w.$$

By Theorem 9.3.2, $\|(Y^0, Z^0)\|_w \leq C I_0$. Then, $\|(Y, Z)\|_w \leq C I_0$. By Theorem 9.3.2 again one can easily get

$$\|\Theta\|^2 = \|\Theta^{Y,Z}\|^2$$
$$\leq C E\left\{|x|^2 + \left(\int_0^T [|b| + |f|](t, 0, Y_t, Z_t) dt\right)^2 \right.$$
$$\left. + \int_0^T |\sigma(t, 0, Y_t, Z_t)|^2 dt + |g(0)|^2\right\}$$
$$\leq C I_0^2 + C E\left\{\int_0^T [|Y_t|^2 + |Z_t|^2] dt\right\} \leq C I_0^2 + C\|(Y, Z)\|_w^2 \leq C I_0^2.$$

It remains to check (11.12) under the five cases in Assumption 11.2.1.

(i) Note that, for $0 \leq s < r \leq T$, we have $(\Gamma_s)^{-1} \Gamma_r \leq e^{(4K + C\varepsilon^{-1})T}$. Without loss of generality we assume $T \leq 1$. By (11.11) we have

$$E\left\{\Gamma_y(t)|\Delta Y_t|^2 + (1 - \varepsilon f_z^2) \int_t^T \Gamma_y(s)|\Delta Z_s|^2 ds\right\}$$
$$\leq E\left\{\int_0^T (\varepsilon + (1 + \varepsilon)(1 + \delta^{-1})) K^2 (K^2 + \varepsilon K^2 T) e^{(4K + C\varepsilon^{-1})T} \Gamma_y(s)|\Delta y_s|^2 ds\right.$$
$$\left. + \int_0^T (\varepsilon K^2 + (1 + \varepsilon)(1 + \delta)\sigma_z^2)(g_x^2 + \varepsilon K^2 T) e^{(4K + C\varepsilon^{-1})T} \Gamma_y(s)|\Delta z_s|^2 ds\right\}$$
$$\leq (\varepsilon + (1 + \varepsilon)(1 + \delta^{-1})) K^2 (K^2 + \varepsilon K^2) e^{(4K + C\varepsilon^{-1})T} T \|(\Delta y, \Delta z)\|_w^2$$
$$+ (\varepsilon K^2 + (1 + \varepsilon)(1 + \delta)\sigma_z^2)(g_x^2 + \varepsilon K^2) e^{(4K + C\varepsilon^{-1})T}$$
$$\times (1 - \varepsilon f_z^2)^{-1} \|(\Delta y, \Delta z)\|_w^2$$
$$=: c(T) \|(\Delta y, \Delta z)\|_w^2.$$

Note that, by (11.4),

$$\lim_{\varepsilon, \delta \to 0} (\varepsilon K^2 + (1 + \varepsilon)(1 + \delta)\sigma_z^2)(g_x^2 + \varepsilon K^2)(1 - \varepsilon f_z^2)^{-1} = \sigma_z^2 g_x^2 < 1.$$

Fix ε, δ small enough so that

$$(\varepsilon K^2 + (1 + \varepsilon)(1 + \delta)\sigma_z^2)(g_x^2 + \varepsilon K^2)(1 - \varepsilon f_z^2)^{-1} < 1.$$

Clearly,

$$\lim_{T \to 0} c(T) = (\varepsilon K^2 + (1 + \varepsilon)(1 + \delta)\sigma_z^2)(g_x^2 + \varepsilon K^2)(1 - \varepsilon f_z^2)^{-1}.$$

Thus, when T is small enough, we have $c(T) < 1$.

(ii) In this case we set $\varepsilon := \frac{1}{2 f_z^2} \wedge 1$ and $\delta := 1$. Then, $1 - \varepsilon f_z^2 \geq \frac{1}{2}$ and thus (11.11) leads to

$$E\left\{\Gamma_y(t)|\Delta Y_t|^2 + (1 - \varepsilon f_z^2) \int_t^T \Gamma_y(s)|\Delta Z_s|^2 ds\right\}$$

11.2 Fixed Point Approach

$$\leq E\left\{\int_0^T C(b_y^2 + \sigma_y^2)(e^{CT} + e^{CT}T)\Gamma_y(s)|\Delta y_s|^2 ds\right.$$
$$\left. + \int_0^T C(b_z^2 + \sigma_z^2)(e^{CT} + e^{CT}T)(1 - \varepsilon f_z^2)\Gamma_y(s)|\Delta z_s|^2 ds\right\}$$
$$\leq Ce^{CT}(1+T)^2[b_y^2 + \sigma_y^2 + b_z^2 + \sigma_z^2]\|(\Delta y, \Delta z)\|_w^2$$
$$=: c\|(\Delta y, \Delta z)\|_w^2.$$

Clearly, when $b_y^2 + \sigma_y^2 + b_z^2 + \sigma_z^2$ is small, we have $c < 1$.

(iii) In this case we also set $\varepsilon := \frac{1}{2f_z^2} \wedge 1$ and $\delta := 1$. Then, $1 - \varepsilon f_z^2 \geq \frac{1}{2}$ and thus (11.11) leads to

$$E\left\{\Gamma_y(t)|\Delta Y_t|^2 + (1 - \varepsilon f_z^2)\int_t^T \Gamma_y(s)|\Delta Z_s|^2 ds\right\}$$
$$\leq E\left\{\int_0^T CK^2 e^{CT}(g_x^2 + Tf_x^2)\Gamma_y(s)|\Delta y_s|^2 ds\right.$$
$$\left. + \int_0^T CK^2(g_x^2 + Tf_x^2)(1 - \varepsilon f_z^2)\Gamma_y(s)|\Delta z_s|^2 ds\right\}$$
$$\leq Ce^{CT}(1+T)^2[g_x^2 + Tf_x^2]\|(\Delta y, \Delta z)\|_w^2$$
$$=: c\|(\Delta y, \Delta z)\|_w^2.$$

Clearly, when f_x and g_x are small, we have $c < 1$.

(iv) and (v). In this case, similar to case (i) we first choose ε and δ small enough so that

$$(\varepsilon K^2 + (1+\varepsilon)(1+\delta)\sigma_z^2)(g_x^2 + \varepsilon f_x^2 T)(1 - \varepsilon f_z^2) < 1.$$

Now, fix ε and δ. Let $k_b + k_f < -C\varepsilon^{-1}$ for the constant C in (11.9). Then, $(\Gamma_s)^{-1}\Gamma_r \leq 1$ for $0 \leq s \leq r \leq T$. Therefore, (11.11) leads to

$$E\left\{\Gamma_y(t)|\Delta Y_t|^2 + (1 - \varepsilon f_z^2)\int_t^T \Gamma_y(s)|\Delta Z_s|^2 ds\right\}$$
$$\leq CK^2\left(K^2 \int_t^T e^{2(k_b + k_f + C\varepsilon^{-1})(T-s)} ds + \int_t^T \int_{s \vee t}^T e^{2(k_b + k_f + C\varepsilon^{-1})(r-s)} dr ds\right)$$
$$\times \|(\Delta y, \Delta z)\|_w^2$$
$$+ (\varepsilon K^2 + (1+\varepsilon)(1+\delta)\sigma_z^2)(g_x^2 + \varepsilon f_x^2 T)(1 - \varepsilon f_z^2)\|(\Delta y, \Delta z)\|_w^2$$
$$\leq \left[-\frac{C(1+T)}{k_b + k_f + C\varepsilon^{-1}} + (\varepsilon K^2 + (1+\varepsilon)(1+\delta)\sigma_z^2)(g_x^2 + \varepsilon f_x^2 T)(1 - \varepsilon f_z^2)\right]$$
$$\times \|(\Delta y, \Delta z)\|_w^2$$
$$=: c\|(\Delta y, \Delta z)\|_w^2.$$

When $k_b + k_f$ is negative enough, we obtain $c < 1$.

11.3 Four-Step Scheme—The Decoupling Approach

This approach uses the connection between Markovian FBSDEs and PDEs. Consider the following FBSDE:

$$\begin{cases} X_t = x + \int_0^t b(s, X_s, Y_s, Z_s)ds + \int_0^t \sigma(s, X_s, Y_s)dB_s; \\ Y_t = g(X_T) + \int_t^T f(s, X_s, Y_s, Z_s)ds - \int_t^T Z_s dB_s, \end{cases} \quad (11.13)$$

where the coefficients b, σ, f, g are deterministic measurable functions. We note that the forward diffusion coefficient σ does not depend on Z. This FBSDE is associated with the following quasi-linear parabolic PDE:

$$\begin{cases} u_t + \frac{1}{2}u_{xx}\sigma^2(t, x, u) + u_x b\big(t, x, u, u_x\sigma(t, x, u)\big) + f\big(t, x, u, u_x\sigma(t, x, u)\big) \\ = 0; \\ u(T, x) = g(x). \end{cases} \quad (11.14)$$

We first have the following result:

Theorem 11.3.1 *Assume b, σ, f, g are uniformly Lipschitz continuous in (x, y, z); σ^2 is uniformly Lipschitz continuous in y; PDE (11.14) has a classical solution u with bounded u_x and u_{xx}; and functions*

$$\tilde{b}(t, x) := b\big(t, x, u(t, x), u_x\sigma(t, x, u(t, x))\big), \qquad \tilde{\sigma}(t, x) := \sigma\big(t, x, u(t, x)\big),$$

are uniformly Lipschitz continuous in x and $\int_0^T [|\tilde{b}(t, 0)| + |\tilde{\sigma}(t, 0)|^2]dt < \infty$. Then, FBSDE (11.13) has a unique solution (X, Y, Z) and we have

$$Y_t = u(t, X_t), \qquad Z_t = u_x(t, X_t)\sigma\big(t, X_t, u(t, X_t)\big). \quad (11.15)$$

Proof We first prove existence. By our assumptions on $\tilde{b}$ and $\tilde{\sigma}$, there is a solution X to the following SDE:

$$X_t = x + \int_0^t \tilde{b}(s, X_s)ds + \int_0^t \tilde{\sigma}(s, X_s)dB_s. \quad (11.16)$$

Let us define (Y, Z) by (11.15). Then, applying Itô's rule one can check straightforwardly that (X, Y, Z) solves FSBDE (11.13) and satisfies $\|(X, Y, Z)\| < \infty$.

To prove uniqueness, let (X, Y, Z) be an arbitrary solution to FBSDE (11.13). Denote

$$\tilde{Y}_t := u(t, X_t), \qquad \tilde{Z}_t := u_x(t, X_t)\sigma\big(t, X_t, u(t, X_t)\big) \quad \text{and} \quad \Delta Y_t := \tilde{Y}_t - Y_t,$$
$$\Delta Z_t := \tilde{Z}_t - Z_t.$$

Applying Itô's rule, we have

11.3 Four-Step Scheme—The Decoupling Approach

$$d\tilde{Y}_t = \left(u_t(t, X_t) + u_x(t, X_t)b(t, X_t, Y_t, Z_t) + \frac{1}{2}u_{xx}(t, X_t)\sigma^2(t, X_t, Y_t)\right)dt$$
$$+ u_x(t, X_t)\sigma(t, X_t, Y_t)dB_t.$$

Then, since u satisfies PDE (11.14),

$$d(\Delta Y_t)$$
$$= \left[u_t(t, X_t) + u_x(t, X_t)b(t, X_t, Y_t, Z_t) + \frac{1}{2}u_{xx}(t, X_t)\sigma^2(t, X_t, Y_t)\right.$$
$$\left. + f(t, X_t, Y_t, Z_t)\right]dt + \left[u_x(t, X_t)\sigma(t, X_t, Y_t) - Z_t\right]dB_t$$
$$= -\left[u_x(t, X_t)b(t, X_t, \tilde{Y}_t, \tilde{Z}_t) + \frac{1}{2}u_{xx}(t, X_t)\sigma^2(t, X_t, \tilde{Y}_t) + f(t, X_t, \tilde{Y}_t, \tilde{Z}_t)\right]dt$$
$$+ \left[u_x(t, X_t)b(t, X_t, Y_t, Z_t) + \frac{1}{2}u_{xx}(t, X_t)\sigma^2(t, X_t, Y_t) + f(t, X_t, Y_t, Z_t)\right]dt$$
$$+ \left[u_x(t, X_t)\sigma(t, X_t, Y_t) - u_x(t, X_t)\sigma(t, X_t, \tilde{Y}_t) + \Delta Z_t\right]dB_t$$
$$= [\alpha_t \Delta Y_t + \beta_t \Delta Z_t]dt + [\gamma_t \Delta Y_t + \Delta Z_t]dB_t$$
$$= \left[(\alpha_t - \beta_t \gamma_t)\Delta Y_t + \beta_t(\gamma_t \Delta Y_t + \Delta Z_t)\right]dt + [\gamma_t \Delta Y_t + \Delta Z_t]dB_t,$$

where α, β, γ are bounded. That is, $(\Delta Y, \gamma_t \Delta Y_t + \Delta Z_t)$ satisfies the above linear BSDE. Note that $\Delta Y_T = u(T, X_T) - g(X_T) = 0$. Then, $\Delta Y = 0$ and $\gamma_t \Delta Y_t + \Delta Z_t = 0$, and thus $\Delta Z = 0$. Therefore,

$$b(t, X_t, Y_t, Z_t) = b\big(t, X_t, u(t, X_t), u_x(t, X_t)\sigma\big(t, X_t, u(t, X_t)\big)\big) = \tilde{b}(t, X_t),$$
$$\sigma(t, X_t, Y_t) = \sigma\big(t, X_t, u(t, X_t)\big) = \tilde{\sigma}(t, X_t).$$

That is, X satisfies (11.16). Since SDE (11.16) has a unique solution, we know X is unique, and then so is (Y, Z). □

The main idea of the above approach is to use the decoupling function u. We now extend it to general FBSDEs with random coefficients:

$$\begin{cases} X_t = x + \int_0^t b(s, \omega, X_s, Y_s, Z_s)ds + \int_0^t \sigma(s, \omega, X_s, Y_s)dB_s; \\ Y_t = g(\omega, X_T) + \int_t^T f(s, \omega, X_s, Y_s, Z_s)ds - \int_t^T Z_s dB_s, \end{cases} \quad (11.17)$$

where σ does not depend on Z. Assume Assumption 11.1.1 holds.

Let $0 \leq t_1 < t_2 \leq T$, η be an $\mathcal{F}_{t_1}$-measurable square integrable random variable, and $\varphi(\omega, x)$ be a random field such that φ is $\mathcal{F}_{t_2}$-measurable for any fixed x and uniformly Lipschitz continuous in x with a Lipschitz constant K. Consider the following FBSDE over $[t_1, t_2]$:

$$\begin{cases} X_t = \eta + \int_{t_1}^t b(s, \omega, \Theta_s)ds + \int_{t_1}^t \sigma(s, \omega, X_s, Y_s)dB_s; \\ Y_t = \varphi(\omega, X_{t_2}) + \int_t^{t_2} f(s, \omega, \Theta_s)ds - \int_t^{t_2} Z_s dB_s. \end{cases} \quad (11.18)$$

Note that the above FBSDE satisfies condition (11.4) automatically. Applying Theorem 11.2.3(i), there exists a constant $\delta(K)$, which depends only on the dimensions, the Lipschitz constants in Assumption 11.1.1, and the Lipschitz constant K of φ, such that whenever $t_2 - t_1 \leq \delta(K)$, FBSDE (11.18) has a unique solution.

Theorem 11.3.2 *Assume Assumption* 11.1.1 *holds, and there exists a random field $u(t, \omega, x)$ such that*

(i) $u(T, \omega, x) = g(\omega, x)$;
(ii) *For each* (t, x), u *is $\mathcal{F}_t$-measurable;*
(iii) u *is uniformly Lipschitz continuous in x with a Lipschitz constant K;*
(iv) *For any $0 \leq t_1 < t_2 \leq T$ such that $t_2 - t_1 \leq \delta(K)$, the constant introduced above, the unique solution to FBSDE (11.18) over $[t_1, t_2]$ with terminal condition $\varphi(\omega, x) := u(t_2, \omega, x)$ satisfies $Y_{t_1} = u(t_1, \omega, X_{t_1})$.*

Then, FBSDE (11.17) *has a unique solution* (X, Y, Z) *on* $[0, T]$ *and we have* $Y_t = u(t, \omega, X_t)$.

Proof We first prove existence. Let $0 = t_0 < \cdots < t_n = T$ be a partition of $[0, T]$ such that $t_i - t_{i-1} \leq \delta(K)$ for $i = 1, \ldots, n$. Denote $X_0^0 := x$. For $i = 1, \ldots, n$, let (X^i, Y^i, Z^i) be the unique solution of the following FBSDE over $[t_{i-1}, t_i]$:

$$\begin{cases} X_s^i = X_{t_{i-1}}^{i-1} + \int_{t_{i-1}}^s b(r, \omega, X_r^i, Y_r^i, Z_r^i)dr + \int_{t_{i-1}}^s \sigma(r, \omega, X_r^i, Y_r^i)dB_r; \\ Y_s^i = u(t_i, \omega, X_{t_i}^i) + \int_s^{t_i} f(r, \omega, X_r^i, Y_r^i, Z_r^i)dr - \int_s^{t_i} Z_r^i dB_r. \end{cases}$$

Define

$$X_t := \sum_{i=1}^n X_t^i \mathbf{1}_{[t_{i-1}, t_i)}(t) + X_T^n \mathbf{1}_{\{T\}}(t),$$

$$Y_t := \sum_{i=1}^n Y_t^i \mathbf{1}_{[t_{i-1}, t_i)}(t) + Y_T^n \mathbf{1}_{\{T\}}(t),$$

$$Z_t := \sum_{i=1}^n Z_t^i \mathbf{1}_{[t_{i-1}, t_i)}(t) + Z_T^n \mathbf{1}_{\{T\}}(t).$$

Note that $X_{t_{i-1}}^i = X_{t_{i-1}}^{i-1}$ and $Y_{t_{i-1}}^i = u(t_{i-1}, X_{t_{i-1}}^i) = u(t_{i-1}, X_{t_{i-1}}^{i-1}) = Y_{t_{i-1}}^{i-1}$. Then, X and Y are continuous, and one can check straightforwardly that (X, Y, Z) solves (11.17).

It remains to prove uniqueness. Let (X, Y, Z) be an arbitrary solution to FBSDE (11.17), and let $0 = t_0 < \cdots < t_n = T$ be as above. Since $g(\omega, x) = u(T, \omega, x)$, then on $[t_{n-1}, t_n]$ we have

$$\begin{cases} X_s = X_{t_{n-1}} + \int_{t_{n-1}}^s b(r, \omega, X_r, Y_r, Z_r)dr + \int_{t_{n-1}}^s \sigma(r, \omega, X_r, Y_r)dB_r; \\ Y_s = u(t_n, \omega, X_{t_n}) + \int_s^{t_n} f(r, \omega, X_r, Y_r, Z_r)dr - \int_s^{t_n} Z_r dB_r. \end{cases}$$

11.3 Four-Step Scheme—The Decoupling Approach

By our assumption, $Y_{t_{n-1}} = u(t_{n-1}, \omega, X_{t_{n-1}})$. By induction, one sees that, for $i = n, \ldots, 1$,

$$\begin{cases} X_s = X_{t_{i-1}} + \int_{t_{i-1}}^{s} b(r, \omega, X_r, Y_r, Z_r) dr + \int_{t_{i-1}}^{s} \sigma(r, \omega, X_r, Y_r) dB_r; \\ Y_s = u(t_i, \omega, X_{t_i}) + \int_{s}^{t_i} f(r, \omega, X_r, Y_r, Z_r) dr - \int_{s}^{t_i} Z_r dB_r; \\ s \in [t_{i-1}, t_i]. \end{cases}$$

Now, since $X_{t_0} = x$, for $i = 1, \ldots, n$, by forward induction one sees that (X, Y, Z) is unique on $[t_{i-1}, t_i]$.

Finally, for any $t \in [t_{i-1}, t_i]$, considering the FBSDE on $[t, t_i]$ we see that $Y_t = u(t, \omega, X_t)$. □

We conclude this subsection with two sufficient conditions for the existence of the decoupling random field u.

Theorem 11.3.3 *Assume Assumption* 11.1.1 *holds; b, σ, f, g are deterministic and σ does not depend on z; and $\sigma \geq c > 0$. Then, there exists a deterministic function u satisfying the conditions in Theorem* 11.3.2, *and consequently FBSDE* (11.17) *admits a unique solution.*

The proof is quite lengthy, and we refer to Delarue (2002) which proves the theorem under slightly weaker conditions.

Theorem 11.3.4 *Assume Assumption* 11.1.1 *holds; b, σ, f are continuously differentiable in (x, y, z); all the processes are one-dimensional; and there exists a constant $c > 0$ such that*

$$\partial_y \sigma \partial_z b \leq -c |\partial_y b + \partial_x \sigma \partial_z b + \partial_y \sigma \partial_z f|. \tag{11.19}$$

Then, there exists a random field u satisfying the conditions in Theorem 11.3.2, *and consequently FBSDE* (11.17) *admits a unique solution.*

Remark 11.3.5 (i) If $\partial_y \sigma \leq -c_1$ and $\partial_z b \geq c_2$ or $\partial_y \sigma \geq c_1$ and $\partial_z b \leq -c_2$ for some $c_1, c_2 > 0$, then (11.19) holds.

(ii) The following three classes of FBSDEs satisfy condition (11.19) with both sides equal to 0:

$$\begin{cases} X_t = X_0 + \int_0^t b(s, \omega, X_s) ds + \int_0^t \sigma(s, \omega, X_s) dB_s; \\ Y_t = g(\omega, X_T) + \int_t^T f(s, \omega, X_s, Y_s, Z_s) ds - \int_t^T Z_s dB_s; \end{cases}$$

$$\begin{cases} X_t = X_0 + \int_0^t b(s, \omega, X_s, Z_s) ds + \int_0^t \sigma(s, \omega) dB_s; \\ Y_t = g(\omega, X_T) + \int_t^T f(s, \omega, X_s, Y_s, Z_s) ds - \int_t^T Z_s dB_s; \end{cases}$$

$$\begin{cases} X_t = X_0 + \int_0^t b(s,\omega,X_s)ds + \int_0^t \sigma(s,\omega,X_s,Y_s)dB_s; \\ Y_t = g(\omega,X_T) + \int_t^T f(s,\omega,X_s,Y_s)ds - \int_t^T Z_s dB_s. \end{cases}$$

Also, instead of differentiability, it suffices to assume uniform Lipschitz continuity in these cases.

Proof of Theorem 11.3.4 We proceed in several steps.

Step 1. Let K denote the Lipschitz constant of b, σ, f with respect to (x,y,z), and K_0 the Lipschitz constant of g with respect to x. Denote

$$K^* := e^{\bar{K}T}(1+K_0) - 1 \quad \text{where } \bar{K} := 2K + K^2 + \frac{K+2K^2}{2c} \qquad (11.20)$$

and let $\delta^* := \delta(K^*)$ be the constant introduced in Theorem 11.3.2. Fix a partition $0 = t_0 < \cdots < t_n = T$ such that $\Delta t_i \le \delta^*$, $i = 1,\ldots,n$. For each $(t,x) \in [t_{n-1},t_n] \times \mathbb{R}^d$, consider the following FBSDE on $[t,t_n]$:

$$\begin{cases} X_s^{t,x} = x + \int_t^s b(r,\omega,X_r^{t,x},Y_r^{t,x},Z_r^{t,x})dr + \int_t^s \sigma(r,\omega,X_r^{t,x},Y_r^{t,x})dB_r; \\ Y_s^{t,x} = u(t_n,\omega,X_{t_n}^{t,x}) + \int_s^{t_n} f(r,\omega,X_r^{t,x},Y_r^{t,x},Z_r^{t,x})dr - \int_s^{t_n} Z_r^{t,x}dB_r. \end{cases} \qquad (11.21)$$

Note that $K^* \ge (1+K_0) - 1 = K_0$. Then, by Theorem 11.2.3(i), FBSDE (11.21) has a unique solution. Define

$$u(t,x) := Y_t^{t,x}, \quad t \in [t_{n-1},t_n]. \qquad (11.22)$$

Step 2. Given x_1, x_2 and $t \in [t_{n-1},t_n]$, denote $\Delta x := x_1 - x_2$, $\Delta\Theta := \Theta^{t,x_1} - \Theta^{t,x_2}$. Then, $\Delta\Theta$ satisfies the following FBSDE:

$$\begin{cases} \Delta X_s = \Delta x + \int_t^s [\alpha_r^1 \Delta X_r + \beta_r^1 \Delta Y_r + \gamma_r^1 \Delta Z_r]ds \\ \qquad + \int_t^s [\alpha_r^2 \Delta X_r + \beta_r^2 \Delta Y_r]dB_r; \\ \Delta Y_s = \lambda \Delta X_{t_n} + \int_s^{t_n} [\alpha_r^3 \Delta X_r + \beta^3 \Delta Y_r + \gamma^3 \Delta Z_r]dr - \int_s^{t_n} \Delta Z_r dB_r, \end{cases} \qquad (11.23)$$

where $|\alpha^j|, |\beta^j|, |\gamma^j| \le K$, $j = 1,2,3$, and $|\lambda| \le K_0$. Clearly, FBSDE (11.23) also satisfies Assumption 11.1.1, and by (11.6),

$$\|\Theta\|^2 \le C|\Delta x|^2.$$

We note that FBSDE (11.23) is on $[t,t_n]$, and all the arguments in Theorem 11.2.3 hold by replacing the expectation with conditional expectation E_t. Thus, by (11.6) we have

$$E_t \left\{ \sup_{t \le s \le t_n} [|\Delta X_s|^2 + |\Delta Y_s|^2] + \int_t^{t_n} |\Delta Z_s|^2 ds \right\} \le C|\Delta x|^2.$$

11.3 Four-Step Scheme—The Decoupling Approach

In particular, this implies
$$|u(t,x_1) - u(t,x_2)| = |\Delta Y_t| \le E_t[|\Delta Y_t|] \le C|\Delta x|. \tag{11.24}$$
That is, for $t \in [t_{n-1}, t_n]$, $u(t,x)$ is uniformly Lipschitz continuous in x.

Step 3. Let $\eta = \sum_{j=1}^m x_j 1_{E_j}$ where $x_1, \ldots, x_m \in \mathbb{R}^d$ and $E_1, \ldots, E_m \in \mathcal{F}_t$ form a partition of Ω. One can check straightforwardly that $\Theta^{t,\eta} := \sum_{j=1}^m \Theta^{t,x_j} 1_{E_j}$ satisfy FBSDE (11.21) with initial value $X_t^{t,\eta} := \eta$. In particular, this implies that
$$Y_t^{t,\eta} = \sum_{j=1}^m Y_t^{t,x_j} 1_{E_j} = \sum_{j=1}^m u(t,\omega,x_j) 1_{E_j} = u\left(t,\omega, \sum_{j=1}^m x_j 1_{E_j}\right) = u(t,\omega,\eta).$$

Moreover, for general $\mathcal{F}_t$-measurable square integrable η, there exist $\{\eta_m, m \ge 1\}$ taking the above form such that $\lim_{m \to \infty} E\{|\eta_m - \eta|^2\} = 0$. Denote again by $\Theta^{t,\eta}$ the unique solution to FBSDE (11.21) with initial value $X_t^{t,\eta} := \eta$. By the arguments in Step 2, we know that
$$E\{|Y^{t,\eta} - Y^{t,\eta_m}|^2\} \le CE\{|\eta - \eta_m|^2\} \to 0.$$
Since $u(t,x)$ is Lipschitz continuous in x, we get
$$Y_t^{t,\eta} = \lim_{m \to \infty} Y_t^{t,\eta_m} = \lim_{m \to \infty} u(t, \eta_m) = u\left(t, \lim_{m \to \infty} \eta_m\right) = u(t,\eta). \tag{11.25}$$

Step 4. We now get a more precise estimate for constant C in (11.24). Since b, σ, f, g are continuously differentiable, by standard arguments one can easily see that $u(t,x)$ is differentiable in x with $u_x(t,x) = \nabla Y_t^{t,x}$, where

$$\begin{cases} \nabla X_s = 1 + \int_t^s \left[b_x(r,\omega,\Theta_r)\nabla X_r + b_y(r,\omega,\Theta_r)\nabla Y_r + b_z(r,\omega,\Theta_r)\nabla Z_r\right]ds \\ \qquad + \int_t^s \left[\sigma_x(r,\omega,\Theta_r)\nabla X_r + \sigma_y(r,\omega,\Theta_r)\nabla Y_r\right]dB_r; \\ \nabla Y_s = g_x(\omega, X_{t_n})\nabla X_{t_n} + \int_s^{t_n}\left[f_x(r,\omega,\Theta_r)\nabla X_r + f_y(r,\omega,\Theta_r)\nabla Y_r \right. \\ \qquad \left. + f_z(r,\omega,\Theta_r)\nabla Z_r\right]dr - \int_s^{t_n} \nabla Z_r dB_r. \end{cases} \tag{11.26}$$

Denote
$$\tilde{Y}_s := \nabla Y_s (\nabla X_s)^{-1}, \qquad \bar{Z}_s := \nabla Z_s (\nabla X_s)^{-1}, \qquad \tilde{Z}_s := \bar{Z}_s - \tilde{Y}_s[\sigma_x + \sigma_y \tilde{Y}_s].$$
Applying Itô's rule we obtain:
$$\begin{aligned} d(\nabla X_s)^{-1} &= -(\nabla X_s)^{-1}[b_x + b_y \tilde{Y}_s + b_z \bar{Z}_s]ds + (\nabla X_s)^{-1}[\sigma_x + \sigma_y \tilde{Y}_s]^2 ds \\ &\quad - (\nabla X_s)^{-1}[\sigma_x + \sigma_y \tilde{Y}_s]dB_s; \\ d(\tilde{Y}_s) &= -[f_x + f_y \tilde{Y}_s + f_z \bar{Z}_s]ds - \tilde{Y}_s[b_x + b_y \tilde{Y}_s + b_z \bar{Z}_s]ds \\ &\quad + \tilde{Y}_s[\sigma_x + \sigma_y \tilde{Y}_s]^2 ds - \bar{Z}_s[\sigma_x + \sigma_y \tilde{Y}_s]ds + \tilde{Z}_s dB_s. \end{aligned}$$
Note that $\bar{Z}_s = \tilde{Z}_s + \tilde{Y}_s[\sigma_x + \sigma_y \tilde{Y}_s]$. Then,

$$d(\tilde{Y}_s) = -[f_x + f_y\tilde{Y}_s + f_z\tilde{Z}_s]ds - \tilde{Y}_s[b_x + b_y\tilde{Y}_s + b_z\tilde{Z}_s]ds$$
$$- \tilde{Z}_s[\sigma_x + \sigma_y\tilde{Y}_s]ds + \tilde{Z}_s dB_s$$
$$= -\big[f_x + f_y\tilde{Y}_s + b_x\tilde{Y}_s + b_y\tilde{Y}_s^2\big]ds - [f_z + b_z\tilde{Y}_s]\big[\tilde{Z}_s + \tilde{Y}_s[\sigma_x + \sigma_y\tilde{Y}_s]\big]ds$$
$$- \tilde{Z}_s[\sigma_x + \sigma_y\tilde{Y}_s]ds + \tilde{Z}_s dB_s$$
$$= -\big[f_x + (f_y + b_x + f_z\sigma_x)\tilde{Y}_s + (b_y + f_z\sigma_y + b_z\sigma_x)\tilde{Y}_s^2 + b_z\sigma_y\tilde{Y}_s^3\big]ds$$
$$- [f_z + b_z\tilde{Y}_s + \sigma_x + \tilde{Y}_s]\tilde{Z}_s + \tilde{Z}_s dB_s.$$

We note that $\tilde{Y}_s = u_x(s, X_s)$ and is bounded for $s \in [t, t_n] \subset [t_{n-1}, t_n]$. Denote

$$\Gamma_s := \exp\left(\int_t^s [f_y + b_x + f_z\sigma_x + (b_y + f_z\sigma_y + b_z\sigma_x)\tilde{Y}_s + b_z\sigma_y\tilde{Y}_s^2]dr\right)$$
$$d\tilde{P} := \exp\left(\int_t^{t_n} [f_z + b_z\tilde{Y}_s + \sigma_x + \tilde{Y}_s]dB_s - \frac{1}{2}\int_t^{t_n} [f_z + b_z\tilde{Y}_s + \sigma_x + \tilde{Y}_s]^2 ds\right)dP.$$

Then,

$$\tilde{Y}_t = E_t^{\tilde{P}}\left\{\Gamma_{t_n}\tilde{Y}_{t_n} + \int_t^{t_n} \Gamma_s f_x ds\right\}.$$

Note that $|\tilde{Y}_{t_n}| = |g_x(X_{t_n})| \leq K_0$, and by (11.19),

$$f_y + b_x + f_z\sigma_x + (b_y + f_z\sigma_y + b_z\sigma_x)\tilde{Y}_s + b_z\sigma_y\tilde{Y}_s^2$$
$$\leq f_y + b_x + f_z\sigma_x + (b_y + f_z\sigma_y + b_z\sigma_x)\tilde{Y}_s - c|b_y + f_z\sigma_y + b_z\sigma_x|\tilde{Y}_s^2$$
$$\leq f_y + b_x + f_z\sigma_x + \frac{1}{4c}|b_y + f_z\sigma_y + b_z\sigma_x| \leq 2K + K^2 + \frac{K + 2K^2}{2c} = \bar{K},$$

and thus

$$\Gamma_s \leq e^{\bar{K}(s-t)}.$$

Then,

$$|\tilde{Y}_t| \leq e^{\bar{K}(t_n-t)}K_0 + \int_t^{t_n} e^{\bar{K}(s-t)}K ds$$
$$\leq e^{\bar{K}(t_n-t)}K_0 + e^{\bar{K}(t_n-t)} - 1 = e^{\bar{K}(t_n-t)}(K_0 + 1) - 1.$$

This implies that

$$|u_x(t, x)| \leq K_1 := e^{\bar{K}(t_n-t_{n-1})}(K_0 + 1) - 1, \quad t \in [t_{n-1}, t_n]. \quad (11.27)$$

Step 5. Note that $K_1 \leq K^*$. Assume u is defined on $[t_i, t_n]$ with $|u_x(t_i, x)| \leq K_i \leq K^*$. Define $u(t, x) := Y_t^{t,x}$ for $t \in [t_{i-1}, t_i]$, where

$$\begin{cases} X_s^{t,x} = x + \int_t^s b(r, \omega, X_r^{t,x}, Y_r^{t,x}, Z_r^{t,x})dr + \int_t^s \sigma(r, \omega, X_r^{t,x}, Y_r^{t,x})dB_r; \\ Y_s^{t,x} = u(t_i, \omega, X_{t_i}^{t,x}) + \int_s^{t_i} f(r, \omega, X_r^{t,x}, Y_r^{t,x}, Z_r^{t,x})dr - \int_s^{t_i} Z_r^{t,x}dB_r. \end{cases}$$

11.4 Method of Continuation

As in Step 4, we can prove
$$|u_x(t,x)| \le K_{i-1} := e^{\bar{K}(t_i - t_{i-1})}(K_i + 1) - 1, \quad t \in [t_{i-1}, t_i].$$
By induction we get
$$K_i = e^{\bar{K}(t_n - t_i)}(K_0 + 1) - 1 \le K^*, \quad i = 1, \ldots, n.$$
So, the backward induction can continue until $i = 1$, and thus, we may define u on $[0, T]$ and $|u_x(t,x)| \le K^*$ for all $t \in [0, T]$. Finally, it is clear that u satisfies the other requirements of Theorem 11.3.2. □

11.4 Method of Continuation

In this section we consider again general FBSDE (11.1) with random coefficients, and for notational simplicity we omit ω in the coefficients. Denote $\theta := (x, y, z)$ and $\Delta\theta := \theta_1 - \theta_2$. We adopt the following assumptions.

Assumption 11.4.1 There exists a constant $c > 0$ such that, for any θ_1, θ_2,
$$[b(t, \theta_1) - b(t, \theta_2)]\Delta y + [\sigma(t, \theta_1) - \sigma(t, \theta_2)]\Delta z - [f(t, \theta_1) - f(t, \theta_2)]\Delta x$$
$$\le -c[|\Delta x|^2 + |\Delta y|^2 + |\Delta z|^2], \quad a.s.;$$
$$[g(x_1) - g(x_2)]\Delta x \ge 0, \quad a.s.$$

Theorem 11.4.2 *Assume Assumptions 11.1.1 and 11.4.1 hold. Then, FBSDE (11.1) admits a unique solution.*

Proof of uniqueness Assume Θ^i, $i = 1, 2$ are two solutions. Denote, for $\varphi = b, \sigma, f$,
$$\Delta\Theta := \Theta^1 - \Theta^2; \qquad \Delta\varphi_t := \varphi(t, \Theta_t^1) - \varphi(t, \Theta_t^2); \qquad \Delta g := g(X_T^1) - g(X_T^2).$$
Then,
$$\begin{cases} \Delta X_t = \int_0^t \Delta b_s ds + \int_0^t \Delta \sigma_s dB_s; \\ \Delta Y_t = \Delta g + \int_t^T \Delta f_s ds - \int_t^T \Delta Z_s dB_s. \end{cases} \quad (11.28)$$
Applying Itô's rule on $\Delta X_t \Delta Y_t$ we have
$$d(\Delta X_t \Delta Y_t) = \Delta X_t d\Delta Y_t + \Delta Y_t d\Delta X_t + \Delta \sigma_t \Delta Z_t dt$$
$$= [-\Delta f_t \Delta X_t + \Delta b_t \Delta Y_t + \Delta \sigma_t \Delta Z_t]dt + [\Delta X_t \Delta Z_t + \Delta \sigma_t \Delta Y_t]dB_t.$$
Note that $\Delta X_0 = 0$ and $\Delta Y_T = \Delta g$. Thus,
$$E\{\Delta g \Delta X_T\} = E\{\Delta Y_T \Delta X_T - \Delta Y_0 \Delta X_0\}$$
$$= E\left\{\int_0^T [-\Delta f_t \Delta X_t + \Delta b_t \Delta Y_t + \Delta \sigma_t \Delta Z_t]dt\right\}.$$

By Assumption 11.4.1 we get
$$0 \le -cE\left\{\int_0^T [|\Delta X_t|^2 + |\Delta Y_t|^2 + |\Delta Z_t|^2]dt\right\}.$$
Then, obviously we have $\Delta X_t = \Delta Y_t = \Delta Z_t = 0$. □

The existence is first proved for a linear FSBDE.

Lemma 11.4.3 *Assume $b_0, f_0 \in L^{1,2}(\mathbb{F})$, $\sigma_0 \in L^2(\mathbb{F})$, and $g_0 \in L^2(\mathcal{F}_T)$. Then, the following linear FBSDE admits a (unique) solution:*

$$\begin{cases} X_t = x + \int_0^t [-Y_s + b_0(s)]ds + \int_0^t [-Z_s + \sigma_0(s)]dB_s; \\ Y_t = X_T + g_0 + \int_t^T [X_s + f_0(s)]ds - \int_t^T Z_s dB_s. \end{cases} \quad (11.29)$$

We note that, although we will not use it in the following proof, FBSDE (11.29) satisfies the monotonicity conditions with $c = 1$.

Proof Assume (X, Y, Z) is a solution. Denote $\bar{Y}_t := Y_t - X_t$. Note that
$$X_t = X_T + \int_t^T [Y_s - b_0(s)]ds + \int_t^T [Z_s - \sigma_0(s)]dB_s.$$
Then, $\bar{Y}$ satisfies
$$\bar{Y}_t = g_0 + \int_t^T [-\bar{Y}_s + f_0(s) + b_0(s)]ds - \int_t^T [2Z_s + \sigma_0(s)]dB_s.$$
We now solve (11.29) as follows. First, solve the following linear BSDE:
$$\bar{Y}_t = g_0 + \int_t^T [-\bar{Y}_s + f_0(s) + b_0(s)]ds - \int_t^T \bar{Z}_s dB_s.$$
Second, let $Z_t := 2[\bar{Z}_t - \sigma_0(t)]$. Third, solve the following linear FSDE:
$$X_t = x + \int_0^t [-X_s - \bar{Y}_s + b_0(s)]ds + \int_0^t \left[-\frac{1}{2}Z_s + \frac{3}{2}\sigma_0(s)\right]dB_s.$$
Finally, let $Y_t := \bar{Y}_t + X_t$. Then, one can easily check that (X, Y, Z) is a solution to FBSDE (11.29). □

Now, we fix (b, σ, f, g) satisfying Assumptions 11.1.1 and 11.4.1. The Method of Continuation consists in building a bridge between FBSDEs (11.1) and (11.29). Namely, for $\alpha \in [0, 1]$, let

$$b^\alpha(t, \theta) := \alpha b(t, \theta) - (1 - \alpha)y; \qquad \sigma^\alpha(t, \theta) := \alpha \sigma(t, \theta) - (1 - \alpha)z;$$
$$f^\alpha(t, \theta) := \alpha f(t, \theta) + (1 - \alpha)x; \qquad g^\alpha(x) := \alpha g(x) + (1 - \alpha)x.$$

11.4 Method of Continuation   245

We note that $(b^\alpha, \sigma^\alpha, f^\alpha, g^\alpha)$ satisfies Assumptions 11.1.1 and 11.4.1 with constant
$$c_\alpha := \alpha c + 1 - \alpha \geq \min(c, 1). \tag{11.30}$$
Let FBSDE(α) denote the class of FBSDEs taking the following form with some $(b_0, \sigma_0, f_0, g_0)$:
$$\begin{cases} X_t = x + \int_0^t [b^\alpha(s, \Theta_s) + b_0(s)]ds + \int_0^t [\sigma^\alpha(s, \Theta_s) + \sigma_0(s)]dB_s; \\ Y_t = g^\alpha(X_T) + g_0 + \int_t^T [f^\alpha(s, \Theta_s) + f_0(s)]ds - \int_t^T Z_s dB_s. \end{cases}$$
Then, FBSDE (11.29) is in class FBSDE(0), and FBSDE (11.1) is in class FBSDE(1) (with $b_0 = \sigma_0 = f_0 = g_0 = 0$). We say FBSDE($\alpha$) is solvable if the FBSDE has a solution for any $b_0, \sigma_0, f_0 \in L^2(\mathbb{F})$ and $g_0 \in L^2(\mathcal{F}_T)$. The following lemma plays a crucial role.

Lemma 11.4.4 *Assume Assumptions 11.1.1 and 11.4.1 hold. If FBSDE(α_0) is solvable, then there exists $\delta_0 > 0$, depending only on the Lipschitz constants of (b, σ, f, g) and the constant c in Assumption 11.4.1, such that FBSDE(α) is solvable for any $\alpha \in [\alpha_0, \alpha_0 + \delta_0]$.*

Before we prove this lemma, we use it to prove the existence part of Theorem 11.4.2.

Proof of Existence in Theorem 11.4.2. By Lemma 11.4.3 FBSDE(0) is solvable. Assume $(n-1)\delta_0 < T \leq n\delta_0$. Applying Lemma 11.4.4 n times we know FBSDE(1) is also solvable. Therefore, FBSDE (11.1) admits a solution. □

Proof of Lemma 11.4.4 For any $\alpha \in [\alpha_0, \alpha_0 + \delta_0]$ where δ_0 will be determined later, denote $\delta := \alpha - \alpha_0 \leq \delta_0$. For any $b_0, \sigma_0, f_0 \in L^2(\mathbb{F})$ and $g_0 \in L^2(\mathcal{F}_T)$, denote $\Theta^0 := (0, 0, 0)$ and for $n = 0, 1, \ldots,$
$$b_0^n(t) := \delta[Y_t^n + b(t, \Theta_t^n)] + b_0(t); \quad \sigma_0^n(t) := \delta[Z_t^n + \sigma(t, \Theta_t^n)] + \sigma_0(t);$$
$$f_0^n(t) := \delta[X_t^n + f(t, \Theta_t^n)] + f_0(t); \quad g_0^n(t) := \delta[-X_T^n + g(X_T^n)] + g_0,$$
and let Θ^{n+1} be the solution to the following FBSDE:
$$\begin{cases} X_t^{n+1} = x + \int_0^t [b^{\alpha_0}(s, \Theta_s^{n+1}) + b_0^n(s)]ds + \int_0^t [\sigma^{\alpha_0}(s, \Theta_s^{n+1}) + \sigma_0^n(s)]dB_s; \\ Y_t^{n+1} = g^{\alpha_0}(X_T^{n+1}) + g_0^n + \int_t^T [f^{\alpha_0}(s, \Theta_s^{n+1}) + f_0^n(s)]ds - \int_t^T Z_s^{n+1} dB_s. \end{cases}$$
By our assumption FBSDE(α_0) is solvable and thus Θ^n are well defined. Denote $\Delta\Theta^n := \Theta^{n+1} - \Theta^n$. Then,
$$d\Delta X_t^n = [[b^{\alpha_0}(t, \Theta_t^{n+1}) - b^{\alpha_0}(t, \Theta_t^n)] + \delta[\Delta Y_t^{n-1} + b(t, \Theta_t^n) - b(t, \Theta_t^{n-1})]]dt$$
$$+ [[\sigma^{\alpha_0}(t, \Theta_t^{n+1}) - \sigma^{\alpha_0}(t, \Theta_t^n)]$$

$$+\delta[\Delta Z_t^{n-1}+\sigma(t,\Theta_t^n)-\sigma(t,\Theta_t^{n-1})]]dB_t;$$
$$d\Delta Y_t^n = -[[f^{\alpha_0}(t,\Theta_t^{n+1})-f^{\alpha_0}(t,\Theta_t^n)]$$
$$+\delta[\Delta X_t^{n-1}+f(t,\Theta_t^n)-f(t,\Theta_t^{n-1})]]dt$$
$$+\Delta Z_t^n dB_t.$$

Applying Itô's rule we have

$$d(\Delta X_t^n \Delta Y_t^n) = [\cdots]dB_t$$
$$+[-[f^{\alpha_0}(t,\Theta_t^{n+1})-f^{\alpha_0}(t,\Theta_t^n)]\Delta X_t^n$$
$$+[b^{\alpha_0}(t,\Theta_t^{n+1})-b^{\alpha_0}(t,\Theta_t^n)]\Delta Y_t^n$$
$$+[\sigma^{\alpha_0}(t,\Theta_t^{n+1})-\sigma^{\alpha_0}(t,\Theta_t^n)]\Delta Z_t^n]dt$$
$$+\delta[-[\Delta X_t^{n-1}+f(t,\Theta_t^n)-f(t,\Theta_t^{n-1})]\Delta X_t^n$$
$$+[\Delta Y_t^{n-1}+b(t,\Theta_t^n)-b(t,\Theta_t^{n-1})]\Delta Y_t^n$$
$$+[\Delta Z_t^{n-1}+\sigma(t,\Theta_t^n)-\sigma(t,\Theta_t^{n-1})]\Delta Z^n]dt.$$

Recall (11.30). We have

$$d(\Delta X_t^n \Delta Y_t^n) \leq [\cdots]dB_t$$
$$-c_{\alpha_0}|\Delta \Theta_t^n|^2 dt + C\delta[|\Delta X_t^{n-1}||\Delta X_t^n|+|\Delta Y_t^{n-1}||\Delta Y_t^n|$$
$$+|\Delta Z_t^{n-1}||\Delta Z_t^n|]dt$$
$$\leq [\cdots]dB_t + [(\delta-c_{\alpha_0})|\Delta \Theta_t^n|^2 + C\delta|\Delta \Theta_t^{n-1}|^2]dt.$$

Note that $\Delta X_0^n = 0$ and

$$\Delta X_T^n \Delta Y_T^n = \Delta X_T^n [g^{\alpha_0}(X_T^{n+1})-g^{\alpha_0}(X_T^n)] \geq 0.$$

Then,

$$E\left\{(c_{\alpha_0}-\delta)\int_0^T |\Delta \Theta_t^n|^2 dt\right\} \leq C\delta E\left\{\int_0^T |\Delta \Theta_t^{n-1}|^2 dt\right\}.$$

Without loss of generality we assume $c \leq 1$. Then, $c_{\alpha_0} \geq c$ and thus

$$(c-\delta)E\left\{\int_0^T |\Delta \Theta_t^n|^2 dt\right\} \leq C_1 \delta E\left\{\int_0^T |\Delta \Theta_t^{n-1}|^2 dt\right\}.$$

Choose $\delta_0 := \frac{c}{1+4C_1} > 0$. Then, for any $d \leq d_0$, $C_1\delta \leq \frac{1}{4}(c-\delta)$. Therefore,

$$E\left\{\int_0^T |\Delta \Theta_t^n|^2 dt\right\} \leq \frac{1}{4}E\left\{\int_0^T |\Delta \Theta_t^{n-1}|^2 dt\right\}.$$

By induction we get

$$E\left\{\int_0^T |\Delta \Theta_t^n|^2 dt\right\} \leq \frac{C}{4^n}.$$

Thus,
$$E\left\{\int_0^T |\Delta\Theta_t^n|^2 dt\right\}^{\frac{1}{2}} \le \frac{C}{2^n}.$$

Then, for any $n > m$,
$$E\left\{\int_0^T |\Theta_t^n - \Theta_t^m|^2 dt\right\}^{\frac{1}{2}} \le \sum_{i=m}^{n-1} E\left\{\int_0^T |\Delta\Theta_t^i|^2 dt\right\}^{\frac{1}{2}} \le C\sum_{i=m}^{n-1} \frac{1}{2^i} \le \frac{C}{2^m} \to 0,$$

as $m \to \infty$. So, there exists Θ such that
$$\lim_{n\to\infty} E\left\{\int_0^T |\Theta_t^n - \Theta_t|^2 dt\right\} = 0.$$

Note that
$$b^{\alpha_0}(t, \Theta_t^{n+1}) + b_0^n(t) = b^{\alpha_0}(t, \Theta_t^{n+1}) + \delta[Y_t^n + b(t, \Theta_t^n)] + b_0(t)$$
$$\to b^{\alpha_0}(t, \Theta_t) + \delta[Y_t + b(t, \Theta_t)] + b_0(t)$$
$$= b^{\alpha_0+\delta}(t, \Theta_t) + b_0(t) = b^{\alpha}(t, \Theta_t) + b_0(t).$$

Similar results hold for the other terms. Thus, Θ satisfies FBSDE(α) for any $\alpha \in [\alpha_0, \alpha_0 + \delta_0]$. Finally, it is straightforward to check that $\|\Theta\| < \infty$. □

Remark 11.4.5 The monotonicity conditions in Assumption 11.4.1 can be replaced by
$$[b(t, \theta_1) - b(t, \theta_2)]\Delta y + [\sigma(t, \theta_1) - \sigma(t, \theta_2)]\Delta z - [f(t, \theta_1) - f(t, \theta_2)]\Delta x$$
$$\ge -c[|\Delta x|^2 + |\Delta y|^2 + |\Delta z|^2], \quad a.s.;$$
$$[g(x_1) - g(x_2)]\Delta x \le 0, \quad a.s.$$

and we can still obtain existence and uniqueness results in a similar way.

11.5 Further Reading

The first paper on coupled FBSDE is Antonelli (1993), studying the case in which T is small. That work was extended by Pardoux and Tang (1999) following the fixed point approach. The four-step scheme was proposed by Ma et al. (1994), whose main idea is the decoupling strategy. This strategy was further exploited by Delarue (2002) in the Markovian framework, and by Zhang (2006) and Ma et al. (2011) in non-Markovian frameworks. The method of continuation is due to Hu and Peng (1995), Peng and Wu (1999), and Yong (1997). The book Ma and Yong (1999) is a common reference for the theory.

References

Antonelli, F.: Backward-forward stochastic differential equations. Ann. Appl. Probab. **3**, 777–793 (1993)

Delarue, F.: On the existence and Uniqueness of solutions to FBSDEs in a non-degenerate case. Stoch. Process. Appl. **99**, 209–286 (2002)

Hu, Y., Peng, S.: Solution of forward–backward stochastic differential equations. Probab. Theory Relat. Fields **103**, 273–283 (1995)

Ma, J., Yong, J.: Forward-Backward Stochastic Differential Equations and Their Applications. Lecture Notes in Mathematics, vol. 1702. Springer, Berlin (1999)

Ma, J., Protter, P., Yong, J.: Solving forward-backward stochastic differential equations explicitly—a four step scheme. Probab. Theory Relat. Fields **98**, 339–359 (1994)

Ma, J., Wu, Z., Zhang, D., Zhang, J.: On well-posedness of forward-backward SDEs—a unified approach. Working paper, University of Southern California (2011)

Pardoux, E., Tang, S.: Forward-backward stochastic differential equations and quasilinear parabolic PDEs. Probab. Theory Relat. Fields **114**, 123–150 (1999)

Peng, S., Wu, Z.: Fully coupled forward-backward stochastic differential equations and applications to optimal control. SIAM J. Control Optim. **37**, 825–843 (1999)

Yong, J.: Finding adapted solutions of forward-backward stochastic differential equations: method of continuation. Probab. Theory Relat. Fields **107**, 537–572 (1997)

Zhang, J.: The well-posedness of FBSDEs. Discrete Contin. Dyn. Syst., Ser. B **6**, 927–940 (2006)

References

Adrian, T., Westerfield, M.: Disagreement and learning in a dynamic contracting model. Rev. Financ. Stud. **22**, 3839–3871 (2009)

Antonelli, F.: Backward-forward stochastic differential equations. Ann. Appl. Probab. **3**, 777–793 (1993)

Bender, C., Denk, R.: A forward scheme for backward SDEs. Stoch. Process. Appl. **117**, 1793–1812 (2007)

Biais, B., Mariotti, T., Plantin, G., Rochet, J.-C.: Dynamic security design: convergence to continuous time and asset pricing implications. Rev. Econ. Stud. **74**, 345–390 (2007)

Biais, B., Mariotti, T., Rochet, J.-C., Villeneuve, S.: Large risks, limited liability, and dynamic moral hazard. Econometrica **78**, 73–118 (2010)

Bismut, J.M.: Conjugate convex functions in optimal stochastic control. J. Math. Anal. Appl. **44**, 384–404 (1973)

Bolton, P., Dewatripont, M.: Contract Theory. MIT Press, Cambridge (2005)

Borch, K.: Equilibrium in a reinsurance market. Econometrica **30**, 424–444 (1962)

Bouchard, B., Touzi, N.: Discrete-time approximation and Monte-Carlo simulation of backward stochastic differential equations. Stoch. Process. Appl. **111**, 175–206 (2004)

Briand, P., Hu, Y.: BSDE with quadratic growth and unbounded terminal value. Probab. Theory Relat. Fields **136**, 604–618 (2006)

Briand, P., Hu, Y.: Quadratic BSDEs with convex generators and unbounded terminal conditions. Probab. Theory Relat. Fields **141**, 543–567 (2008)

Cadenillas, A., Cvitanić, J., Zapatero, F.: Optimal risk-sharing with effort and project choice. J. Econ. Theory **133**, 403–440 (2007)

Crisan, D., Manolarakis, K.: Solving backward stochastic differential equations using the cubature method: application to nonlinear pricing. SIAM J. Financ. Math. **3**, 534–571 (2012)

Cvitanić, J., Zhang, J.: Optimal compensation with adverse selection and dynamic actions. Math. Financ. Econ. **1**, 21–55 (2007)

Cvitanić, J., Wan, X., Zhang, J.: Optimal contracts in continuous-time models. J. Appl. Math. Stoch. Anal. **2006**, 95203 (2006)

Cvitanić, J., Wan, X., Zhang, J.: Optimal compensation with hidden action and lump-sum payment in a continuous-time model. Appl. Math. Optim. **59**, 99–146 (2009)

Cvitanić, J., Wan, X., Yang, H.: Dynamics of contract design with screening. Manag. Sci. (2012, forthcoming)

Delarue, F.: On the existence and Uniqueness of solutions to FBSDEs in a non-degenerate case. Stoch. Process. Appl. **99**, 209–286 (2002)

DeMarzo, P.M., Fishman, M.: Optimal long-term financial contracting. Rev. Financ. Stud. **20**, 2079–2128 (2007a)

DeMarzo, P.M., Fishman, M.: Agency and optimal investment dynamics. Rev. Financ. Stud. **20**, 151–188 (2007b)

DeMarzo, P.M., Sannikov, Y.: Optimal security design and dynamic capital structure in a continuous-time agency model. J. Finance **61**, 2681–2724 (2006)

DeMarzo, P.M., Sannikov, Y.: Learning, termination and payout policy in dynamic incentive contracts. Working paper, Princeton University (2011)

Duffie, D., Epstein, L.G.: Stochastic differential utility. Econometrica **60**, 353–394 (1992)

Duffie, D., Geoffard, P.Y., Skiadas, C.: Efficient and equilibrium allocations with stochastic differential utility. J. Math. Econ. **23**, 133–146 (1994)

Dumas, B., Uppal, R., Wang, T.: Efficient intertemporal allocations with recursive utility. J. Econ. Theory **93**, 240–259 (2000)

Ekeland, I.: On the variational principle. J. Math. Anal. Appl. **47**, 324–353 (1974)

El Karoui, N., Peng, S., Quenez, M.C.: Backward stochastic differential equations in finance. Math. Finance **7**, 1–71 (1997)

El Karoui, N., Peng, S., Quenez, M.C.: A dynamic maximum principle for the optimization of recursive utilities under constraints. Ann. Appl. Probab. **11**, 664–693 (2001)

Fleming, W., Soner, H.M.: Controlled Markov Processes and Viscosity Solutions, 2nd edn. Springer, New York (2006)

Fong, K.G.: Evaluating skilled experts: optimal scoring rules for surgeons. Working paper, Stanford University (2009)

Giat, Y., Subramanian, A.: Dynamic contracting under imperfect public information and asymmetric beliefs. Working paper, Georgia State University (2009)

Giat, Y., Hackman, S.T., Subramanian, A.: Investment under uncertainty, heterogeneous beliefs and agency conflicts. Review of Financial Studies **23**(4), 1360–1404 (2011)

Gobet, E., Lemor, J.-P., Warin, X.: A regression-based Monte-Carlo method to solve backward stochastic differential equations. Ann. Appl. Probab. **15**, 2172–2202 (2005)

He, Z.: Optimal executive compensation when firm size follows geometric Brownian motion. Rev. Financ. Stud. **22**, 859–892 (2009)

He, Z., Wei, B., Yu, J.: Permanent risk and dynamic incentives. Working paper, Baruch College (2010)

Holmström, B., Milgrom, P.: Aggregation and linearity in the provision of intertemporal incentives. Econometrica **55**, 303–328 (1987)

Hu, Y., Peng, S.: Solution of forward–backward stochastic differential equations. Probab. Theory Relat. Fields **103**, 273–283 (1995)

Kamien, M.I., Schwartz, N.L.: Dynamic Optimization. Elsevier, Amsterdam (1991)

Karatzas, I., Shreve, S.E.: Methods of Mathematical Finance. Springer, New York (1998)

Kobylanski, M.: Backward stochastic differential equations and partial differential equations with quadratic growth. Ann. Probab. **28**, 558–602 (2000)

Laffont, J.J., Martimort, D.: The Theory of Incentives: The Principal-Agent Model. Princeton University Press, Princeton (2001)

Larsen, K.: Optimal portfolio delegation when parties have different coefficients of risk aversion. Quant. Finance **5**, 503–512 (2005)

Leland, H.E.: Agency costs, risk management, and capital structure. J. Finance **53**, 1213–1243 (1998)

Ma, J., Yong, J.: Forward-Backward Stochastic Differential Equations and Their Applications. Lecture Notes in Mathematics, vol. 1702. Springer, Berlin (1999)

Ma, J., Protter, P., Yong, J.: Solving forward-backward stochastic differential equations explicitly—a four step scheme. Probab. Theory Relat. Fields **98**, 339–359 (1994)

Ma, J., Wu, Z., Zhang, D., Zhang, J.: On well-posedness of forward-backward SDEs—a unified approach. Working paper, University of Southern California (2011)

Mirrlees, J.A.: The theory of moral hazard and unobservable behaviour. Rev. Econ. Stud. **66**, 3–21 (1999) (Original version 1975)

Ou-Yang, H.: Optimal contracts in a continuous-time delegated portfolio management problem. Rev. Financ. Stud. **16**, 173–208 (2003)

References

Ou-Yang, H.: An equilibrium model of asset pricing and moral hazard. Rev. Financ. Stud. **18**, 1219–1251 (2005)

Pardoux, E., Peng, S.: Adapted solution of a backward stochastic differential equation. Syst. Control Lett. **14**, 55–61 (1990)

Pardoux, E., Peng, S.: Backward Stochastic Differential Equations and Quasilinear Parabolic Partial Differential Equations. Lecture Notes in Control and Inform. Sci., vol. 176, pp. 200–217. Springer, New York (1992)

Pardoux, E., Tang, S.: Forward-backward stochastic differential equations and quasilinear parabolic PDEs. Probab. Theory Relat. Fields **114**, 123–150 (1999)

Peng, S.: A general stochastic maximum principle for optimal control problems. SIAM J. Control **28**, 966–979 (1990)

Peng, S.: Probabilistic interpretation for systems of quasilinear parabolic partial differential equations. Stochastics **37**, 61–74 (1991)

Peng, S., Wu, Z.: Fully coupled forward-backward stochastic differential equations and applications to optimal control. SIAM J. Control Optim. **37**, 825–843 (1999)

Piskorski, T., Tchistyi, A.: Optimal mortgage design. Rev. Financ. Stud. **23**, 3098–3140 (2010)

Prat, J., Jovanovic, B.: Dynamic incentive contracts under parameter uncertainty. Working paper, NYU (2010)

Quadrini, V.: Investment and liquidation in renegotiation-proof contracts with moral hazard. J. Monet. Econ. **51**, 713–751 (2004)

Ross, S.A.: The economic theory of agency: the principal's problem. Am. Econ. Rev. **63**, 134–139 (1973). Papers and Proceedings of the Eighty-fifth Annual Meeting of the American Economic Association

Salanie, B.: The Economics of Contracts: A Primer, 2nd edn. MIT Press, Cambridge (2005)

Sannikov, Y.: Agency problems, screening and increasing credit lines. Working paper, Princeton University (2007)

Sannikov, Y.: A continuous-time version of the principal-agent problem. Rev. Econ. Stud. **75**, 957–984 (2008)

Sannikov, Y.: Contracts: the theory of dynamic principal-agent relationships and the continuous-time approach. Working paper, Princeton University (2012)

Schattler, H., Sung, J.: The first-order approach to continuous-time principal-agent problem with exponential utility. J. Econ. Theory **61**, 331–371 (1993)

Spear, S., Srivastrava, S.: On repeated moral hazard with discounting. Rev. Econ. Stud. **53**, 599–617 (1987)

Sung, J.: Linearity with project selection and controllable diffusion rate in continuous-time principal-agent problems. Rand J. Econ. **26**, 720–743 (1995)

Sung, J.: Lectures on the Theory of Contracts in Corporate Finance: From Discrete-Time to Continuous-Time Models. Com2Mac Lecture Note Series, vol. 4. Pohang University of Science and Technology, Pohang (2001)

Sung, J.: Optimal contracts under adverse selection and moral hazard: a continuous-time approach. Rev. Financ. Stud. **18**, 1121–1173 (2005)

Williams, N.: On dynamic principal-agent problems in continuous time. Working paper, University of Wisconsin-Madison (2009)

Wilson, R.: The theory of syndicates. Econometrica **36**, 119–132 (1968)

Wu, Z.: Maximum principle for optimal control problem of fully coupled forward-backward stochastic systems. J. Syst. Sci. Math. Sci. **11**, 249–259 (1998)

Yong, J.: Finding adapted solutions of forward-backward stochastic differential equations: method of continuation. Probab. Theory Relat. Fields **107**, 537–572 (1997)

Yong, J.: Optimality variational principle for optimal controls of forward-backward stochastic differential equations. SIAM J. Control Optim. **48**, 4119–4156 (2010)

Yong, J., Zhou, X.Y.: Stochastic Controls: Hamiltonian Systems and HJB Equations. Springer, New York (1999)
Zhang, J.: A numerical scheme for BSDEs. Ann. Appl. Probab. **14**, 459–488 (2004)
Zhang, J.: The well-posedness of FBSDEs. Discrete Contin. Dyn. Syst., Ser. B **6**, 927–940 (2006)
Zhang, Y.: Dynamic contracting with persistent shocks. J. Econ. Theory **144**, 635–675 (2009)
Zhang, J.: Backward Stochastic Differential Equations. Book manuscript, University of Southern California (2011, in preparation)

Index

A
Adjoint process, 28, 159
Adrian, 113
Adverse selection, 3, 5, 137
Agent, 3
Antonelli, 248

B
Backward Stochastic Differential Equation (BSDE), 28, 157
 linear, 159
 Markovian, 170
 PDE, 170
 with quadratic growth, 173
 stochastic control of, 183
Bargaining power, 4
Bender, 181
Black–Scholes formula, 158, 171
Biais, 113, 115
Bismut, 227
Black, 158, 171
Bolton, 6, 153
Borch, 14
Borch rule, 7, 20, 26, 29, 36, 57
Bouchard, 181
Briand, 90
BSDE, *see* Backward Stochastic Differential Equation
Budget constraint, 18

C
Cadenillas, 24, 43
Call option, 23
Certainty equivalent, 13, 90
Commitment, 5
Comparison principle/theorem, 90, 165
Cost function, 7

Crisan, 181
Cvitanić, J., 14, 24, 43, 84, 113, 153

D
Delarue, 239
DeMarzo, 113, 115
Denk, 181
Dewatripont, 6, 153
Dual problem, 38
Duffie, 24, 43, 183
Dumas, 24, 43
Dynamic Programming Principle, 107, 124

E
Ekeland, 202
El Karoui, 181, 227
Epstein, 43
Euler equation, 147

F
FBSDE, *see* Forward-Backward Stochastic Differential Equation
Feynman, 131, 170
First best, 3, 8, 18
Fishman, 134
Fong, 113
Forward-Backward Stochastic Differential Equation (FBSDE), 158, 229
 stochastic control of, 188

G
Geoffard, 24, 43
Giat, 113
Gobet, 181

H
Hackman, 113
Hamilton–Jacobi–Bellman PDE, 76, 106
Hamiltonian, 30, 63, 72, 183, 192, 199, 200
Hazard function, 144, 149
He, 113
Hedging, 158
Hidden action, 3, 4, 8, 137
Hidden type, 3, 5, 10, 135
HJB, 76, 92, 106
Holmström, 14, 83, 93
Hu, 90, 209

I
Implementable, 31, 49
Incentive compatible, 31
Individual rationality constraint, 3, 49
Informational rent, 14, 137, 147
IR constraint, 4
ISCC, 33

J
Jovanovic, 113

K
Kac, 131, 170
Karatzas, 43
Kmien, 147
Kobylanski, 182

L
Laffont, 6, 153
Larsen, 24
Leland, 134
Lemor, 181
Limited liability, 150

M
Ma, 248
Malliavin derivative, 53
Manolarakis, 181
Mariotti, 113, 134
Martimort, 6, 153
Martingale representation theorem, 35
Menu of contracts, 5, 11, 138
Merton's problem, 20
Milgrom, 14, 84, 113
Mirrlees, 9
Misreporting, 115
Moral hazard, 3, 4, 47

N
Non-separable utility, 26

O
Option pricing, 158
Ou-Yang, 24, 43, 113
Output process, 25

P
PA problem, 6
Pardoux, 182, 248
Pareto optimal, 4
Participation constraint, 3
Peng, 181, 182, 227, 248
Piskorski, 113
Plantin, 113, 134
Prat, 113
Principal, 3
Principal–Agent problem, 3, 51, 230
Protter, 248

Q
Quadrini, 134
Quenez, 181, 227

R
Ratio of marginal utilities process, 47
Recursive utility, 40, 183
Reflection process, 122, 124
Rent, 13, 137, 147
Revelation principle, 5, 138
Risk Sharing, 3, 4, 7, 18, 25
Rochet, 113, 134
Ross, 24

S
Salanie, 6, 153
Sannikov, 84, 103, 130, 153
Schattler, 84
Schwartz, 147
Second best, 3, 4
Self-financing, 158
Sensitivity, 8, 10
Separable utility, 26, 72
Shreve, 43
Skiadas, 24, 43
Spear, 134
Spence, 144
Srivastrava, 134
Stochastic Differential Utility, 40
Stochastic Maximum Principle, 26, 84, 110, 183
Subramanian, 113
Sung, 84, 153

T
Tang, 248
Target action, 50, 57, 67, 79, 87, 91, 96
Tchistyi, 113
Terminal condition, 157, 192
Third best, 3, 5, 135
Touzi, 181
Truth-telling, 11, 120, 127, 130, 143, 144, 148

U
Uppal, 24, 43

W
Wan, 43, 84, 153
Wang, 24, 43
Warin, 181
Weak formulation, 47, 50, 51, 183, 203
Weak solution, 48
Wei, 113
Westerfield, 113
Williams, 84
Wilson, 14
Wu, 227, 248

Y
Yang, 43, 153
Yong, 108, 227, 247, 248
Yu, 113

Z
Zapatero, 24, 43
Zhang, J., 14, 84, 113, 153, 181, 182, 247, 248
Zhang, Y., 113
Zhou, 108